CRC Press / Balkema
Taylor & Francis Group

T0187751

Errata

Information, Communication and Environment
Marine Navigation and Safety of Sea Transportation
Editors: Adam Weintrit & Tomasz Neumann
ISBN: 978-1-138-02857-9

Pages 56, 227:
Due to an error during the typesetting stage of this volume, the following
corrections should be applied to the equations:
Page 56:

Equation in figure caption 2 should be $N = N_1 \cdot N_2 \cdot N_3 \cdot N_4 = 4 \cdot 5 \cdot 7 \cdot 13$

Page 227:

Eqn (2) $S = 51{,}432 \times V^{0{,}728}$

We sincerely apologize for these errors.

The publishers

Information, Communication and Environment

Marine Navigation and Safety of Sea Transportation

Editors

Adam Weintrit & Tomasz Neumann
Gdynia Maritime University, Gdynia, Poland

CRC Press
Taylor & Francis Group
Boca Raton London New York

CRC Press is an imprint of the
Taylor & Francis Group, an **informa** business

A BALKEMA BOOK

Published by:
CRC Press/Balkema
P.O. Box 447, 2300 AK Leiden, The Netherlands
e-mail: Pub.NL@taylorandfrancis.com
www.crcpress.com – www.taylorandfrancis.com

First issued in paperback 2020

Typeset by V Publishing Solutions Pvt Ltd., Chennai, India

ISBN 13: 978-0-367-73822-8 (pbk)
ISBN 13: 978-1-138-02857-9 (hbk)

Visit the Taylor & Francis Web site at
http://www.taylorandfrancis.com

and the CRC Press Web site at
http://www.crcpress.com

Contents

List of reviewers 7

Information, Communication and Environment. Introduction... 9
A. Weintrit & T. Neumann

Chapter 1. *Maritime Communications* 11

1.1. Hidden Communication in the Terrestrial and Satellite Radiotelephone Channels of Maritime Mobile Services 13
O. Shyshkin & V. Koshevyy

1.2. Introduction to Inmarsat GEO Space and Ground Segments... 21
D.S. Ilcev

1.3. Integration of Radio and Satellite Automatic Identification System for Maritime Applications............. 33
D.S. Ilcev

1.4. Satellite Antenna Infrastructure Onboard Inmarsat Spacecraft for Maritime and Other Mobile Applications 45
D.S. Ilcev

1.5. Synthesis of Composite Biphasic Signals for Continuous Wave Radar 55
V.M. Koshevyy, I.V. Koshevyy & D.O. Dolzhenko

1.6. Zero Levels Formation of Radiation Pattern Linear Antennas Array with Minimum Quiantity of Controlling Coefficients Weights.. 61
V.M. Koshevyy & A.A. Shershnova

1.7. Radio Refractivity and Rain-Rate Estimations over Northwest Aegean Archipelagos for Electromagnetic Wave Attenuation Modelling 67
E.A. Karagianni, A.P. Mitropoulos, N.G. Drolias, A.D. Sarantopoulos & A.A. Charantonis

1.8. Concepts of the GMDSS Modernization 75
K. Korcz

Chapter 2. *Decision Support System*.. 83

2.1. Supporting Situation Awareness on the Bridge: Testing Route Exchange in a Practical e-Navigation Study 85
T. Porathe, A. Brodje, R. Weber, D. Camre & O. Borup

2.2. PARK Model and Decision Support System based on Ship Operator's Consciousness 93
S.W. Park, Y.S. Park, J.S. Park & N.X. Thanh

2.3. Multi-objective Route Optimization for Onboard Decision Support System 99
R. Vettor & C. Guedes Soares

2.4. Simulation-Augmented Methods for Manoeuvring Support – On-Board Ships and from the Shore 107
K. Benedict, M. Kirchhoff, M. Gluch, S. Fischer, M. Schaub & M. Baldauf

2.5. 3D Navigator Decision Support System Using the Smartglasses Technology 117
A. Łebkowski

2.6. Neuroevolutionary Ship Maneuvering Prediction System.. 123
M. Łącki

Chapter 3. *Deoinformation Systems and Maritime Spatian Planning* 129

3.1. Information and Communication Technologies in the Area with a Complex Spatial Structure................. 131
A. Kuśmińska-Fijałkowska & Z. Łukasik

3.2. Establishing a Framework for Maritime Spatial Planning in Europe 135
A. Kuśmińska-Fijałkowska & Z. Łukasik

3.3. Application of Intelligent Geoinformation Systems for Integrated Safety Assessment of Marine Activities 139
V.V. Popovich, O.V. Smirnova, M.V. Tsvetkov & R.P. Sorokin

Chapter 4. *Hydrometeorological Aspects* 145

4.1. Design Tide and Wave for Santos Offshore Port (Brazil) Considering Extreme Events in a Climate Changing Scenario.... 147
P. Alfredini, E. Arasaki & A.S. Moreira

4.2. Mathematical Modeling of Wave Situation for Creation of Protective Hydrotechnical Constructions in Port Kulevi 153
A. Gegenava, I. Sharabidze & A. Kakhidze

4.3. The Northerly Summer Wind off the West Coast of the Iberian Peninsula.. 157
N. Rijo, A. Semedo, D.C.A. Lima, P. Miranda, R.M. Cardoso & P.M.M. Soares

Chapter 5. *Inland Shipping* 163

5.1. Emergency Group Decision-Making with Multidivisional Cooperation for Inland Maritime Accident 165
B. Wu, X.P. Yan, Y. Wang & J.F. Zhang

5.2. The Concept of Emergency Notification System for Inland Navigation 173
T. Perzyński, A. Lewiński & Z. Łukasik

5.3. Ship Design Optimization Applied for Urban Regular Transport on Guadalquivir River (GuadaMAR) 179
A. Querol, R. Jiménez-Castañeda & F. Piniella

5.4. Ship Emission Study Under Traffic Control in Inland Waterway Network Based on Traffic Simulation Data....................185
X. Chen, J. Mou, L. Chen & X. Yue

5.5. The Using of Risk to Determination of Safety Navigation in Inland Waters 195
W. Galor

5.6. Inland Water Transport and its Impact on Seaports and Seaport Cities Development 201
A.S. Grzelakowski

Chapter 6. *Maritime Pollution and Environment Protection*..209

6.1. Determination of Marine Pollution Caused by Ship Operations Using the DEMATEL Method...211
Ü. Özdemir, H. Yılmaz & E. Başar

6.2. Joint-Task Force Management in Cross-Border Emergency Response. Managerial Roles and Structuring Mechanisms in High Complexity-High Volatility Environments..217
O.J. Borch & N. Andreassen

6.3. Environmental Risk Assessment for the Aegean Sea 225
I. Koromila, Z. Nivolianitou, S. Perantonis, T. Giannakopoulos, E. Charou, S. Gyftakis & K. Spyrou

6.4. Probabilistic Meta-models Evaluating Accidental Oil Spill Size from Tankers 231
J. Montewka, F. Goerlandt & X. Zheng

6.5. Negative Impact of Cruise Tourism Development on Local Community and the Environment 243
J. Kizielewicz & T. Luković

Chapter 7. *Vessel Traffic Service (VTS)* 251

7.1. Improving Safety of Navigation by Implementing VTS/VTMIS: Experiences from Montenegro 253
S. Bauk   & N. Kapidani

7.2. Evolutionary Methods in the Management of Vessel Traffic..259
A. Łebkowski

7.3. Supporting Voice Communication Between Navigator and VTS by Visual Solutions – Exploring the Use of the "Route Suggestion" Functionality within VTS 267
A. Brodje, R. Weber, D. Camre, O. Borup & T. Porathe

7.4. 4M Overturned Pyramid (MOP) Model: Case Studies on Indonesian and Japanese Maritime Traffic Systems (MTS)........275
W. Mutmainnah & M. Furusho

List of reviewers

Prof. Paolo **Alfredini**, University of São Paulo, Polytechnic School, São Paulo, Brazil
Prof. Michael **Baldauf**, Word Maritime University, Malmö, Sweden
Prof. Angelica **Baylon**, Maritime Academy of Asia & the Pacific, Philippines
Prof. Neil **Bose**, Australian Maritime College, University of Tasmania, Launceston, Australia
Prof. Zbigniew **Burciu**, Gdynia Maritime University, Poland
Sr. Jesus **Carbajosa Menendez**, President of Spanish Institute of Navigation, Spain
Prof. Ruizhi **Chen**, Texas A&M University, Corpus Christi, US
Prof. Andrzej **Chudzikiewicz**, Warsaw University of Technology, Poland
Prof. Frank **Coolen**, Durham University, UK
Prof. Jerzy **Czajkowski**, Gdynia Maritime University, Poland
Prof. Krzysztof **Czaplewski**, Gdynia Maritime University, Poland
Prof. German **de Melo Rodriguez**, Polytechnic University of Catalonia, Barcelona, Spain
Prof. Bolesław **Domański**, Jagiellonian University, Kraków, Poland
Prof. Eamonn **Doyle**, National Maritime College of Ireland, Cork Institute of Technology, Cork, Ireland
Prof. Branislav **Dragović**, University of Montenegro, Kotor, Montenegro
Prof. Daniel **Duda**, Polish Naval Academy, Polish Nautological Society, Poland
Prof. Akram **Elentably**, King Abdulaziz University (KAU), Jeddah, Saudi Arabia
Prof. Włodzimierz **Filipowicz**, Gdynia Maritime University, Gdynia, Poland
Prof. Masao **Furusho**, Kobe University, Japan
Prof. Wiesław **Galor**, Maritime University of Szczecin, Poland
Prof. Yang **Gao**, University of Calgary, Canada
Prof. Georg **Gartner**, Vienna University of Technology, Wien, Austria
Prof. Péter **Gáspár**, Computer and Automation Research Institute, Hungarian Academy of Sciences, Budapest, Hungary
Prof. Jerzy **Gaździcki**, President of the Polish Association for Spatial Information; Warsaw, Poland
Prof. Witold **Gierusz**, Gdynia Maritime University, Poland
Prof. Dariusz **Gotlib**, Warsaw University of Technology, Warsaw, Poland
Prof. Andrzej **Grzelakowski**, Gdynia Maritime University, Poland
Prof. Vladimir **Hahanov**, Kharkov National University of Radio Electronics, Kharkov, Ukraine
Prof. Jerzy **Hajduk**, Maritime University of Szczecin, Poland
Prof. Michał **Holec**, Gdynia Maritime University, Poland
Prof. Qinyou **Hu**, Shanghai Maritime University, Shanghai, China
Prof. Stojce Dimov **Ilcev**, Durban University of Technology, South Africa
Prof. Marianna **Jacyna**, Warsaw University of Technology, Poland
Prof. Ales **Janota**, University of Žilina, Slovakia
Prof. Jacek **Januszewski**, Gdynia Maritime University, Poland
Prof. Piotr **Jędrzejowicz**, Gdynia Maritime University, Poland
Prof. Jung Sik **Jeong**, Mokpo National Maritime University, South Korea
Prof. Kalin **Kalinov**, Nikola Y. Vaptsarov Naval Academy, Varna, Bulgaria
Prof. John **Kemp**, Royal Institute of Navigation, London, UK
Prof. Eiichi **Kobayashi**, Kobe University, Japan
Prof. Lech **Kobyliński**, Polish Academy of Sciences, Gdansk University of Technology, Poland
Prof. Andrzej **Królikowski**, Maritime Office in Gdynia; Gdynia Maritime University, Poland
Prof. Pentti **Kujala**, Helsinki University of Technology, Helsinki, Finland
Prof. Jan **Kulczyk**, Wroclaw University of Technology, Poland
Prof. Krzysztof **Kulpa**, Warsaw University of Technology, Warsaw, Poland
Prof. Shashi **Kumar**, U.S. Merchant Marine Academy, New York
Prof. Alexander **Kuznetsov**, Admiral Makarov State Maritime Academy, St. Petersburg, Russia
Prof. Bogumił **Łączyński**, Gdynia Maritime University, Poland
Prof. Andrzej **Lewiński**, University of Technology and Humanities in Radom, Poland
Prof. Dieter **Lompe**, Hochschule Bremerhaven, Germany
Prof. Mirosław **Luft**, University of Technology and Humanities in Radom, Poland
Prof. Zbigniew **Łukasik**, University of Technology and Humanities in Radom, Poland
Prof. Evgeniy **Lushnikov**, Maritime University of Szczecin, Poland
Prof. Prabhat K. **Mahanti**, University of New Brunswick, Saint John, Canada
Prof. Artur **Makar**, Polish Naval Academy, Gdynia, Poland
Prof. Francesc Xavier **Martinez de Oses**, Polytechnical University of Catalonia, Barcelona, Spain
Prof. Doyan **Mednikarov**, Nikola Y. Vaptsarov Naval Academy, Varna, Bulgaria
Prof. Jerzy **Mikulski**, University of Economics in Katowice, Poland
Prof. Sergey **Moiseenko**, Kaliningrad State Technical University, Kaliningrad, Russian Federation
Prof. Junmin **Mou**, Wuhan University of Technology, Wuhan, China
Prof. Reinhard **Mueller**-Demuth, Hochschule Wismar, Germany
Prof. Janusz **Narkiewicz**, Warsaw University of Technology, Poland
Prof. Rudy R. **Negenborn**, Delft University of Technology, Delft, The Netherlands
Prof. Nikitas **Nikitakos**, University of the Aegean, Chios, Greece
Prof. Tomasz **Nowakowski**, Wroclaw University of Technology, Poland
Prof. Gyei-Kark **Park**, Mokpo National Maritime University, Mokpo, Korea
Mr. David **Patraiko**, The Nautical Institute, UK

Information, Communication and Environment. Introduction

A. Weintrit & T. Neumann
Gdynia Maritime University, Gdynia, Poland
Polish Branch of the Nautical Institute

The contents of the book are partitioned into seven separate chapters: Maritime communications (covering the chapters 1.1 through 1.8), Decision Support Systems (covering the chapters 2.1 through 2.6), Geoinformation systems and maritime spatial planning (covering the chapters 3.1 through 3.3), Hydrometeorological aspects (covering the chapters 4.1 through 4.3), Inland shipping (covering the chapters 5.1 through 5.6), Maritime pollution and environment protection (covering the chapters 6.1 through 6.5), and VTS - Vessel Traffic Service (covering the chapters 7.1 through 7.4).

In each of them readers can find a few sub-chapters. Sub-chapters collected in the first chapter, titled 'Maritime communications', concerning hidden communication in the terrestrial and satellite radiotelephone channels of maritime mobile services, introduction to Inmarsat GEO space and ground segments, integration of radio and satellite automatic identification system for maritime applications, satellite antenna infrastructure onboard Inmarsat spacecraft for maritime and other mobile applications, synthesis of composite biphasic signals for continuous wave radar, zero levels formation of radiation pattern linear antennas array with minimum quantity of controlling coefficients weights, radio refractivity and rain-rate estimations over northwest Aegean archipelagos for electromagnetic wave attenuation modelling and concepts of the GMDSS modernization.

In the second chapter there are described problems related to decision support systems: supporting situation awareness on the bridge, testing route exchange in a practical e-Navigation study, Park model and decision support system based on ship operator's consciousness, multi-objective route optimization for onboard decision support system, simulation-augmented methods for manoeuvring support – on-board ships and from the shore, 3D navigator decision support system using the smartglasses technology, and neuroevolutionary ship manoeuvring prediction system.

Third chapter concerns geoinformation systems and maritime spatial planning. The readers can find some information about information and communication technologies in the area with a complex spatial structure, establishing a framework for maritime spatial planning in Europe, and application of intelligent geoinformation systems for integrated safety assessment of marine activities.

The fourth chapter deals with hydrometeorological aspects. The contents of the fourth chapter are partitioned into three subchapters: design tide and wave for Santos Offshore Port, considering extreme events in a climate changing scenario, mathematical nodeling of wave situation for creation of protective hydrotechnical constructions in Port Kulevi, and the northerly summer wind off the West coast of the Iberian Peninsula.

The fifth chapter deals with inland shipping problems. The contents of the fifth chapter are partitioned into six subchapters: emergency group decision-making with multidivisional cooperation for inland maritime accident, the concept of emergency notification system for inland navigation, ship design optimization applied for urban regular transport on Guadalquivir River (Guadamar), ship emission study under traffic control in inland waterway network based on traffic simulation data, the using of risk to determination of safety navigation in inland waters, and inland water transport and its impact on seaports and seaport cities development.

In the sixth chapter there are described problems related to maritime pollution and environment protection: determination of marine pollution caused by ship operations using the Dematel method, joint-task force management in cross-border emergency response, managerial roles and structuring mechanisms in high complexity-high volatility environments, environmental risk assessment for the Aegean Sea, probabilistic meta-models evaluating accidental oil spill size from tankers, and negative impact of cruise tourism development on local community and the environment.

Seventh chapter concerns Vessel Traffic Service (VTS). The readers can find some information about improving safety of navigation by implementing VTS/VTMIS: experiences from Montenegro, evolutionary methods in the management of vessel traffic, supporting voice communication between navigator and VTS by visual solutions – exploring the use of the "route suggestion" functionality within VTS, and 4M Overturned Pyramid (MOP) model: case studies on Indonesian and Japanese Maritime Traffic Systems (MTS).

Each subchapter was reviewed at least by three independent reviewers. The Editor would like to express his gratitude to distinguished authors and reviewers of chapters for their great contribution for expected success of the publication. He congratulates the authors and reviewers for their excellent work.

Maritime Communications

Hidden Communication in the Terrestrial and Satellite Radiotelephone Channels of Maritime Mobile Services

O. Shyshkin & V. Koshevyy

Odessa National Maritime Academy, Ukraine

ABSTRACT: Additional imperceptible data transmission by means of audio watermarking (AW) in analog and digital voice channels of traditional and satellite radiocommunication system is presented. High AW performances of the designed system are obtained by: (a) realizing asynchronous embedding regime, in which the rigorous sliding researches of transmitting/receiving frame for data embedding/detecting are conducted, (b) watermark for every bit allocation in the narrow frequency bands while preserving the total spectral power of every band and (c) application of error correcting BCH code. Detection of watermarks doesn't require any synchronizing or marker data expenses. AW system allows embedding from 100 to 400 bps under watermark-to-signal ratio -20 dB and less and robustness against resampling, companding, additive noise, filtering, quantization, flat fading and MP3 compression. It may be better implemented in analog and digital radiotelephone channels for automatic identification of transmitting stations and hidden data transmissions to enhance maritime safety and security.

1 INTRODUCTION

Voice communication is a main mode of communication with all priorities in the satellite and traditional terrestrial maritime channels. At present radiotelephone systems in Very High Frequency (VHF), Medium Frequency (MF) and High Frequency (HF) bands use analog modulations. One of the basic drawbacks of maritime radiotelephony is the lack of automatic identification (AI) of radio transmissions. Meanwhile validity, reliability and timeliness of identification plays an important role in navigation, port control and search and rescue operations. In existing radiotelephony the responsibility of identification entirely lies on a navigator and his adequate actions. An absence, delay or incorrectly understood voice vessel identification directly affects the safety of navigation. Relation of identification to human factor provokes to anonymous transmissions that is of great harm to distress and safety VHF channel 16.

Importance of AI issue was expressed particularly in proposition [1]. It is known also "keying phenomenon", relating to push-to-transmit (PTT) button falling back in a VHF transceiver because of various reasons [2]. This phenomenon brings the communication blackout of other stations near the ship or very poor communication state in relevant

areas around the ship, which is especially harmful when the ship is in the area of Vessel Traffic Services (VTS).

To settle the identification problem Automatic Transmitter Identification System (ATIS) [3] is implemented on inland European waterways. In ATIS the identity of the vessel is sent digitally immediately after the ship's radio operator has finished talking and releases PTT button. But because of above mentioned "keying phenomenon" the problem cannot be solved in full.

An innovative approach to AI on the base of speech watermarking technology applied in air traffic control system is considered in article [4]. The authors proposed replacing some frames of voice signal, which correspond to non-voiced phonemes, by a white Gaussian data signal that carries the watermark information to solve the speech watermarking problem. The obvious limitation of proposed method is that only the speech is subject to watermarking and not audio as a whole. If the speech transmission is absent, the method fails.

In articles [5], [6] audio watermarking (AW) schemes for AI in the VHF maritime radiotelephony are proposed.

In this paper we present novel advances that considerably improve characteristics of AW system in the frame of a fundamental trade-off between watermark data payload, perceptual fidelity and robustness against channel attacks. Moreover, application of AW in analog channel is spread to digital telephone satellite channels. Such technology implements abilities of hidden and protected transmission of additional information for security purpose in a certain circumstances.

In the designed AW system improved capabilities are achieved by the following innovations:
- application of basic embedding algorithm with preserving the signal norm and minimizing distortions;
- concentration of vector coordinates in narrow frequency band and application of OFDM-like multichannel embedding algorithm;
- selection a suitable audio frame for embedding in asynchronous watermarking regime and
- using error correcting BCH codes.

2 EMBEDDING ALGORITHM

2.1 Basic concept of audio watermarking

Audio watermarking implies techniques that are used to imperceptible information convey by a certain embedding it into the virgin audio signal. AW doesn't call for any additional time or frequency channel resources. The basic principle of AW is shown in the Fig. 1. AW encoder is placed in the circuit break immediately after microphone (points 1 – 2) and before standard channel. Embedded data present additional information for identification or covert messaging. Under standard channel we mean a complex of communication facilities, including properly analog or digital radio channel. AW decoder is connected to audio output of receiver (point 3). The extracted data are directed for the proper utilization (displaying, printing, Electronic Chart Display and Information System, etc.).

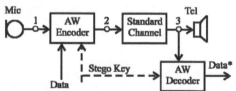

Figure 1. General scheme of watermarked communication

Both encoder and decoder may utilize some secret stego key to protect from unauthorized access to embedded information.

Such philosophy demands no alterations neither in operating radio equipment nor operational procedures. Besides it provides the full compatibility with the standard radios without AW function. The only side effects of AW function consist in:

1 appearing some limited distortions in watermarked signal that are practically inaudible and
2 time delay (64 msec) at transmission of audio signal that is quite tolerable for real time conversation.

Generally AW system has the following characteristics: (a) Perceptual fidelity of watermarks. As usual perceptual fidelity is evaluated by Watermark-to-Signal Ratio (WSR) in dB. Acceptable level of signal degradation according to proposals by the International Federation for the Phonographic Industry (IFPI) is taken equal to -20 dB or less [7]; (b) Data payload – the amount data that can be embedded into the host audio per time unit and measured in bits per second (bps); (c) Robustness – ability of watermarks to resist against channel attacks. It can be evaluated by the probability of successive watermark extraction.

These characteristics are mutually contradictive and an optimal trade-off should be chosen for the best capabilities of AW system.

In the present paper the AW process is considered as a classical communication problem with side information about channel state [8]. Channel state here represents the host audio signal, the influence of which on watermarking signal may be eliminated in the case it is known at the transmitter.

2.2 The main one channel AW algorithm

The core idea of AW algorithm lies in quantization of correlation coefficient between signal vector $x = (x_1, x_2, ..., x_L)$ and a certain random vector $u = (u_1, u_2, ..., u_L)$, and generation a watermarked vector $s = (s_1, s_2, ..., s_L)$ with the same norm. In the article [6] it is shown that the present algorithm provides minimization of introduced distortions $\|x - s\| = \min$ while keeping the signal norm $\|s\| = \|x\|$. Preserving norm is necessary for applying normalization and therefore providing watermarked signal immunity against amplitude scaling.

Random vector u plays as a stego key in Fig. 1 and is considered known at the transmission and receiving parties for secret communication.

The amplitude of correlation coefficient $\tilde{x} = (x, u) / \|x\| \|u\|$ undergoes the quantization process on the numerical lattice

$$\Lambda_m = \{\Delta \mathbf{N} + \frac{\Delta}{2} m + \delta\} \cap [0,1) \tag{1}$$

where Δ – quantization step, $\mathbf{N}$ – set of natural numbers, including zero, $m = \{0,1\}$ – embedded information bit, δ – lattice offset.

For instance, assuming $\Delta = 0.5$, $\delta = 0.1$, the numerical lattices become: $\Lambda_0 = \{0.1, 0.6\}$, $\Lambda_1 = \{0.35, 0.85\}$. Parameters Δ, δ may be included in the stego key as well.

Processing procedures are summarized in the Table 1. Random vector u is assumed uniformly distributed at the interval $[a,b]$.

Table 1. AW encoding

Step	Calculations
0	$u \sim U[a,b]$, $u' = \dfrac{u}{\|u\|}$
1	Correlation coefficient (or inner product) $$\tilde{x} = (x', u'), \quad x' = \frac{x}{\|x\|},$$ $\|x\| = \sqrt{x_1^2 + x_2^2 + ... + x_L^2}$ – Euclidian norm
2	Quantization $$A_s = \arg\min_{\lambda \in \Lambda_m} d\left(\lambda, |\tilde{x}|\right), \quad \varphi_s = \varphi_x$$ $\tilde{s} = A_s \exp(j\varphi_s)$, $\tilde{s} \in \mathbf{C}$, complex numbers
3	Transformation coefficients $$\alpha_{1,2} = \pm\sqrt{\frac{1 - \tilde{s}\,\tilde{s}^*}{1 - \tilde{x}\,\tilde{x}^*}}, \quad \beta = \tilde{s} - \alpha\tilde{x}$$
4	Watermarked vector $$s' = \alpha x' + \beta u', \quad s = s'\|x\|$$

2.3 OFDM-like multichannel algorithm

Orthogonal Frequency Division Multiplexing (OFDM) is well approved communication technology for various channels, especially with multipath propagation. Multipath propagation leads to intersymbol interference (ISI), which greatly corrupts the transmitting signal. In the maritime radiotelephony ISI appears due to band limited channel. Frequency response of audio channel is essentially nonuniform and limited by the frequencies (300 – 3000) kHz. As a result of this the different frequency signal components undergo the different amplitude variations. Hence normalization in the general frequency band as a measure against amplitude scaling is not efficient.

In the proposed algorithm signal vector is composed from the frequency coefficients of Fourier transform. It is fundamentally important that the every vector is composed from the adjacent coefficients so that they are placed in the narrow band and exposed to the same amplitude distortions. By means of normalization the influence of amplitude scaling is eliminated.

The vector forming principle is demonstrated by the Fig. 2. Firstly, the audio frame of N samples is transforming from time to frequency domain using Fast Fourier Transform (FFT). Then adjacent frequency coefficients compose sub channel vectors X_i, $i = \overline{1,n}$. The zeroth FFT coefficient is remaining free. Sub channel band comes to value of $\Delta f = F_s L / N$ Hz. For $F_s = 8$ kHz, $L = 2$ (as in the

Fig. 2) and $N = 512$ $\Delta f = 31.25$ Hz. One embedded bit occupies one sub channel per frame. The overall number of sub channel comes to n. After watermarking (WM) vector X_i is transforming into vectors S_i in such a way that their total power is preserved $\|S_i\| = \|X_i\|$. Only the powers of harmonics are varying within narrow band Δf while the total power balance and harmonic phases remain unchangeable, that provides watermarks inaudibility. Modified coefficients and samples are shown gray. At last, modified coefficients are transformed in the time domain by means inverse FFT (IFFT).

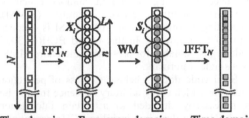

Time domain   Frequency domain   Time domain

Figure 2. Multichannel OFDM-like watermarking

2.4 Estimation of suitable frame

Information embedding is achieved by a certain distortions of the host signal. These distortions may appear depending less or more on current frame characteristic and embedded information. "Suitable" frame is considered by less distortion, ideally no distortion at all is needed. Therefore using the suitable frame for embedding is reasonable for minimizing introduced distortion.

As a measure of introduced distortion in i-th sub channel we assume the normalized distance between vectors X_i и S_i $d_i = \|X_i - S_i\| / \|X_i\|$. The total distortion is estimated as $D = \sum_{i=1}^{n} d_i$. Then the decision on watermarking the current frame may be accepted by comparing D with a certain threshold ρ_{enc}. If $D \leq \rho_{enc}$ then embedding is possible, otherwise giving up modification the current frame.

To stabilize payload and distortions to loudness of audio signal the floating threshold is proposed in the form

$$\rho_{enc} = h\Delta\frac{\sigma_{long}}{\sigma_{frame}}, \qquad (2)$$

where $h = 0.5...2$ – a certain threshold factor that specifies data payload; Δ – quantization step; σ_{long} – long term root mean square (rms) of the host signal in the interval of (3 – 5) sec; σ_{frame} – short term rms of the current frame.

The floating threshold implements asynchronous mode of watermarked frame transfer and provides approximately constant data payload subject to peak-factor of watermarking signal.

2.5 *Error-correction encoding*

To enhance interference immunity the Bose-Chaudhuri-Hocquenghem error-correcting code BCH(63,30,6) was used. It includes $k = 30$ information bits for total block length of $n = 63$ bits and is capable to correct up to $t = 6$ errors per block [9].

In the maritime mobile service with digital selective calling (DSC) the identification is produced by using of 9-digit number called as maritime mobile service identity (MMSI). According recommendation [10] MMSI is encoded into 30 bit sequence. In this regard BCH(63,30,6) code is just suitable for presenting the entire MMSI by a single watermark.

BCH code doesn't belong to class of perfection codes for which every arbitrary sequence from n bit is necessarily decoded to a certain information block. Having decoding inability BCH decoder sets a rejection decoding flag. The probability of decoding failure (or not correcting) one can obtain from the formula:

$$P_{nc} = \frac{2^n - 2^k (1 + \sum_{i=1}^{t} C_n^i)}{2^n},$$ (3)

where $C_n^i = \dfrac{n!}{i!(n-i)!}$ – binomial coefficients.

For BCH(63,30,6) code $P_{nc} = 0.9912$. Utilization the state of decoding flag provides decreasing of false-negative probability in the detecting algorithm.

3 WATERMARK DETECTION AND DECODING

Watermark detection is produced by sliding analysis of the received frame of length N samples in the frequency FFT domain. It is essential that detection and decoding processes do not require any starting synchronization and marking.

Detection algorithm may be summarized as follows. On every frame AW decoder tries detecting and decoding watermark. The same lattices Λ_0, Λ_1 and random sequence u as in AW encoder are used. The basic formulas for single frame processing are presented in the Table 2.

Table 2. Detection and decoding

Step	Calculations
1	Correlation coefficient of the received signal $\tilde{y}_i = (Y_i', u')$, $Y_i' = \dfrac{Y_i}{\|Y_i\|}$, $i = \overline{1, n}$
2	$d_i = \min\limits_{\lambda \in \Lambda_{10}} d(|\tilde{y}_i|, \lambda)$, $\Lambda_{10} = \Lambda_0 \bigcup \Lambda_1$
3	$D = \sum_{i=1}^{n} d_i$, if $D \begin{cases} \le \rho_{det} - \text{WM detected} \\ > \rho_{det} - \text{WM not detected} \end{cases}$
4	Estimation of received bit in в i-th channel $\hat{m}_i = \arg\min\limits_{\lambda \in \Lambda_m} d(|\tilde{y}_i|, \lambda)$

In every frequency sub channel correlation coefficient $\tilde{y}_i$ is computed. Then distances from number $|\tilde{y}_i|$ to the nearest lattice node in every sub channel and the total distance D along all channel are calculated. AW decoder approves decision on watermark detection by comparing D to a certain threshold ρ_{det} as shown on step 3 in the Table 2.

The value ρ_{det} must be set assuming the following conditions. Distance D presents a random value. Its distribution may be approximated by gamma distribution. In the case of watermark absence within analyzed frame D is expressed by a sum of n independent uniformly distributed values in the interval $[0, \Delta / 4]$ with equal means and root mean square. Mean and rms of D are evidently written taking into account $m_D = m_d n$ and $\sigma_D = \sigma_d \sqrt{n}$ by the next relations:

$$m_D = \frac{\Delta n}{8}, \ \sigma_D = \frac{\Delta \sqrt{3n}}{24}.$$

In the watermark presence deflections d_i are caused by external interferences. In the absence of any interferences absolute values of correlation coefficients are placed exactly in the lattice nodes $|\tilde{y}_i| \in \Lambda_{10}$. Distributions of D for cases watermark present/absent are shown in Fig. 3.

In watermark application for automatic identification it is necessary to minimize false-negative probability (shown black). Herewith false-positive errors (shading) are not critical because the same watermark is permanently transmitted in the form of MMSI.

We set decision boundary to $\rho_{det} = m_D - 5\sigma_D$. On account of central limit theorem distribution of D in the watermark presence is satisfactory approximated by normal law $D \sim N(m_D, \sigma_D)$. Therefore false-negative probability may be estimated through error function that gives $P_{FN} \approx 2 \times 10^{-7}$ for parameters $\Delta = 1, n = 63$.

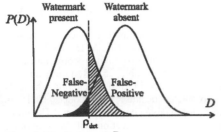

Figure 3. Distributions of D

$$WSR = 10\lg\frac{\sum_{j=1}^{T_{SP}}\sum_{i=1}^{N_{SP}}(X_{ij}-\hat{S}_{ij})^2}{\sum_{j=1}^{T_{SP}}\sum_{i=1}^{N_{SP}}X_{ij}^2},$$

where X_{ij}, S_{ij} – amplitude of spectrograms of dimension [Time, Frequency] $= [T_{SP}, N_{SP}]$.

Table 3. Data payload and fidelity versus step size and threshold factor

Step size	Threshold factor	Data payload, bps	Fidelity, dB
0.25	0.5	102	-49.04
	1	136	-41.36
	2	255	-31.57
	4	356	-25.12
0.5	0.5	102	-42.16
	1	136	-34.82
	2	272	-24.56
	4	365	-18.87
1	0.5	102	-36.56
	1	144	-28.5
	2	297	-18.71
	4	399	-12.17
2	0.5	110	-28.97
	1	178	-20.19
	2	322	-11.85
	4	450	-6.05

BCH decoder decreases the total false-negative probability on account of rejection of decoding with the previously computed probability $P_{nc} = 0.9912$. Hence the resulting false-negative probability assuming formula (3) comes to the value

$$P_{false} = P_{FN}(1-P_{nc}) \approx 2\times10^{-9}.$$

4 RESULTS OF SIMULATION AND EXPERIMENT

Proposed AW algorithms was thoroughly investigated in conditions of varying independent parameters and influence the whole spectrum of interferences which can take place in the real channel. Computer simulation was carried out in MatLab environment for speech wav-files with sampling frequency $F_s = 8$ kHz.

Simulation results that refer to data payload and fidelity subjected to step size Δ and threshold factor h from formula (2) are shown in the Table 3.

Data payload is considered as the total number of pure bits related to the file duration and calculated by the formula:

$$DP = \frac{N_{TX}k}{l_{wav}F_s},$$

where N_{TX} – number of transmitted watermarked frames; l_{wav} – length of the tested file in number of samples per file; k – number of information bits per block of BCH code.

Fidelity was evaluated as Watermark-to-Signal Ratio in the frequency domain:

Spectrograms were computed in accordance of ITU Recommendation [11] by mapping the time signals to the time-frequency domain using a short-term FFT with a Hann window of size 256 samples.

Considering an acceptable level of speech signal degradation under proposals to WSR from [7] of magnitude -20 dB or less the independent parameters Δ and h may be chosen from Table 3 for practical implementation.

Other AW system parameters were taken as follows: vectors length $L = 2$, FFT dimension $N = 512$, number of embedded bits per frame $n = 63$, parameters of BCH code $(n,k,t) = (63,30,6)$.

AW robustness was evaluated as true-positive probability in the form of relation of received and correctly decoded watermarked frames to the total number of transmitted frames $P_{cor} = N_{RX}/N_{TX}$. Influence of channel attacks on is presented in the Table 4.

Table 4. Probability of frame correct detection versus step size subjected to various channel attacks

Step size Δ	Companding, μ-law	Filtering	Additive noise, dB -40	-30	-20	Quantization by 2^q levels $q=8$	$q=7$	$q=6$	Clipping 50%**	Flat fading	Complex interf.*
0.125	0.76	0.13	0.23	0	0	0.5	0.27	0	0.9 / 1	0.13	0
0.25	1	0.26	0.6	0.1	0	0.63	0.56	0.23	0.93 / 1	0.37	0
0.5	1	0.72	0.72	0.37	0	0.84	0.75	0.62	0.94 / 1	0.5	0.03
1	1	0.94	0.86	0.71	0.14	1	0.83	0.66	0.97 / 1	1	0.43
2	1	1	1	0.74	0.37	1	0.97	0.74	1	1	0.68

* Complex interference: Filtering + additive noise -30 dB + flat fading
** For the thresholds h = 2 / 0.5 accordingly

Specifications of variety channel attacks, the watermarked signal is subjected to, are shown below:

1. Resampling: The watermarked signal, originally sampled at $F_s = 8\,\text{kHz}$, was interpolated at 16 kHz, shifted by one sample and then restored back at 8 kHz. Appropriate procedure `interp(s,4)`

2. MP3 compression 9.6 kbps: The watermarked signal was compressed at the bitrate of 9.6 kbps and then decompressed back to wav format.

3. Compressing – Expanding (companding): The watermarked signal undergoes standard companding procedure according μ-law. `lin2mu(s), mu2lin(y)`

4. Filtering: The watermarked signal was passed through the passband Butterworth filter of order 2 and cutting frequencies 400 Hz and 2 kHz. `butter(1,[.1 0.5],filter(bb,aa,s)`

5. Additive White Gaussian Noise (AWGN): White Gaussian noise was added at the level measured for the whole signal fragment. `awgn(s,SNR,'measured')`

6. Quantization: The watermarked signal was quantized by Q levels. `1/Q*round(s*Q)`

7. Clipping: The watermarked signal was cut with saturation at levels $\pm 0.5\max|s|$

8. Flat fading: The watermarked signal undergoes sinusoidal amplitude modulation with a modulation frequency of 3 Hz and a modulation depth of 0.5.

9. Complex interference: A simultaneous influence of attacks (4) + (5) -30 dB + (8)

For channel attacks (1), (2) no robustness degradation was registered.

In all experiments no false-negative errors were registered.

Reasonably high level of false-positive errors is not meaningful for watermarking in AI application, because of repeatedly transmitted same watermark in the view of MMSI. Single true-positive watermark detection within (1 – 3) sec to consider identification successful is sufficient.

Signalograms that illustrate asynchronous watermarking and influence channel attacks on its detection are shown in Fig. 4 for parameters $\Delta = 1$, $h = 1$. Peak-factor for the tested voice signal comes to 12.7 dB. The total number of watermarked frames amounted to 17, at that the data payload amounted to 63 watermarked (including 30 pure) bits per frame. An evident tendency of watermarked frames is placing them within pauses (intervals 1.8 – 2.1 sec and 3.2 – 3.5 sec) and unvoiced phonemes (all the rest) as shown in Fig. 4 a). Companding attack No 3 from attack specification destroyed no watermarks and all 17 data packets were detected and correctly decoded. For more complicated situation against complex attack No 9: Filtering + additive noise (-30 dB) + flat fading only 5 data blocks were restored.

For AI implementation such outcome is quite accessible.

Recommended parameters and related characteristics separately for VHF analog R/T identification and satellite hidden communication are shown in the Table 5. Requirements to watermark robustness in digital satellite channels may be reduced because the codec itself is responsible for stable communication. Therefore proposed hidden communication will be supported as long as the main link is possible. In particular the value of step size Δ and correcting strength t of BCH may be decreased in favor of data payload.

Table 5. Recommended independent parameters and related characteristics

	VHF analog R/T identification	Satellite hidden communication
Independent parameters:		
Step size Δ	1; 2	0.25; 0.5
Threshold factor h	0.5; 1	2; 4
FFT dimension N	512	512
BCH code (n,k,t)	(63,30,6)	(63,30,6); (63,36,5); (63,39,4)
Vector length L	2	2
Related characteristics:		
Data payload, bps	102 … 178	255* … 365*
Fidelity, dB	-36 … -20	-25 … -19
Delay, msec	64	64

* For BCH code (63,30,6)

5 CONCLUSION

Presented philosophy of watermarked radiotelephone terrestrial and satellite maritime communication gives new possibilities of enhancing navigational safety and security. The comprehensive design of AW system makes possible its application in analog and digital communication channel irrespective of radio installation already involved. A single hardware upgrading includes two electronic chips. One of them is placed directly into the handset at the transmitting part and another one is switched to standard audio output at the receiving part. No other invasions in operating radios and operational procedures are needed. Besides, the designed system is fully compatible with the installation without AW function.

AW for VHF radiotelephone identification is exclusively appropriate for smart and addressed ship-to-ship and ship-to-shore with all priorities radiotelephony, especially taking into account obligatory watch on channel 16 [12]. It is also helpful to monitor intentional and accidental violations of radiocommunication.

AW application in digital satellite voice communication gives additional chances for hidden data communication for the sake of security.

The proposed AW system has rather easy implementation because embedding/detection algorithms are based on standard FFT and BCH encoding/decoding procedures. All calculations are also suitable to vectorization.

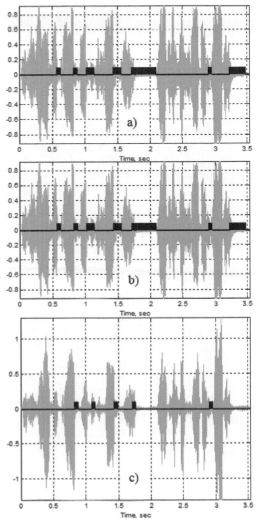

Figure 4. Watermarked audio signals and disposition of watermarked frames (black): a) before channel attacks and after channel attacks b) μ-law companding c) complex interference: Filtering + additive noise -30 dB + flat fading

REFERENCE

[1] Automatic Transmission of the Identification of the Radiotelephone Station. COMSAR 16/7, Submitted by Poland, December 2011.
[2] Proposals to Amend the Performance Standards for Shipborne VHF Radiotelephone Facilities. COMSAR 14/7/5, Submitted by the Republic of Korea, December 2009.
[3] ETSI EN 300698-1. Radio telephone transmitters and receivers for the maritime mobile service operating in the VHF bands used on inland waterways; Part 1: Technical characteristics and methods of measurement. 50 p.
[4] K. Hofbauer, G. Kubin, W.Kleijn "Speech Watermarking for Analog Flat-Fading Bandpass Channels", *IEEE Transactions on Audio, Speech and Language Processing,* November 2008, Vol. 17, No. 8, pp. 1624 – 1637
[5] Shishkin A. V. "Identification of radiotelephony transmissions in VHF band of maritime radio communications", *Radioelectronics and Communication Systems,* November 2012, Vol. 55, No. 11, pp. 482-489
[6] Shishkin A. V., Koshevoy V.M. "Stealthy Information Transmission in the Terrestrial GMDSS Radiotelephone Communication", TransNav – *International Journal on Marine Navigation and Safety of Sea Transportation,* Vol. 7, No. 4, pp. 541 – 548, 2013.
[7] Vivekananda Bhat K, Indranil Sengupta, Abhijit Das "An adaptive audio watermarking based on the singular value decomposition in the wavelet domain". *Digital Signal Processing,* Volume 20, Issue 6, December 2010, Pages 1547-1558. Available at http://202.114.89.42/resource/pdf/5566.pdf
[8] Gel'fand S. I., Pinsker M. S. "Coding for channel with random parameters" / S. I. Gel'fand, M. S. Pinsker // *Problems of Control and Information Theory.* – 1980. – Vol. 9 (1), pp. 19 – 31.
[9] Morelos-Zaragoza R.H. The art of error correcting codes. John Wiley, 2006, 263 p.
[10] Operational procedures for the use of digital selective-calling equipment in the maritime mobile service. RECOMMENDATION ITU-R M.541.
[11] *ITU-T Recommendation P.862: Perceptual evaluation of speech quality (PESQ),* International Telecommunication Union, 02/2001.
[12] Brzoska S. "Advantages of Preservation of Obligatory Voice Communication on the VHF Radio Channel 16", TransNav – *International Journal on Marine Navigation and Safety of Sea Transportation,* Vol. 4, No. 2, pp. 137 – 141, 2010.

Introduction to Inmarsat GEO Space and Ground Segments

D.S. Ilcev
Durban University of Technology (DUT), South Africa

ABSTRACT: This article presents retrospective of development Inmarsat space and ground segments as a Geostationary Earth Orbit (GEO) satellite system with significant contribution to the modern maritime and other mobile satellite applications. The technical parameters and comparison of previous and new generations of Inmarsat GEO satellite constellations with their advantages and disadvantages are described. The choice and upgrading of a particular satellite configuration depends mainly on its mission objectives, development trends of satellite mobile service and applications, modernization of onboard equipment and characteristics of the satellite payloads. Each Inmarsat satellite provides global beam spacecraft antenna for global coverage covering approximately one-third of the Earth's surface (including land and sea) and spot beam coverages from an GEO nearly 36,000 kilometers above the Equator. In this orbit each satellite moves at exactly the same rate as the Earth rotates, so remaining in the same position relative to the Earth's surface. Fife generations of Inmarsat GEO satellite constellations, Inmarsat link budget, ground segment, ground networks, coordination and different control centres are introduced.

1 INTRODUCTION

Inmarsat was established in 1979 known as the International Maritime Satellite Organization (Inmarsat) not-profit international organization, set up at the behest of the International Maritime Organization (IMO) and an UN body, for the purpose of establishing a satellite communications network for the commercial maritime community, ship management including, emergency, distress and safety applications. It began trading in 1982 and was used the acronym "Inmarsat", which operates and maintains the Inmarsat Ground Network (IGN) constellation of nine GEO satellites and many Mobile Earth Station (MES) and Land Earth Station (LES) terminals. The company has over 33 years experience in designing, implementing and operating innovative satellite communication networks and Mobile Satellite Communication (MSC) configurations.

Inmarsat delivers its services through an IGN of about 260 partners in over 80 countries including some of the world's largest telecommunications companies and offers a portfolio of visionary Mobile Satellite Service including Fixed Satellite Service (FSS) via portable units for governments and enterprises requiring reliable voice, data and video communication on land, at sea or in the air, with over 98% of the Earth's surface (except the polar regions). Inmarsat also provides service in regions of developed and developing countries were there is no available service of Terrestrial Telecommunication Networks (TTN).

The intent was to create a self-financing body, which would improve communications and other safety service according to the Safety of Life at Sea (SOLAS) Convention and IMO regulations and recommendations.

The name was changed to International Mobile Satellite Organization (IMSO) when it began to provide services to road, rails, aircraft and portable users, but the acronym "Inmarsat" was kept. When the organization was converted into a private company in 1999, the business was split into two parts: the bulk of the organization was converted into the commercial company, Inmarsat Plc, and a small group became the regulatory body, IMSO. In 2005 Apax Partners and Permira bought shares and the Company was also first listed on the London Stock Exchange in that year. In 2008 the USA-based Harbinger Capital owned 28% of the Company and one year later Inmarsat completed the acquisition of a 19% stake in SkyWave manufacturer.

Thus, Inmarsat was the world's first international and nongovernmental Global Mobile Satellite Communication (GMSC) operator offering a mature range of modern communications services to maritime, land, aeronautical and semi-fixed users, such as Broadband Global Area Network (BGAN). The Inmarsat satellites are controlled from Inmarsat's headquarters in London, which is also home to Inmarsat Ventures and small IGO created to supervise the company's public service for the maritime community, known as Global Maritime Distress and Safety System (GMDSS), implemented with IMO, and aviation Air Traffic Control ATC/CNS implemented with ICAO. The keystone of the Inmarsat strategy is the current I-4 advanced satellite constellation, which from 2004 started to support the Inmarsat BGAN, mobile satellite data communications at up to 432 Kb/s for Internet access, mobile multimedia via satellite and many other advanced applications [1, 2, 3].

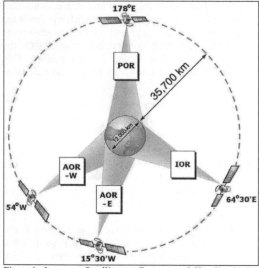
Figure 1. Inmarsat Satellites – **Courtesy of Handbook: by Inmarsat** [1]

2 INMARSAT SPACE SEGMENT

For the first decade of Inmarsat's operations, the space constellation has been leased from Comsat (series of three Marisat satellites F1, F2 and F3), from ESA (series of two Marecs satellites A and B2) and from Intelsat (series of three Intelsat V-MCS A, B and D). These satellites were initially configured in three ocean regions: Atlantic Ocean region (AOR), Indian Ocean Region (IOR) and Pacific Ocean region (POR), each containing an operational satellite and at least one spare in-orbit. This satellite constellation is known as the first generation of the Inmarsat network. Inmarsat was not responsible for TT&C, but operations were controlled by Inmarsat Network Control Centre (NCC) in London.

2.1 Second and Third Generation of Inmarsat Satellite Constellation

The previous space segment of Inmarsat network consisted in total four GEO I-3 and I-2 satellites.

The main global beam of each satellite provides overlapping coverage of the whole surface of the Earth apart from the Poles and in this way it is possible to extend the reach of terrestrial wired and cellular networks to almost anywhere on Earth. Thus, a call from an Inmarsat Mobile Earth Station (MES) goes directly to the satellite overhead, which routes it back down to a Gateway on the ground called a Land Earth Station (LES). From there, the calls/messages are passed into the public phone of Terrestrial Telecommunication Network (TTN), data and Integrated Services Digital Network (ISDN) networks.

The Inmarsat-3 satellites are backed up by a five Inmarsat-3 and four Inmarsat-2. A key advantage of the Inmarsat-3 over their predecessors is their ability to generate a number of spot-beams as well as large global beams. Spot-beams concentrate extra power in areas of high demand and making it possible to supply standard services to smaller and simpler mobile terminals. In Table 1 is shown a list of the Inmarsat spacecraft of 1998/99 deployed for global coverage except the Poles.

Table 1. Ocean Regions and Satellite Longitudes

Satellite Status	Atlantic (West and East), Indian and Pacific Ocean Regions			
	Atlantic/AOR-W	Atlantic/AOR-E	Indian/IOR	Pacific/POR
1st Operational Position	Inmarsat-2 F4 54° W	Inmarsat-3 F2 15.5° W	Inmarsat-3 F1 63.9° E	Inmarsat-3 F3 178.1° E
2nd Operational Position	Inmarsat-3 F4 54° W	– –	– –	– –
In-Orbit Spare Position	Inmarsat-2 F2 55° W	–	Inmarsat-2 F3 65° E	Inmarsat-2 F1 179° E

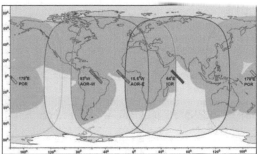

Figure 2. Inmarsat-3 Constellation of four Satellite Ocean Regions – Courtesy of Prospect: by Inmarsat [4]

Moreover, the information list of current Inmarsat GEO satellite constellations is presented in the Inmarsat Maritime Communications Handbook (Issue 4 of 2002) and gives the same 1st positions for operational satellites but in the event of a satellite failure the values change as follows: AOR-W at 98° W; AOR-E at 25° E; POR at 179° E; IOR (For Inmarsat-A, B, C and M) at 109° E and IOR (For Inmarsat-C, mini-M and Fleet) at 25° E. The Inmarsat organization bases its Earth coverage on a constellation of four prime GEO satellites covering four ocean regions with four overlapping, illustrated in Figure 1.

The coverage area for any satellite is defined as the area on the Earth's surface (sea, land or air), within which Line-of-Sight (LOS) communication can be made with the visible satellite. That means, if an Inmarsat MES terminal is located anywhere within a particular satellite coverage area and an antenna of MES is directed towards that satellite, it will be possible to communicate through that satellite with any LES that is also pointed at the same particular satellite.

In Figure 2 is illustrated the footprints of global beam coverages projected onto the surface of the Earth from the four Inmarsat-3 GEO satellites in use for all mobile application. Thus, it should be noted that the recommended limit of latitudinal beam coverage is within the area between 75° North and South [2, 3, 4, 5].

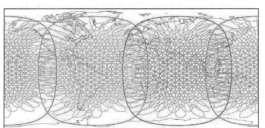

Figure 3. Inmarsat I-4 Spot-Beam Coverage for Digital Mobile Broadband – Courtesy of Prospect: by Inmarsat [6]

The previous map presented in Figure 2 depicts coverage following completion of the satellite repositioning process. In the AOR-W region, Classic services were delivered via the I-4 satellite. Now the I-4 satellite over the Americas is activated, Classic services traffic in AOR-W has been transferred back to an I-3 satellite. As a result, there is reduced spot beam coverage in AOR-W, but Inmarsat will make best commercial efforts to increase the I-3 spot beam coverage so that it exceeds the original I-3 coverage.

In Figure 3 is shown the global narrow spot-beam coverage of each Inmarsat-4 satellite for three ocean regions. Thus, I-4 satellites are covering three oceans with overlappings, which narrow beams vary in size with tend to be several hundred kilometers across. In Table 2 are presented the characteristics of 3 Inmarsat's generations.

Table 2. Comparison of Three Generations Inmarsat Satellites

Inmarsat Satellites	Inmarsat-2	Inmarsat-3	Inmarsat-4
No. Satellites	4	5	2 + 1
Coverage	1 Global Beam	7 Wide Spots 1 Global Beam	228 Narrow Spots 19 Wide Spots 1 Global Beam
Mobile Link EIRP	39 dBW	49 dBW	67 dBW
Channelisation	4 channels (4.5 to 7.3 MHz)	46 channels (0.9 to 2.2 MHz)	588 channels (EOL) (200 KHz)
S/C Dry Mass	700 kg	1000 kg	3340 kg
Solar Array Span	14.5m	20.7m	45m
Voice (4.8kbps)	250	1000	18000
M4 (64 kbps)	N/A	200	2250
PMC (384 kbps)	N/A	N/A	588

Therefore, I-4 satellites are covering Maritime service such as FleetPhone, Fleet 33/55/77 and FleetBroadband, Land service such as BGAN, IsatPhone, LandPhone, GAN and Mini-M, while for Aeronautical service are covering Aero-C, I, M, mini-M, Swift64 and SwiftBroadband. As a large percentage of the Earth's MSC requirements lies within this roaming area, the system is considered to possess a global coverage pattern. As well as a "global" beam covering a complete hemisphere, each satellite generates up to seven spot beams designed to increase the amount of communications capacity available in areas of high demand. Namely, the MSC services are delivered to ships, land vehicles (road and rail), aircraft, transportable and semi-fixed mobile terminals through the spot beams of the three Inmarsat-4 satellites [3, 5, 6, 7].

In Figure 4 is illustrated Inmarsat I-4 global-beam network coverage following completion of the satellite-repositioning program, which ended on 24 February 2009. In such a way, to reflect the geographic locations covered by the satellites, Inmarsat refers to its three I-4 satellite regions as follows: I-4 Americas at position 98° W, I-4 EMEA (Europe, Middle East, and Africa) at 25° E and I-4 Asia-Pacific at 143.5° E.

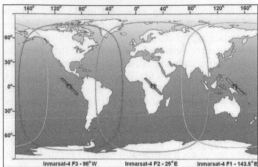

Figure 4. Inmarsat I-4 Global Coverage for GMSC – Courtesy of Prospect: by Inmarsat [6]

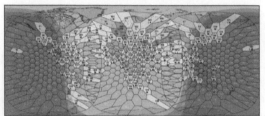

Figure 5. Inmarsat I-5 Global Spot-Beam for Digital Mobile Broadband – Courtesy of Prospect: by Inmarsat [8]

In Figure 4 is illustrated Inmarsat I-4 global-beam network coverage following completion of the satellite-repositioning program, which ended on 24 February 2009. In such a way, to reflect the geographic locations covered by the satellites, Inmarsat refers to its three I-4 satellite regions as follows: I-4 Americas at position 98° W, I-4 EMEA (Europe, Middle East, and Africa) at 25° E and I-4 Asia-Pacific at 143.5° E.

In this sense, with its own three GEO satellite constellation the Inmarsat network provides the GMSC service, which in Article-3 of the Inmarsat convention is stated as: "The purpose of Inmarsat is to make provision for the space segment necessary for improving mobile maritime and, as practicable, aeronautical and land communication and as well as MSC service on other waters not part of the marine environment, thereby assisting in improving GMSC for distress and safety of life, communications for Air Traffic Service (ATS), the efficiency and management of transportation by sea, on land and in air and other mobile public correspondence services and radio determination capabilities".

In such a way, the present spot-beam coverage concentrates extra power in areas of high demand as well as making it possible to supply standard service to smaller, simpler and less powerful mobile terminals. At this point, a key advantage of the Inmarsat-4 satellites over their predecessors is their ability to generate a great number of spot-beams as well as single large global beams.

The coverage map of new forthcoming fifth generation of Inmarsat I-5 spacecraft is shown in Figure 5, known as Global Xpress new Inmarsat I-5 satellite constellation, which offers seamless global coverage with mobile broadband speeds of up to 50 MB/s for users in the government, maritime, land, energy and aeronautical sectors. Inmarsat is investing an estimated amount of 1.2 billion US$ in the Global Xpress program, which includes launch costs [5, 7, 8, 9].

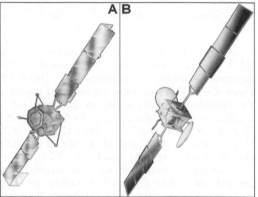

Figure 6. I-2 and I-3 Spacecraft – Courtesy of Book: by Ilcev [7]

2.2 Second Generation of Inmarsat-2 Satellites

Inmarsat operates a total of four 2nd generation Inmarsat-2 birds launched in 1990/92 with a capacity equivalent to about 250 Inmarsat-A voice circuits. In fact, these four Inmarsat-2 satellites were built according to Inmarsat specifications by an international consortium headed by the firm known as Space and Communications Division of British Aerospace (now Matra Marconi Space).

The Inmarsat-2 satellite design is based on the Eurostar three-axis-stabilized satellite platform with 10 years lifetime, shown in Figure 6 (A). At launch, each satellite weighed is 1,300 kg and had an initial in-orbit mass of 800 kg and 1,200 watts of available power. Each communications payload provides two satellite transponders, which realize outbound (C to L-band) and inbound (L to C-band) links with MES in the 6.4/1.5 and 1.6/3.6 GHz bands, respectively. So, the L-band EIRP is a minimum of 39 dBW, while G/T for L-band is about –6 dB/K for global satellite coverage. Each satellite's global beam covers roughly one-third of the Earth's surface.

2.3 Third Generation of Inmarsat-3 Satellites

The Lockheed Martin Astro Space is US-based Company that built the new spacecraft bus for next Inmarsat-3 generation, which is based on the GE Astro Space Series 4000, 2.5 m high and with a 3.2

radial envelope centered on a thrust cone. Matra Marconi Space built the communications payload, spacecraft antenna systems, repeater and other communications electronics.

Thus, payload and solar arrays are mounted on N and S-facing panels. The L-band Receiver (Rx) and Transmitter (Tx) reflector arrays are mounted on E and W panels and are fed by an array with cup-shaped elements. Furthermore, the navigation antenna is located on the Earth-facing panel.

Table 3. Orbital Parameters of Inmarsat-3 Spacecraft

Background	
Owner/Operator:	Inmarsat Organization
Present status:	Operational
Orbital location:	64o East
Altitude:	About 36,000 km
Type of orbit:	Inclined GEO
Inclination angle:	±2.7o
Number of satellites:	1 operational & 1 spare
Number of spot beams:	5
Coverage:	IOR
Additional information: Other Ocean regions POR, AOR-W and AOR-E have 1 operational and 1 or 2 spare Inmarsat-2 or 3 satellites	

Spacecraft	
Name of satellite:	Inmarsat-3 F1
Launch date:	4 April, 1996
Launch vehicle:	Atlas IIA
Typical users:	Maritime, Land and Aeronautical
Cost/Lease information:	Nil
Prime contractors:	Lockheed Martin
Other contractors:	Matra Marconi
Type of satellite: GE Astro Series 4000 Stabilization: 3-Axis	
Design lifetime:	13 years
Mass in orbit:	860 kg
Launch weight:	2,066 kg
Dimensions deployed:	2 x 7 x 20 m,
Electric power:	2.8 kW
SSPA power:	C-band 1 @ 15 W; L-band 1 @ 490 W

Communications Payload	
Frequency bands:	
a) Communications:	
L-band (Service Link)	1.6/1.5 GHz
C-band (Feeder Link)	6.4/3.6 GHz
b) Navigation:	L1 1.5 & C-band 6.4/3.6 GHz
Multiple access:	TDM/TDMA
Modulation:	BPSK, O-QPSK, FEC
Transponder type:	L-C/C-L & L1-C-band
Number of transponders:	1 L & C-band
Channel bit rate:	From 600 b/s to 24 Kb/s
Channel capacity:	About 2000 voice circuits
Channel bandwidth: L-C/C-L 34 MHz; Navigation 2.2 MHz; L-L 1 MHz; C-C 9 MHz	
Channel polarization: L-band RHCP; C-band LHCP & RHCP	
EIRP:	L-band Global 44 dBW & Spot 48 dBW; C-band 27.5 dBW
G/T:	L-band Global –6.5 & Spot –2.5 dB/K

The tremendous advantage of the Inmarsat-3 over previous satellites is their ability to concentrate power on particular areas of high traffic within the footprint. Each satellite utilizes a maximum of seven spot beams and one global beam. In such a way, the number of spot beams will be chosen according to traffic demands. In addition, these new satellites can re-use portions of the L-band frequency for non-adjacent spot beams coverage, effectively doubling the capacity of the satellite. Each satellite weighs about 2,066 kg at launch, compared to 1,300 kg for an Inmarsat-2 satellite. However, the Inmarsat I-3 satellite produces up to 48 dBW of EIRP, a measure of how much signal strength a satellite can concentrate on its service area coverage. All other parameters of Inmarsat-3 F1 satellite are presented in Table 2 and 3.

The main mission spacecraft onboard payloads of Inmarsat-3 are the communication transponders on both C and L-band and are the frequency-translated by the transponders for the downlink within the same band, shown in Figure 6 (B). The uplink signal is rebroadcast to users within each ocean coverage and spot beam footprints. As a secondary payload, Inmarsat-3 has the navigation transponders that provide the WAAS capability of Augmentation CNS capabilities. Two frequencies, L1 on 1.57542 GHz and C-band on 3.6 GHz, are used to allow correction of ionospheric delay.

The WAAS signal will be broadcast to the users at L1 frequency, the same as GPS or GLONASS. For additional integrity purposes and for checking the data received by the satellite and data being broadcast to mobile users is also down linked back to the control site in the C-band. In fact, the 6.4 GHz L-band repeaters is power-limited to ensure that the navigation signal can never interfere with the GPS or GLONASS signals.

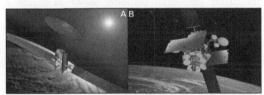

Figure 7. Inmarsat-4 and Forthcoming Inmarsat-5 Spacecraft – Courtesy of Book: by Inmarsat [7]

2.4 Fourth Generation of Inmarsat-4 Satellites

Responding to the growing demands from all corporate mobile satellite users of high-speed Internet access and multimedia connectivity, Inmarsat built fourth generation of satellites as a Gateway for the new satellite mobile broadband network. Inmarsat has awarded European Astrium a 700 million US$ contract to build three Inmarsat I-4 satellites, which will support the new BGAN, shown in Figure 7 (A). The BGAN solution was introduced in 2004 to deliver Internet and Intranet content and all multimedia solutions, video on demand, videoconferencing, fax, E-mail, phone and mobile office LAN access at speeds up to 432 Kb/s worldwide and it will be compatible with third generation (3G/4G) cellular systems.

Three Inmarsat I-4 F1 satellite launched March 11th 2005, I-4 F2 launched November 8th 2005, and I-4 F3 anticipated launch in 2007 (POR) subject to business case and successful service using IOR and AOR. All three spacecraft have the advanced technology to reduce service costs by 75%, compared to existing Inmarsat-M4 charges. They will be 100 times more powerful than the present generation and BGAN will provide at least 10 times as much capacity as today's network. The spacecraft will be built in the UK; the bus will be assembled in Stevenage and the payload in Portsmouth. The two sections will be united in France, together with the US-built antenna and German-built solar arrays.

The Inmarsat BGAN was later used as an example in order to be designed new maritime broadband known as a FleetBroadband and new aeronautical broadband known as a SwiftBroadband.

However, Inmarsat GMSC service using current Inmarsat I-4 spacecraft is trusted with Maritime and Aeronautical commercial distress and safety services through compliance with IMO and ICAO requirements overseen by IMSO [5, 7, 10, 11].

2.5 *Fifth Generation of Inmarsat-5 Satellites*

Inmarsat has contracted Boeing, the US aerospace company to build a new Inmarsat-5 constellation, for 1.2 billion US$, to provide global broadband network called Inmarsat Global Xpress. Boeing will construct three Inmarsat-5 (I-5) satellites based on its 702HP spacecraft platform, shown in Figure 7 (B). The first satellite was scheduled for 2013, with full global coverage expected by the end of 2014. Thus, this spacecraft will break new ground by transmitting in a portion of the radio spectrum never before utilized by the commercial operator of a global satellite system, which will be the extremely high radio frequency Ka-band.

Each I-5 beard will carry a payload of 89 Ka-band coverage beams capable of flexing capacity across the globe and enabling Inmarsat network to adapt to shifting subscriber usage patterns over their projected lifetime of 15 years. The Inmarsat future Global Xpress network will take advantage of the additional bandwidth available in the Ka-band to offer download rates up to 50 Mb/s and upload speeds of 5Mb/s from all mobile user terminals as small as 60 centimeters. New I-5 spacecraft will operate independently from the L-band satellites offering complementary services for a wide range of mobile and fixed solutions. The new satellites will join Inmarsat's fleet of 11 GEO satellites in total that provide a wide range of voice, data and video (VDV) transmissions through an established global distributors and service providers.

Leveraging Boeing's big expertise in government applications I-5 satellites will provide an array of secure voice and high-speed communications applications between land, sea and air services and multinational coalitions [5, 7, 12, 13].

3 INMARSAT MSC LINK BUDGET

The Inmarsat satellites link budget analysis forms the cornerstone of the space system design. Link budgets are performed in order to analyze the critical factors in the transmission chain and to optimize the performance characteristics, such as transmission power, bit rate and so on, in order to ensure that a given target quality of service can be achieved. The sample a of maritime link budget for the link MES-to-GEO at 1.64 GHz and from the LES-to-GEO at 6.42 GHz and in the reverse direction for the link GEO-to-LES at 4.2 GHz and GEO-to-MES at 1.5 GHz is shown in Table 4.

Table 4. Maritime Mobile Link Budget

Parameter	MES-to-GEO	LES-to-GEO
MES/LES EIRP Carrier	36 dBW	58 dBW
Absorption & FSL at 1.6/ 6.42 GHz of 5° Elevation	189.4 dB	201.3 dB
Satellite Rx G/T	–13.0 dBK	–14.0 dBK
Uplink C/N$_o$	62.2 dBHz	17.3 dBHz
Total Satellite EIRP	16.0 dBW	33.0 dBW
Intermodulation Noise Power Ratio	15.0 dB	9.0 dB
Transponder Bandwidth (7.5 MHz)	68.8 dBHz	68.8 dBHz
Satellite EIRP/Carrier	–5.0 dBW	18.0 dBW
Satellite C/N$_o$	62.8 dBHz	62.8 dBHz

Parameter	GEO-to-LES	GEO-to-MES
GEO Satellite EIRP Carrier	–5.0 dBW	18.0 dBW
Atmospheric & FSL at 4.2/ 1.5 GHz of 5° Elevation	197.6 dB	188.9 dB
LES/SES Rx G/T	32.0 dBK	–4.0 dBK
Downlink C/N$_o$	58.0 dBHz	53.7 dBHz
Satellite Link C/N$_o$ (Up/ Intermodulation/Downlink)	55.7 dBHz	53.1dBHz
Intersystem Interference C/I$_o$	64.4 dBHz	61.8 dBHz
Overall C/N$_o$	55.2 dBHz	52.6 dBHz
Required C/N$_o$	52.5 dBHz	52.5 dBHz
Margin	2.7 dB	0.1 dB

The MES terminal is considered as any Inmarsat installation, which G/T is –4 dB/K. The up and downlink budget for C/N$_o$ are fairly standard except can be noted that: C/N$_o$ = EIRP – FSL – L$_f$ + G/T – K, where FSL = Free Space Loss; L$_f$ = fixed losses made up of antenna misalignment of the receiver; and K = Boltzmann Constant in logarithmic form (–228.6 dBW/Hz/K).

The intermodulation on board the spacecraft is given as follows: the total intermodulation noise in the 7.5 MHz (68.8 dBHz) transponder is 24 dBW. The subtraction 24 – 68.8 = –44.8 dBW/Hz, so the carrier to intermodulation noise $(C/N_0)_{IM} = 18$ dBW -44.8 dBW/Hz = 62.8dB-Hz. Thus, the required (C/N_0) has to be 52.5 dB-Hz [1, 2, 5, 14].

4 INMARSAT GROUND SEGMENT AND NETWORKS

The Inmarsat ground segment comprises a network of LES, which are managed by different LES operators, Network Coordination Stations (NCS) and Network Control Centre (NCC). The major part of the ground segment and other networks are mobile subscribers or MES, which can be Ship Earth Station (SES), Vehicle Earth Station (VES) or Aircraft Earth Station (AES).

Each LES satellite operator provides a reliable transmission link between satellite network and TTN, capable of handling many types of calls to and from MES terminals simultaneously over the Inmarsat networks.

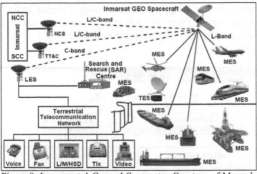

Figure 8. Inmarsat-4 Ground Segment   Courtesy of Manual: by Ilcev [5]

The Inmarsat network is providing many types of Mobile Satellite Services (MSS): maritime, land and aeronautical, such as Voice, Fax, L/M/HSD (Low/Medium/High Speed Data) via TTN, which I-4 ground segment is illustrated in Figure 8. In addition, the ground segment containing LES, Telemetry, Tracking & Command (TT&C), NCS NCC, Satellite Control Centre (SCC), Search & Rescue Centre and different MES terminals.

4.1 *Inmarsat Mobile Earth Station (MES) Solutions*

An MES is an RF device installed on board different mobiles, such as SES, VES, AES and Transportable Earth Station (TES), or it can be installed in a fixed location in maritime or land-based environments, indoor or outdoor, public payphones, on board mobiles and in suburban, rural and remote locations including SCADA applications. Inmarsat does not manufacture such equipment itself but permits manufacturers to produce models, which are type-approved to standards that have been set by Inmarsat and other international bodies, such as IMO, ICAO and the International Electrotechnical Commission (IEC). Therefore, only type-approved terminals are permitted to communicate via Inmarsat's space and ground segments. At this point, all types of MES provide different communication services in both mobile-to-ground and ground-to-mobile direction and inter-mobile communications. The list of MES terminals, types of service and access codes and the countries in which they are registered are given in Inmarsat Operational Handbooks, in the Admiralty List of Radio and Satellite Services, in the ITU list of Ship Stations, and in SITA and ARINC list of Aircraft Stations.

The Maritime MSS (MMSS) is a service in which MES is fitted onboard merchant or military ships, floating objects, rigs or offshore constructions, hovercrafts and survival craft stations providing commercial, logistics, tactical, defense and safety communications. The ship's Emergency Position Indicating Radio Beacon (EPIRB) terminal, either portable or fixed onboard may also participate in this service. The EPIRB is a special MES in the GMDSS, the emission of which is intended to facilitate urgent SAR operation for vessels in distress. The MMSS service enables mobile satellite links between LES and SES, between two or more SES and/or between associated ships and other satellite communications stations in all positions at sea or in ports. The SES is a mobile Earth station in the MMSS capable of surface movement at sea within the geographical limits of a country or continent. Thus, in distinction from conventional maritime communications, a ship fitted with SES in or near a port may operate with LES or other SES terminals in cases of distress and commercial operations [2, 3, 5, 13].

4.2 *Inmarsat Land Earth Stations (LES)*

The LES terminal is a powerful land-based receiving and transmitting station serving in a GMSC system. The LES infrastructure is fixed on the ground, and so it can serve FSS as well, but cannot be part of FSS. In a more precise sense, every LES is a part of MSS network, although it has a fixed location and can provide FSS.

Some LES, such as Goonhilly, provide widely fixed satellite links for FSS and consequently, the Inmarsat MSS network is a small part of the overall LES providing service for maritime and other mobile applications. Each LES in the IGN is owned and operated by an Inmarsat Signatory with the mission to provide a range of services to all types of MES. There are more than 40 LES terminals located in 30 countries around the globe but usually in the Northern Hemisphere, which is a small anomaly. The MES operator and shore subscribers can choose the most suitable LES, as long as they are within the same Ocean Region. The fundamental requirement for each Inmarsat LES terminal with Antenna Control and Signaling Equipment (ACSE) is that it be capable of reliable communications with all MES terminals, which block diagram is shown in Figure 9.

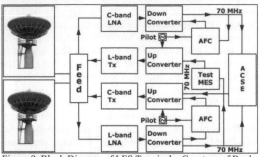

Figure 9. Block Diagram of LES Terminal – Courtesy of Book: by Dalgleish [13]

There are two major types of Inmarsat LES: Coast Earth Station (CES) for maritime and land mobile applications, such as SES and VES standards, and Ground Erath Station (GES) for aeronautical mobile applications providing service to all AES standards. The list of LES terminals, types of service and access codes, the countries in which they are based and Ocean Region of operation are given in Inmarsat Operational Handbooks, in the Admiralty List of Radio and Satellite Services, in the ITU list of Coast Stations and in the SITA and ARINC documentations. The technical side of typical mobile LES (CES and GES) consists in three main features: the antenna system (left), the communication RF equipment (between Feed and ACSE) and ACSE unit (right), as illustrated in Figure 9.

The CES is a maritime Earth station located at a specified fixed point on the coast to provide a feeder link for MMSS. The SES is a maritime Earth station fixed on board ships or other floating objects, which can provide communications links with subscribers onshore via CES and spacecraft. The ship on scene radiocommunications and alert service performs a distress and safety service in the MMSS between one or more SES and CES, or between two or more nearby SES, or between SES and Rescue Coordination Centre (RCC), or between portable or floating EPIRB and ground Local User Terminal (LUT) stations serving in the Cospas-Sarsat SAR satellite network in which alert messages are useful to those concerned with the movement and position of ships and of ships in distress.

Antenna System – A typical LES antenna for an entire IGN would be a Cassegrain structure with a dish reflector of about 14 m diameter. Each LES can have a minimum of one operational and one spare

antenna communication system in order to continue transmissions during maintenance. Some LES terminals have more than two operational antennas, which depends on the Ocean Region covered and the services provided. The antenna operates in both the L and C-band to and from the satellite, with gain requirements of 50.5 dBi and 29.5 dBi, respectively. It is designed to withstand high wind speeds up to 60 m/h in its operational attitude and 120 m/h when stowed at 90o and the parabolic dish is steerable ±135o in azimuth and 0 to 90o in elevation angle. Tracking is either by automatic program control or operator initiated.

An antenna tracking accuracy of 0.01o r.m.s. and a repositioning velocity of 1o s-1 would be typical parameters for such a dish. RF and base band processing hardware design varies greatly with LES design and requirements. In the other words, a single antenna may be used to transmit and receive L-band as well as the C-band signals or the employment of a separate L-band antenna avoids the need for a relatively complex feed system (to combine and separate the outgoing and incoming L-band and C-band signals) but this advantage must be weighed against the cost of procuring and installing a second antenna.

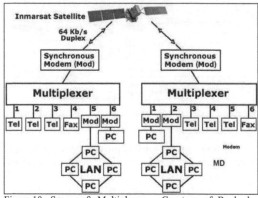

Figure 10. Spacecraft Multiplexer – Courtesy of Book: by Dalgleish [13]

Communication RF Equipment – The radio RF equipment is situated inside the LES building and must be able to operate in Tx and Rx L-band links to monitor the MSC L-band channel and respond to requests for frequency allocations by the NCS; to verify signal performance by loop testing between satellite and LES and to receive the C-to-L Automatic Frequency Control (AFC). The AFC provides Tx and Rx direction control, which helps to keep MES as simple and as cheap as is practicable. A complete test of the LES equipment can be carried out without the cooperation of the MES because a separate test terminal is provided at each LES for this purpose. The Inmarsat system requires AFC to correct for Doppler shift (caused by inclination of the GEO) and errors in frequency translation in the satellite and LES. The total frequency shift from this cause without AFC could be more than 50 kHz and thus is of the same order as the spacing between the Narrow Band Frequency Modulation (NBFM) channels and would be enough to cause failure of the system. The AFC reduces the RF shift to a few hundred hertz by comparing pilot carriers transmitted via the satellite with reference oscillators at the LES and using the difference signals to control the RF of the local oscillator associated with the up and down converter. A pilot transmitted at C and received at L-band is used to control the up converter and thus offset the frequencies of the operational carriers to compensate for Doppler shift and satellite frequency translation errors in the ground-to-mobile direction. Similarly, a pilot transmitted at L and received at C-band controls the down converter and corrects for Doppler shift and errors in translation in the mobile-to-ground direction. All RF errors are corrected except those arising from the frequency instability of the MES up and down converter and Doppler shift resulting from the relative velocities of satellite and MES.

ACSE – This Antenna Control and Signaling Equipment (ACSE) is part of each LES, whose principal purpose is to recognize requests for calls sent by MES, to set and release. This requires response to and initiation of in-band and out-bands signaling over the GEO satellite and the terrestrial path. The next ACSE tasks are to recognize distress calls sent by MES and preempt channels for them when necessary, to check that MES are on the list of authorized users and to bar calls from or to unauthorized MES, to switch voice circuits between TTN circuit and the LES FM channel modem, to switch Telex circuits between TTN channels and the Time Division Multiple/TDM Access (TDM/TDMA) time slots, to determine Tx/Rx frequencies used by the FM channel unit according to the channel allocations made by the NCS, to allocate TDM/TDMA time slots and to collect data statistics for billings, international accountings (for transit calls), traffic analyses, management and maintenance purposes, etc. The ACSE block is sometimes specified to include all the communication equipment of LES terminal other than RF and IF equipment, such as modulators and demodulators, data and voice channels, RF assignments to MES, line control subsystem, system control processor, and etc. The MSC services offered by an LES vary depending upon the complexity of the station selected, such as two-way voice including Fax/Paging, Telex, all data rate, Video, Global Are Network (GAN) – Internet, Web and Mobile Emergency services (Distress, Urgency, Safety and Medical assistance calls). At this point, multiplexing as a number of communication channels onto a single satellite link become possible by using duplex High Speed Data (HSD), such as multiplexing deploying six communication channels onto a single satellite connection through the Inmarsat duplex HSD service, shown in Figure 10 [1, 2, 5, 13].

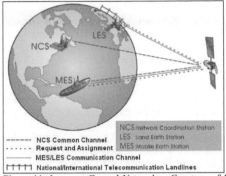

Figure 11. Inmarsat Ground Network – Courtesy of Manual: by Inmarsat [1]

5 INMARSAT GROUND NETWORK (IGN)

The maritime and other mobile configuration of IGN is applied to each of the four or three I-4 Ocean Regions for setting up MSC channels for ship-to-ground calls and vice versa, shore-to-ships calls. The same scheme can be implemented for any other MES configuration of mobile-to-ground calls, such as for land and aeronautical applications. Each MES has always to be tuned to the Common Signaling Channel (CSC) and to listen for assignments Requesting Channel (RC), when is not engaged in passing traffic, namely when MES is an idle state, while each LES also watches the CSC to receive their channel assignments. The CSC is also referred to as TDM0 and is the origin of all traffic. The IGN is interfaced to the TTN as a Gateway to all fixed subscribers, shown in Figure 11.

5.1 Network Coordination Stations (NCS)

The Inmarsat system uses four NCS, one in each Ocean Region separate for each standard, to monitor and control MSC traffic within the region. Usually some LES perform dual services and when required to be specifically identified, the LES serving as an NCS or Standby NCS will be referred as a collocated station. Hence, the NCS is involved in monitoring and control functions and in setting up calls between MES and LES, which is shown in Figure 11. The illustration shows in general terms how the NCS responds to a request from an MES (SES) for a communication channel, by assigning a channel to which both the MES and LES (CES) operator must tune for the call to proceed.

Therefore, an LES serving as the NCS or Standby NCS terminals shall comply with all the technical requirements applicable to any Inmarsat standard LES and shall normally process its own calls in the same manner as a normal LES. In addition, the ground station serving as the NCS terminal shall perform the following necessary conditions for the important functions of IGN system:

1 Transmits continuously signals on a special channel known as an CSC at 6 GHz L-band;
2 Accepts incoming request for Tlg (telegraphy) assignment MSG from all LES and re-broadcasts it to MES on the CSC;
3 Accepts Tel and HSD channel request-for-assignment messages from all LES in IGN and makes Tel and HSD channel assignment via the common TDM channel;
4 Maintains a Tel and HSD channel-activity list that indicates which channels are in use as well as the LES and MES using each channel;
5 Determines if an addressed MES is busy with another call;
6 Clears a telephone call in progress if necessary to service an SOS priority request; and
7 Maintains a record of RC, CSC and satellite Tel channel used for the IGN terminal's analysis purposes.

Furthermore, NCS terminal shall also change to the Alternative Common TDM RF of 6 GHz for Inmarsat I-4 satellite, Tx transponder load control carriers and facilitating measurements of RF signals at both C and L-band from the satellite. For these functions to be performed, an MES Rx must initially be synchronized to the NCS common channel and logged-in to the NCS for its Ocean Region, either automatically or manually at the MES [1, 2, 5, 14].

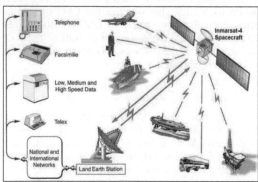

Figure 12. Inmarsat L-band GMSC Network Courtesy of Manual: ALRS [15]

5.2 Network Control Centre (NCC)

The Inmarsat NCC is located in the Inmarsat Headquarters building in London. It monitors, coordinates and controls the operational activities of all satellites (payload and antennas) and makes it possible to transfer operational information throughout the Network and via worldwide TTN routes, data between the NCC in each of the 4 Ocean Regions, see Figure 8. The NCC can send system messages via one or all of the NCS to inform the MES in their Ocean Regions of news relevant to any Inmarsat standards. It controls characteristics of the space segment throughout TT&C stations located in different countries; realizes all plans for new technical solutions and conducts development of the entire system; controls functions of current and newly introduced MES and LES and provides information about all MES, LES, NCS and the working condition of the entire Inmarsat system.

5.3 Satellite Control Centre (SCC)

Whereas the NCC is crucial to the MSC service management, the SCC located in London at Inmarsat House is crucial to spacecraft management and functions of station-keeping and TT&C, illustrated in Figure 8. All data to and from the SCC is routed over worldwide TTN or tracking stations, which also provide backup capacity if required. The TT&C LES terminals are equipped with VHF, C and L-band for controlling spacecraft in all four Ocean Regions. Data on the status of the nine Inmarsat satellites is supplied to the SCC by four TT&C stations located at Fucino (Italy), Beijing (China), Lake Cowichan and Pennant Point in western and eastern Canada and there is also a back-up station at Eik in Norway. Thus, this service provides TT&C, i.e., operational status of spacecraft subsystems and payload, such as transponder signals; decoders and converters; temperature of all equipment and surface; diagnostics on all electrical functions; satellite orientation in space; situation of attitude control fuel; telemetry of process decoder and Rx beacons and provides tracking and control of all parameters during launch of satellite.

5.4 Rescue Coordination Centers (RCC)

As the name implies, RCC are used to assist with SAR in distress situations for maritime and aeronautical applications, shown in Figure 8. Namely, extensive MSC links provide end-to-end connection between the vessel or airplane in distress and competent rescue authorities. Because of the very high priority status accorded to distress alerts and the use of automatic signaling systems, this direct connection linking is rapidly established, usually within only a few seconds. Comprehensive MSC systems link an individual RCC site and can have LES, MCC or LUT. When an RCC receives an original distress alert (SOS or MAYDAY) via one of these stations, it will relay details of the alert to SAR units and to other ships (if the distress is at sea) within the general area of the reported distress.

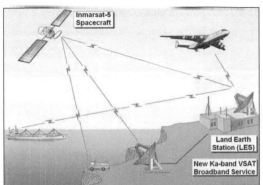

Figure 13. Inmarsat Ka-band GMSC Network – Courtesy of Manual: ALRS [15]

Hence this relayed message should provide the vessel or aircraft in distress with identification, its position and any other relevant information of practical use in rescue operations. The RCC terminal, which initially receives a distress alert, however, appropriately called First RCC, assumes responsibility for all further on-scene coordination of subsequent SAR operations. In such a way, this initial responsibility may be transferred to another RCC terminal, which may be in a better position to coordinate rescue efforts. The RCC stations are also generally involved in subsequent SAR team coordinating communications, such as between the designated On-scene Commander in SAR or Coordinating Surface Search (ship or helicopter), who are onboard SAR units within the general area of the distress incident.

5.5 Terrestrial Telecommunications Network

The Terrestrial Telecommunications Network (TTN) operators are usually Inmarsat Signatories and can be PTT (Telecom) or any government or private TTN providing landline public Tel and Tlx service. The TTN shore operators enable interface of IGN on their landline infrastructure for voice, Fax, Tlx, Low, Medium and High Speed data (LSD/MSD/HSD) and video services, which Inmarsat Mobile Network (IMN) is presented in Figure 12. Each CES or GES as a part of initial L-band IMN has facilities to interface telephone, facsimile, low/medium/high speed data and telex networks of the TTN.

On the other hand, IMN is providing services for maritime, land, aeronautical, personal, military, off-shore and semi-fixed applications. Thus, the new Inmarsat HSD service enables connections to the TTN sites and other infrastructures, such as ISDN, PSDN, PSTN, Leased Lines, Data Network (X.25, X.75 and X.400), Private Data Networks, ATC Network, and etc.

As stated before, the existing L band Inmarsat service is providing digital transmission using mobile broadband, such as FleetBroadband for maritime and SwiftBroadband for aeronautical applications throughout current I-4 and future I-5 spacecraft.

In addition, the new projected Ka-band Inmarsat service will start with mobile VSAT broadband service via I-5 spacecraft, illustrated in Figure 13. The mobile Global Broadband Satellite System (GBSS) is recently introduced as a modern mobile and maritime VSAT networks, equipment and DVB-RCS technique [2, 5, 12, 15].

6 CONCLUSION

The Inmarsat space and ground segments have great significance with regard to satellite use for maritime and other MSS applications. Here are introduced very important context of fundamental contribution in development advanced Inmarsat satellite constellations, coverages, spot networks and the principal parameters that describe the new deployments of modern communication service for maritime and other mobile applications.

REFERENCES

[1] Inmarsat, " Maritime Handbook" London, 2002.
[2] Ilcev D. S. "Global Mobile Satellite Communications for Maritime, Land and Aeronautical Applications", Springer, Boston, 2005.
[3] Zhilin V.A., "Mezhdunarodnaya sputnikova sistema morskoy svyazi - Inmarsat", Sudostroenie, Leningrad, 1988.
[4] Inmarsat, "Inmarsat I-3 Global Coverage Map", Prospect, London, 2000.
[5 Ilcev D.S., "Global Mobile Communications, Navigation and Surveillance (CNS)", Manual, DUT, Durban 2014 [www.dut.ac.za/space_science].
[6] Inmarsat, "Inmarsat I-4 Global Coverage Map for GMSC", Prospect, London, 2010.
[7] Ilcev D. S. "Global Aeronautical Communications, Navigation and Surveillance (CNS)", Volume 1 & 2, AIAA, Reston, 2013.
[8] Inmarsat, "Inmarsat I-5 Spot-Beam Coverage Map for Digital Mobile Broadband", London, 2012.
[9] Inmarsat, "Global Mobile Communications Solutions", London 2001.
[10] Acerov A.M. & Other, "Morskaya radiosvyaz i radionavigaciya", Transport, Moskva, 1987.
[11] Novik L.I. & Other, "Sputnikovaya svyaz na more", Sudostroenie, Leningrad, 1987.
[12] Gallagher B., "Never Beyond Reach", Inmarsat, London, 1989.
[13] Dalgleish D.I., "An Introduction to Satellite Communications", IEE, London, 1989.
[14] Kantor L.Y. & Other, "Sputnikovaya svyaz i problema geostacionarnoy orbiti", Radio i svyaz, Moskva, 1988.
[15] ALRS, "GMDSS" - Volume 5, Admiralty List of Radio Signals, Taunton, 1999.

31

Integration of Radio and Satellite Automatic Identification System for Maritime Applications

D.S. Ilcev

Durban University of Technology (DUT), South Africa

ABSTRACT: This paper introduces the main technical characteristics and integration of current Radio and new Automatic Identification System (AIS) as more reliable solution for enhanced tracking and detecting systems for maritime applications. The technical parameters and comparison of current Radio AIS (R-AIS) and new developed Satellite AIS (S-AIS) with their advantages and disadvantages are described. The possibility in upgrading of a particular AIS configuration depends mainly on its mission objectives, development trends of radio and satellite mobile tracking and surveillance service, modernization of onboard and coastal equipment and characteristics of the satellite payloads for S-AIS. The R-AIS network provides local coverage in coastal waters of VHF-band range, however new S-AIS will be able to provide global beam coverage using Low Earth Orbit (LEO) or even Geostationary Earth Orbit (GEO) spacecraft. In addition, The R-AIS network and its classes, basic technical details of AIS transponder, Communication, Navigation and Surveillance (CNS) Systems for AIS Base Station (BS) and software, the S-AIS Network, enhancements of surveillance and security, improvement of counter piracy and suspicious movement of ships, better facilitated fisheries and environmental monitoring, more Reliable Search and Rescue (SAR) operations and Nano satellite AIS (Nano S-AIS) are discussed.

1 INTRODUCTION

The Global Maritime Distress and Safety System (GMDSS) radio and satellite integrated networks are already developed and implemented for ships safety, distress alert and SAR communications. Namely, this system is not enough effective to provide a real tracking and detecting system of ships for every day navigation aids. The GMDSS radio and satellite networks need an additional integration with new CNS systems as a proposal that has to be developed providing seafarers with global communications and tracking networks introduced in this research.

At this point, the GMDSS has to integrate CNS systems that use radio and satellite technologies for automated distress alerting, rapid tracking and detecting of ships in navigation and distress. Thus, this important integration has also to prevent all emergency situation during navigation employing more sophisticated, modern and reliable technique for improving collision avoidance and enhanced safety and security. On the basis of these daily requirements and necessity for modern maritime radio and satellite tracking and detecting systems onboard ships are already developed R-AIS and S-

AIS working at VHF frequency band, which can be integrated with current and new CNS networks and equipment onboard ships, vehicles and aircraft.

The R-AIS network and equipment are automatic tracking system used onboard ships and by Vessel Traffic Services (VTS) network for identifying and locating vessels by electronically exchanging data with other nearby ships, AIS BS for coastal waters, and via LEO satellites globally. The AIS network and equipment are providing information that supplements marine radar, and in such a way continues to be the primary method of collision avoidance for water transport.

Because of usual problems during extremely bad weather conditions at sea sometimes VHF-band is so limited or interrupted by reduced propagation effects and interference characteristics caused by rainfall, very deep clouds and thunderstorms too. Therefore, the VHF-band AIS, whether radio or satellite, affected by these weather factors will be not able to provide reliable service for tracking and detecting of ships even via additional satellite VHF-band. For that reason will be necessary to deploy GEO or LEO satellite constellations via L or S-band.

Information provided by AIS equipment, such as unique identification, position in real time, course and speed, can be displayed on a screen or an Electronic Chart Display and Information System (ECDIS). The AIS network is intended to assist watchstanding officers onboard ships and allow maritime authorities in each country to track and monitor ships movements in costal navigation. The AIS equipment integrates a standardized VHF transceiver with a positioning system such as an GPS, GLONASS or LORAN-C receiver, and with other electronic navigation sensors, such as a gyrocompass or rate of turn indicator.

Ocean vessels fitted with AIS transponders and transceivers can be tracked by AIS BS located along coast lines or when are out of radio VHF ranges of terrestrial networks, through a growing number of satellites that are fitted with special AIS receivers. In fact, the S-AIS network needs ships to be equipped with other types of devices than R-AIS network and to use special assigned frequencies different from R-AIS network.

However, the integration has to be realized in the Control Centre, which main task will be to collect position data received via both R-AIS and S-AIS for the same ships and distribute to all users.

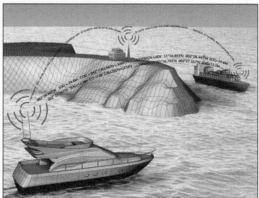

Figure 1. Maritime AIS Network – Courtesy of Brochure: by IMO [1]

2 RADIO VHF AUTOMATIC IDENTIFICATION SYSTEM

The VHF Radio Automatic Identification System (R-AIS), also know as a Radio VHF Data Link (R-VDL) is a most attractive system at present for tracking and detecting of ships for short range service in coastal navigation. Thus, more adequate designation for AIS will be Radio AIS (R-AIS), because recently is developed new Satellite AIS (S-AIS) with similar service for ships and aircraft.

The new Regulation 19 of the SOLAS Chapter V provides requirements for shipborne navigational equipment and sets out this equipment to be carried onboard according to ship type. In 2000, International Maritime Organization (IMO) team adopted a new requirement, as part of a revised new Chapter V that all ships have to carry VHF AIS transponders capable of providing necessary information about the all ship to other ships and to coastal authorities automatically.

The R-AIS is a new maritime surveillance system using the VHF-band to exchange all information between ships and shore Base Stations, including positions, identification, course and speed, which network is shown in Figure 1. It mainly aims is to provide collision avoidance between ships. The link budgets allow receiving transmitted R-AIS signals from space, and consequently a global maritime surveillance can be considered.

However, later arise some challenges, especially regarding message collisions due to the use of a Self Organized Time Division Multiple Access (SOTDMA) protocol (not designed for satellite detecting). Thus, advanced signal processing for separation of received signals is needed.

According to the IMO regulations provided by 31 December 2004 each oceangoing vessel has to install AIS transponder equipment onboard ships, which automatically broadcast regularly to the coast station ships name, call sign and navigation data. This data is programmed soon after the equipment is installed onboard and after that all this information will be transmitted regularly.

The signals are received by R-AIS transponders fitted on other ships or on land based network, such as VMS systems. The received information can be displayed on a screen or a chart plotter, showing the other vessel's positions in much the same manner as a radar display.

Ships fitted with R-AIS onboard equipment shall maintain AIS in operation at all times, except where international agreements or rules provide for the protection of navigational information. The R-AIS standard comprises several substandard called as "types" that specify individual for each product type. The specification for each product's type provides a detailed technical specification, which ensures the overall integrity of the global R-AIS system, within which the entire product types which must operate. Namely, there are two types of R-AIS Transceivers (for transmit and receive) "Class A" and "Class B".

The IMO regulation requires that AIS shall provide information by transmitter including the ship's identity (ID), navigational status and other safety-related information automatically to the appropriately equipped shore stations, other ships and aircraft, than to receive automatically such information from similarly fitted monitor, track ships and exchange data with shore SB facilities.

At its 79th session in December 2004, IMO's Maritime Safety Committee (MSC) agreed that, in relation to the issue of freely available AIS-generated ship data on the worldwide Web. The publication on the worldwide Web or elsewhere of all AIS data transmitted by ships could be detrimental to the safety and security of ships and port facilities. It also was undermining the efforts of the IMO expertise and its Member States to enhance the safety of navigation and security in the international maritime transport sector. In addition, the Committee condemned those who irresponsibly publish AIS data transmitted by ships on the world-wide Web, or elsewhere, particularly if they offer services to the shipping and port industries [1, 2, 3].

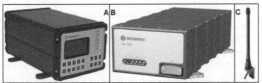

Figure 2. Maritime VDL Classs A R-AIS Stations – Courtesy of Brochure: by CNS Systems [4]

2.1 CNS Systems R-AIS Class A Station

The VDL 6000 Class A ship R-AIS transponder station of the Swedish Company CNS Systems provides IMO SOLAS compliance and also Bundesamt für Seeschiffahrt und Hydrographie (BSH) certification for installation onboard all oceangoing vessels, illustrated in Figure 2 (A). The screen presentation of this transponder is indicating the call signs of other ships and enables a user and other ships to make direct contact by text or voice communication.

Another model of CNS Systems is VDL 6000 Secure Class A shipborne R-AIS transponder that operates in Standard, Silent and Secure Mode, shown in Figure 2 (B). This R-AIS is also serving for naval operation can be configured in "receive only" mode or both "receive and transmit" mode for positive identification and positioning of all ships in the vicinity. The R-AIS units secure system is designed on existing technology and supports a user to receive, schedule and transmit encrypted messages to other users. This service can include transmission of secure text messaging and to receive encrypted range and bearing. The secure R-AIS can also support simulated targets for naval operations.

Both R-AIS transmitters generate output power (adjustable) 1 and 12.5 W and 50 Ohm load. The unit bandwidth is 25 kHz employing TDMA (AIS) protocol with baud rate 9600 b/s (AIS)/1200 b/s Digital Selective Call (DSC) and GMSK (AIS)/FSK (DSC) modulation. Both transceivers are using frequency bands from 156.025 to 162.025 MHz with default channels 87B (161.975 MHz) and 88B (162.025 MHz), 70 (156.525 MHz). Both AIS units of Secure Class A system have 3 (2 AIS TDMA, 1

DSC) number of receivers and Minimum Keyboard and Display (MKD) unit. The R-AIS Class A transponder is easy to install onboard any seagoing ship by connecting it to an GPS and VHF antenna or to the own antenna, illustrated in Figure 2 (C), and is complete after immediate connecting it to the onboard sensors.

To maximize the benefit of the investment, the R-AIS Class A transponder is delivered with an interface to the electronic chart system and/or ARPA radar. Moreover, the system is designed to support long-range reporting via satellite, which will be introduced in the next context below. To maximize the benefit of the functionalities, both R-AIS Class A transponders are delivered with an interface to the chart system and/or ARPA radar.

The data link communication covers identity, position, destination and other required static, data voyage-related and dynamic data, which gives all vessels in an area increased situational awareness and improves safety at sea for the individual ship. Positive identification and positioning of all ships in the vicinity reduces the unnecessary "ship on my port bow" calls. Less information overload greatly enhances safety at sea.

The SOTDMA technology is used in the R-AIS transponder, which transmits and receives useful information on all vessels within VHF coverage. This information includes position, identity (ID), course over ground, heading and rate of turn as well as navigational status and the destination of the ship. Thus, the information received from, and provided to, the ships is easily plotted on any ARPA radar or electronic chart system. This gives the officer of the watch a situational awareness that could never be achieved prior to AIS. In such a way, information on draught, type of cargo and destination could also be used to make decisions related to the ship maneuvering, so at this point is accomplished the maximum awareness.

Targeted at large commercial oceangoing ships, SOTDMA mode requires an AIS transceiver to maintain a constantly updated slot map in its memory such that it has prior knowledge of slots, which are available for it to transmit. Moreover, the SOTDMA transceivers will then pre-announce their transmission and effectively reserving theirs transmit slot. Thus, this transmission is achieved through two receivers in continuous operation. Class A unit has an integrated display, transmit at 12 W, interface capability with multiple ship systems, and offer a sophisticated selection of features and functions. In default transmit rate it sends information every few seconds providing tracking control of vessel. Described system is transmitting all necessary tracking information with help of the VHF transmitter to other nearby ships and to coastal base station (BS) [2, 4, 5].

Figure 3. Maritime Classs B R-AIS – Courtesy of Brochure: by ICOM [6]

2.2 Icom MA500TR Class B AIS Transponder

The MA-500TR is a Class B R-AIS transponder for Non-SOLAS vessels such as pleasure craft, workboat, fishing and small vessels. Transmitter power of this unit is 2 W and not required to have an integrated display.

Class B R-AIS can be connected to most display systems, which the received messages will be displayed in, lists or overlaid on charts. In Figure 3 is presented interface of Icom MA-500TR Class B AIS transponder with antenna, which can be installed usually onboard small vessels, which is not complaining to the necessary requirements of IMO regulations [2, 6, 7].

3 BASIC TECHNICAL DETAILS OF AIS TRANSPONDER

Vessel with mounted onboard both R-AIS type Class A and B transceivers employ Carrier Sense TDMA (SCTDMA) or Self Organizing TDMA (SOTDMA), which AIS network is illustrated in Figure 4 The system is automatically sending and receiving data via the standard vessel's radio units. In such a way, they use two VHF channels simultaneously, such as 87B at 161.975 MHz and channel 88B at 162.025 MHz. Therefore, to order a lot of vessels sending out their data, the AIS transmission protocol works with 2250 slots per minute which can be used by different senders to transmit their information.

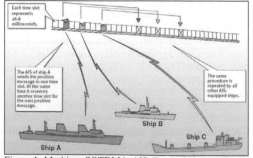

Figure 4. Maritime SOTDMA AIS Transmissions – Courtesy of Brochure: by AMSA [8]

Figure 5. Maritime AIS Base Stations – Courtesy of Brochure: by CNS Systems [4]

The SOTDMA mode is used to autonomously divide the available timeslots between different senders. Due to the limited range of VHF and the different transmission intervals of senders the number of slots is sufficient and collisions hardly occur. An AIS system sends out via Base Station different types of information in varying time intervals. In total, AIS is able to communicate 27 different message types. Message #1 e.g. is the position of the vessel. In such a way, transmitted information by this unit is: Maritime Mobile Service Identity (MMSI) code, Vessel name, Call sign, Type of ship, GPS antenna position, Ship's position, SOG (Speed Over Ground), COG (Course Over Ground), UTC date and time, GPS antenna type, PA (Position Accuracy) and Simple operation. Not all messages are regularly used.

The AIS Plotter display looks like a usual marine radar display. North-up, course-up and range zoom from 0.125 to 24 NM (miles) are supported. The Target list display shows all detected AIS equipped vessels and targets. The Danger list display shows a list of vessels that are within 6 NM of Closest Point of Approach (CPA) and 60 minutes of Time to CPA (TCPA) from own vessel. Therefore, the Danger list can be sorted by CPA or TCPA order. In addition to these display types, the Detail screen shows various information about the selected R-AIS targets, such as CPA and TCPA for collision-risk management. When a vessel comes into the CPA and TCPA range, the unit icon blinks on the Plotter display and emits a beep sound. When connected to external audio equipment installed on the deck tower, the collision alarm function will alert operator even when is away from the AIS transponder [2, 4, 8].

4 CNS SYSTEMS AIS BASE STATION (BS)

Maritime R-AIS Shore Stations are used for surveillance and management of vessel traffic along coastlines, on inland waterways and in ports. These stations are providing all the features required for surveillance and management of vessel traffic at R-AIS VTS.

They are easily configured to the specific needs of necessary service and solutions, from a basic unit to a fully redundant system with an embedded controller providing extensive processing and logging functionality. These R-AIS base stations fulfill the requirements of international AIS standards, provide service for national maritime authority implementing an R-AIS network and are also suitable for stand-alone operation at an VTS or seaport.

The Swedish Company CNS Systems has two designs of Maritime R-AIS Base Stations: VDL 6000/FASS, illustrated in Figure 5 (Left) and VDL 6000/FASS Advanced, shown in Figure 5 (Right).

All AIS ships stations automatically broadcast information on dedicated VHF maritime channels. The AIS broadcasting system consists of static (geostationary), dynamic (non-geostationary) and navigation information, shown in Figure 6. All this information is originating from ships sensors connected to the R-AIS transponder. R-AIS Station sends position reports every 2 seconds to 3 minutes, depending on type of R-AIS transponder, speed and turn rate. Thus, the received R-AIS information at Base Station can be shown on a VTS Operator's screen or an ECDIS display.

4.1 *CNS Systems AIS Base Station VDL 6000/FASS*

The VDL 6000/FASS is configured as an AIS BS terminal illustrated in Figure 5 (Left), which includes two VDL 6000/FASS (Fixed AIS Station System), one base station transponder and one controller in each FASS unit and one Power and Antenna Distribution (PAD) unit. The AIS network with AIS internship reports and reports between ships and AIS BS are presented in Figure 7 (A). In Figure 7 (B) is illustrated the Multifunction Displays for Marine's Integrated Platform Management System (IPMS) for AIS VTS infrastructure, which also can be used as Integrated Bridge System (IBS) display for installation onboard ships [2, 3, 9].

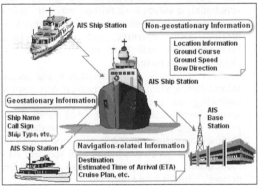

Figure 6. Maritime AIS Network – Courtesy of Brochure: by MRS [9]

This AIS BS configuration is a cross redundant AIS shore terminal with high reliability and full redundancy on all electronic components. On the other hand, this unit exceeds all the requirements of international AIS BS standards and provides all the features required for surveillance, detecting and management of vessel traffic. The AIS BS will automatically switch over to the stand-by FASS unit if the active one goes down (hot stand-by). It can also use the controller and BS terminal crosswise in case of dual failures of hardware.

This solution gives extremely high Mean Time Between Failure (MTBF) and availability and a very low mean time to repair, reducing the need for unscheduled maintenance. The calculated MTBF of the AIS BS is more than 3 million hours, provided faulty units are replaced within 72 hours. The PAD unit contains an VHF antenna switch which can be used when only one VHF antenna is required. If dual VHF antennas are used at the installation site, the VHF switch can be bypassed by connecting directly to the two VDL 6000/FASS units VHF antenna ports. The AIS BS station provides many additional remote functionalities, including configuration, software updates, virtual targets and more.

Other functions are local storage of AIS messages, local target filtering and Simple Network Management Protocol (SNMP). Power management is also integrated in the PAD unit. This feature makes it possible for the user to remotely switch the FASS units power on and off, what is giving the operator full control. Therefore, the AIS BS is designed for operation in secure mode by handling encrypted AIS data. The secure mode enables secure communication between ship-to-shore station and the control centre. Thus, by adding the secure modules to the network software, encryption and decryption can be managed between the users in a secure AIS network. The AIS BS transponder can be supplied with a third AIS channel as well as increased receiver sensitivity.

The AIS BS is employing Power IP relay IF TCP/IP and RJ45 Ethernet with transmitter and receiver tuning range at 156.025 –162.025 MHz and with channel spacing of 12.5 and 25 KHz. Modulation scheme is 25 KHz GMSK (AIS TDMA), 2.5 KHz GFSK (AIS TDMA) and 25 KHz FSK (DSC). The AIS BS station is containing GNSS Receiver as well with GPS L1, 16 parallel channels and DGNSS support.

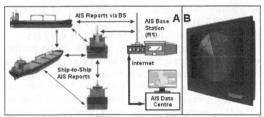

Figure 7. Maritime AIS Network and Display – Courtesy of Manual: by Ilcev [2]

Figure 8. Radar vs AIS – Courtesy of Manual: by Ilcev [2]

4.2 CNS Systems AIS Base Station VDL 6000/FASS Advanced

The Advanced configuration has one transponder and embedded controller hardware running unique software from CNS Systems, shown in Figure 5 (Right). This BS terminal can be installed as a single equipment providing AIS message logging, remote configuration and software update, virtual and synthetic targets, SNMP, local target filtering, remote power on/off and more. Two shore BS terminals can be locally connected to form a fully redundant installation, where their hot-standby operation including mutual cross-wise redundancy transponder and controller between the units provides very high availability.

This AIS shore BS is delivered with the Monitor and Control Tool (MCT) and the Power Supply Management Tool (PSMT). Therefore, the MCT is a graphical interface that provides monitoring and control capabilities such as; change of status of the base station and its subunits, change of operational mode, enabling and disabling of services, configuration of the base station and its subunits, display of number of transmitted and received messages as well as software update.

The PSMT allows a user to control the power supply to the base station and its subsystems. It can be factory configured for the following base station types: AIS BS, Limited AIS BS, Repeater BS and Aids to Navigation (AtoN) transmitting only BS.

With the Repeater station this BS is the perfect choice when implementing an AIS network where extended coverage at remote locations is required. This AIS BS is equipped with dual VHF antennas provides a number of functions, including filtering of AIS targets by selection of MSG type and/or filtering of a defined area. Otherwise, the dual antenna configuration allows the Repeater station to receive and transmit on directional antennas.

This greatly improves coverage and distance in an AIS network, so for example in Figure 8 is shown advantage of AIS vs Radar. Both ships do not have radar contact due to the difficult terrain, but can see each other by AIS. However, sometimes in more mountainous shore line there is not LOS between ships or between ships and base station, so can be used Satellite AIS instead.

In order to simplify the functionality of AIS network it is necessary to construct the network model basing on a LAN. The system can use two ways to construct such a network. The first way is based on computing node, which communicate and exchange data or message each other via computer. The second is based on switching node, which contains data switches and equipment for controlling, formatting, transmitting, routing and receiving data packets [2, 4, 8].

5 CNS SYSTEMS AIS BS AND SHIP STATION SOFTWARE

The CNS Systems company is also supplier of important software for use by the AIS base and ship stations. There are two CNS Systems software for supporting onboard AIS stations, such as Aldebaran II and Sentinel. However, there are four software solutions used at Base stations, such as Horizon, DataStore, DataSwitch and Maestro.

1 Aldebaran II – This is Electronic Charting System (ECS) software for use onboard ships. In fact, it is designed with advanced navigation and communication features and is used by AIS operators worldwide. However, the ECS solutions improve efficiency and safety in today's fast paced computer-aided navigation and situational awareness environments. It offers complete AIS units integration with the ability to display static, dynamic and voyage related information in real time on a multitude of electronic chart formats.

2 Sentinel – This is AIS surveillance and secure information system built on proven AIS display technology for use by operators in the need for secure communication. It offers AIS standard and private communications capabilities to deliver a common operating picture to all users. It permits the simultaneous covert tracking of standard AIS participants, allowing operators to monitor the network without being detected.

3 Horizon – This is shore-based (BS) vessel monitoring solution designed for vessel traffic and monitoring centres. Horizon is ideal for vessel traffic monitoring in national waters, and has been proven to increase safety, security and efficiency. It provides a complete AIS interface

that includes the ability to view and track all vessels, display specific vessel information and send and receive safety related and text messages. Its interface and display of AIS related data offers a substantial leap forward in the ability to communicate and interact with vessels. Since Horizon is fully configurable, operators can adjust the display of information panels, customize color patterns for AIS targets and set entry and exit alarms.

4 DataStore – This is a real-time data logging and playback software solution for National Marine Electronics Association (NMEA) and NMEA-formatted data, specifically AIS data. The "back-end" of the solution is a service that interfaces with a database, and the "front-end" interfaces with an application that allows a user to control logging and playback of data. DataStore can be configured to store all data or a user defined subset of data. Data stored in the AIS database can be queried and played back in the ECS system or an external application.

5 DataSwitch – This is a special data routing and management software application that provides a reliable flow of data to environment. DataSwitch supports the functionality defined in International Association of Lighthouse Authorities (IALA) Recommendation A-124, LSS Layer and Part IV. It is ideal for the collection, filtering, logging and sharing of AIS data over networks. For shore BS networks, it enables the flow of information from one or more AIS base stations and/or receivers to a Vessel Traffic Service (VTS) centre. Similarly, a VTS center can send vital data to a regional headquarters, and then on to a national entity via another DataSwitch. For a vessel-based network, DataSwitch can send information from attached NMEA devices (sensors) to a number of stations on the vessel. The distribution of both standard and proprietary data messages from one central location to shared locations makes DataSwitch essential in many diverse environments.

6 Maestro – This software provides top layer control for the AIS network, supporting the functionality defined in IALA Recommendation A-124, Functionality of the AIS Service Management (ASM) and Part V. It is a graphical display interface and configuration utility for the AIS network.

7 Maestro monitors all AIS network components including status and failure of all components, warnings about failover and backup systems, user account status, and all other relevant events. From the single interface accessed via a web browser, Maestro users can monitor, maintain, and manage all elements of the AIS network. Maestro is an independent process that runs without affecting the other AIS services [2, 3, 4, 5].

6 SATELLITE AUTOMATIC IDENTIFICATION SYSTEMS (S-AIS) NETWORK

The recently developed VHF Radio - Automatic Identification System (R-AIS) network, being an RF-based communications system was never designed for reception of signals from space, however Satellite AIS (S-AIS) greatly extends the range of the original system and creates new application possibilities for competent maritime and aeronautical authorities. Visibility scope is significantly enhanced using S-AIS and in such a way this solution provides global coverage and creates increased CNS situational awareness well beyond the 50 NM range from shore. Similar to the VHF R-AIS ship terminal, the VHF S-AIS unit is easy to install onboard any seagoing ship by connecting it to GPS receiver, gyro and the Pilot Plug interface.

Therefore, most terrestrial-based R-AIS networks provide only limited shore-based coverage via VHF-band to track and monitor vessels, and thus are not able to provide global, open ocean coverage. The Orbcomm and other LEO satellite network including Nano (CubeSat) satellites overcomes many of these challenges with unique S-AIS data service that can monitor a vessel's location and daily status well beyond coastal waters and ports to assist in navigation and improve maritime safety and security.

More importantly, the Orbcomm and other LEO networks can do this cost-effectively and in near-real-time transmissions. The Orbcomm system was the first commercial satellite network providing S-AIS data services. At this point, Orbcomm has licensed S-AIS service to over 100 different customers in a variety of government and commercial organizations.

The Orbcomm S-AIS network and service offers valuable data for applications that maximize global maritime safety and security, enabling the maritime industry to know where nearly every vessel is located, where it is going and when it will arrive at its destination. Orbcomm has already provided access to tracking and detecting information from well over 120,000 unique vessels daily by leveraging proven expertise and existing worldwide ground infrastructure. In addition, by partnering with some of the most trusted maritime information providers in the world, such as Inmarsat, the Orbcomm system may offer the most complete situational picture of global vessel activity.

In May 2004, the US Coast Guard awarded Orbcomm a contract to develop and supply new S-AIS service and network to meet their national security requirements. By utilizing Orbcomm S-AIS service, security and intelligence departments around the world can know where nearly every vessel is located, where it is going and when it will get there. In fact, these valuable data can be used to quickly react to anomalies at sea such as piracy and suspicious movements, contraband and route deviation, SAR, fishery and environmental monitoring and other unusual behavior.

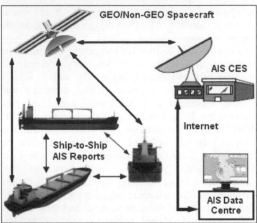

Figure 9. Satellite AIS Network – Courtesy of Manual: by Ilcev [2]

Except Orbcomm satellite system, to establish S-AIS system can be used existing GEO and Non-GEO satellite networks on L or UHF-band, shown in Figure 9.

In fact, it will be necessary to provide adequate satellite constellation, CES terminals, mission operation, data processing centre, operation centers and customer delivery of S-AIS service. Onboard ships S-AIS equipment can send two types of messages, the first type is inter ship communication of ship-to-ship AIS reports, and second type is direct transmission of AIS messages via SES, Internet or terrestrial communication line to the AIS Data Centre. The shore AIS Data centre provides processing of all receiving data and is forwarding S-AIS messages instantly to the customer facilities via Internet.

6.1 *Enhancements of Surveillance and Security*

The proposed S-AIS system and network will improve monitoring, surveillance and security in navigation of ships at deep sea and coastal waters where R-AIS has not coverage. Namely, the new S-AIS network maximizes maritime safety and security, which service is thought to be the most significant development in maritime navigation safety and security since the introduction of radar.

This system is a shipboard broadcast system that also transmits a vessel's identification, position and other critical information to provide the most complete and timely situational picture of vessel activity worldwide. Thus, the benefits of S-AIS in helping maritime authorities and shippers globally are to enhance maritime domain awareness and surveillance through detecting and identification of all ship's sailing route deviation and suspicious movements. In fact, by utilizing S-AIS service, security and intelligence departments around the world can find out where nearly every vessel is located, where it is going and when it will get there. These agencies can use this valuable data to quickly react to any anomalies at sea, such as suspicious movements, route deviation and other unusual behavior of pirate and contraband boats.

The use of data fusion by merging S-AIS data with sensors such as electro optical imaging and Satellite Synthetic Aperture Radar (SSAR) enables the rapid and reliable identification of S-AIS-emitting vessels and highlights non-S-AIS-emitting vessels. This data has proven to be beneficial for government authorities responsible for security, fisheries, exclusive economic zones and environmental monitoring in improving security and safety efforts. From more efficient management of port traffic, to support of national surveillance initiatives, to collision avoidance and other benefits, S-AIS service is helping to keep global waterways more safe and secure.

Therefore, received data from these agencies can be used just for commercial purposes, however for collision avoidance new S-AIS system cannot work without establishment of special ground tracking centres, which have to provide positions of all adjacent ships to the certain ship requesting these data. In such a way, ships captain getting this electronically information with positions of all surrounded ships can easily and safe navigate even in the extremely bad weather conditions.

6.2 *Improvement of Counter Piracy and Suspicious Movement*

Piracy in high-risk areas has become a major threat to regional trade and maritime security in recent years. As the frequency and aggressiveness of pirate attacks has increased, the need for S-AIS service in mitigating the risk of piracy is more important than ever before. As stated before, it can be used to locate approach of pirate boats, suspicious movements and to identify ships of interest, such as terrorist attacks, contraband and so on. Namely, the S-AIS can help alert vessels to a potential threat so fleet operators can avoid, deter or delay piracy attacks and greatly reduce risks to the vessel and crew. This system reduces the escalating threat and dangerous impacts of piracy and improves vessel safety and security is more important than ever before.

The S-AIS system and network enhance detecting of ships movement at sea, identifies vessels of interest and filter out "friendly" ships, especially in high-danger areas. If there is a data anomaly with regard to ID, location or speed, maritime authorities can use S-AIS data to determine the variance in reported versus actual status and take immediate action. This system can also be used to reduce the need for routine patrol missions at sea and help dispatch the appropriate authorities quickly when security incidents occur and in such a way significantly to improve the efficiency of all maritime operations. In addition, by providing proven and reliable real global coverage, S-AIS delivers uninterrupted satellite data service even when the traditional shipboard communications systems and alarm transponders may have been compromised.

Post-piracy tracking and reporting via S-AIS can be also used to identify typical ships traffic patterns and activities in high-risk sea areas, which can significantly help local and national security organizations proactively counter piracy hijackings, kidnappings and extortion.

6.3 *Better Facilitated Fisheries and Environmental Monitoring*

The new S-AIS network is providing access to timely, accurate vessel data as an instrumental in supporting fisheries management, environmental protection, all pollution preventions and modern transport operational compliance programs in global waterways. From the prevention of marine daily pollution incidents to the enforcement of fishery regulations to vessels traffic management, S-AIS service provides the maritime industry with complete and reliable visibility over ships activity worldwide. The S-AIS information service is also cost-effective and reliable resource for shippers, ship operators, seaport authorities and government agencies to help prevent environmental disasters, enforce fishery regulations and ensure the safety of mariners.

The S-AIS service can be used to manage fishing activities, quotas and harvesting limits by alerting authorities of vessels entering closed or protected environmental zones. Namely, the best result is a reduction in illegal fishing and the preservation of depleting ocean resources. Thus, this system is also able to track historical traffic patterns and identify violators within these protected areas.

Environmental organizations leverage S-AIS data to determine if a vessel has been in an area where oil, hazardous waste or ships ballast has been deliberately discharged or leaked to determine who is responsible for polluting the water. This service can also provide valuable data to all maritime authorities regarding the activity of vessels around and within such environmental or navigation hazards at sea, enhancing maritime safety and enabling those authorities to take immediate necessary actions. Besides, in support of fisheries control and environmental protection, relevant maritime authorities around the world have to detect and identify illegal, unreported and unregulated fishing activities. As stated earlier, the S-AIS system can integrate its satellite data with sensors such as surveillance radar and deliver comprehensive monitoring reports to authorities responsible for fishery ships, exclusive economic zones and environmental monitoring.

6.4 *More Reliable Search and Rescue Operations*

In an emergency incident when loss of life or the watercraft is imminent or threatened by grave danger, timing is becoming everything. The faster that information can be communicated to the first responders such as coast guards, maritime or other competent authorities, the faster that help can be sent on the way.

The most effective solution for improving ship safety including the precision and efficiency of international SAR operations is Cospas-Sarsat, LRIT and after that global S-AIS data service. This service is a new shipboard broadcast system that transmits a vessel's identification, position, speed and detecting other critical data to provide a complete and reliable situational picture of near-real-time vessel activity worldwide. For vessels in distress alert and during SAR operations the S-AIS service can identify exactly where a vessel is located anywhere in the world even if it continues to drift from the distress location.

Access to accurate, reliable and timely data about the position and status of a vessel and its crew can greatly improve response time by focusing SAR resources to a specific area and enhancing overall rescue coordination. Most importantly, S-AIS data can help minimize damage to and loss of the vessel, potentially saving lives. This valuable positioning service is also helpful in tracking the status of the rescue team and reducing risk to the rescuers, especially in treacherous heavy weather or water conditions.

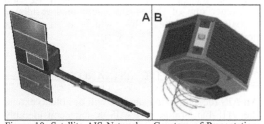

Figure 10. Satellite AIS Network – Courtesy of Presentation: by Ilcev [5]

Orbcomm and other Non-GEO satellite operators today are providing efforts to develop VHF or UHF S-AIS network for all mobile applications and especially for maritime commercial and SAR on scene operations. In such a way, Orbcomm Little LEO satellite operator is the first mobile commercial satellite network who is developing and implementing S-AIS Data Service.

In 2008, Orbcomm team launched the first LEO satellites specially equipped with the capability to collect AIS data and has plans to include these capabilities on all future satellites for ongoing support of global safety and security initiatives. Orbcomm mobile satellite operator recently has successfully launched six satellites with S-AIS equipped payloads. Orbcomm's next launches started in 2011, which AIS satellite is shown in Figure 10 (A). Following the development of Orbcomm system is designed sample of the M3MSat Maritime AIS CubeSat (Nano) satellite for the Canadian government, shown in Figure 10 (B), which was launched in 2012.

The future S-AIS network is ideal for global satellite ships tracking and surveillance useful by maritime and government administration, but as stated earlier, is not dedicated for a real collision avoidance of ships. Thus, to enhance R-AIS and S-AIS service has to be provided the similar system as proposed Global Ship Tracking (GST).

The GST system is augmented LRIT network, which except ship tracking provides tracking of captured ships by pirates and enhanced collision avoidance. Its Tracking Control Stations (TCS) is collecting positioning data of all ships in any sea area, indicating them on the radar like display and distributing them on any ship request sailing in certain sea area.

The disadvantages of Nano and LEO satellites is that they need several dozen of spacecraft to build complete coverage, could interfere with satellite operations and space missions, short equipment lifespan, enhanced atmospheric drag, increasing danger from space debris from many non-working satellites, handover problems and short visibility over horizon. To get more reliable S-AIS network has to be employed GEO satellites, as the best solutions for S-AIS. In Figure 11 is shown by Cain & Meger one of the S-AIS operational results to globally track the ships, which is useful for commercial purposes [2, 3, 5, 10].

7 NANO SATELLITE AIS (NANO S-AIS)

An AIS receiver using satellite will be able to extend the VHF-band range of R-AIS systems considerably and make it easier to monitor ship traffic and fishing in the High North areas.

Figure 11. Operational results of S-AIS Network – Courtesy of Paper: by Cain [10]

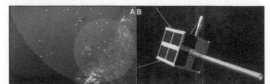

Figure 12. Coverage of AIS Networks and Nano-AIS – Courtesy of Presentation: by Ilcev [5]

Recently many countries developed cost effective small satellites in LEO constellations. They are categorized by their weight, such as Pico less than 1 – 5 Kg, Nano less than 10 – 50 Kg and Micro satellites less than 100 – 200 Kg, which measure less than 0.5 meter The ITU regulations are not geared for these smaller satellites, but recently were proposed their frequencies in the range of 137 MHz to 2,450 MHz.

No matter what smaller satellites have limited lifetime, short lifetime of batteries and orbit control capabilities, they can be used as cost-effective missions for tracking, detecting and remote sensing. Based on different studies and examples in which Nano satellites were exposed to the different levels of radiation, some CubeSat electronics Secure Digital (SD) cards especially are susceptible to any errors from radiation. In addition, space radiation can be mitigated to some degree through shielding and material choice or via special software by clever use of watchdog timers and special "self-aware" coding as well, where self-verification is consistently monitored.

With regards to the smallest satellite life time, CubeSat at orbit lower than 300 Km will be 0-100 days, these from 300 to 400 Km are a danger of collision with the International Space Station (ISS), because that the orbit of ISS is usually maintained between 335 km perigee and 400 km apogee. A CubeSat in this altitude band could last for 0.5 to 2 years. In contrast, the lifetimes of higher altitude satellites than 600 Km could be theoretically up to two decades.

In Figure 12 (A) is illustrated larger coverage area using S-AIS or even AIS via High Altitude Platforms (HAP) versus using smaller coverage of conventional system of VHF R-AIS. The altitude of the LEO satellite or HAP station gives the AIS receiver a large range of coverage and both can therefore make observations over extended sea areas. The VHF AIS signals are strong enough to be received by an SCP or satellite.

The new developed AISSat-1 is a Nano satellite in LEO measuring 20x20x20 cm, weight is 6 Kg and is shaped like a cube (CubeSat), shown in Figure 5.24 (B). The satellite payload is designed by the Kongsberg-Seatex AS Company and the purpose of the satellite is to improve ships surveillance of maritime activities in the High North sea areas. It is believed that the low traffic density in the High North requires one receiver and antenna only to handle the expected volume of AIS messages. So, the AISSat-1 satellite is being launched in order to test these presumptions and if is successful to be used for S-AIS facilities.

The AISSat-1 satellite will operate in a Polar orbit at an altitude of about 600 Km and is proposed to be launched by the PSLV rocket of the Indian Space Research Organization (ISRO). Thus, the Norwegian Space Centre is project owner and the Norwegian Coastal Administration (FFI) will receive the data, while the Norwegian Defense Research Establishment is responsible for the technical implementation. Additional information about this project is possible to see at Web pages of Kongsberg Maritime.

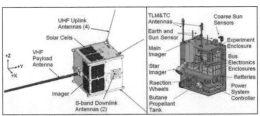

Figure 13. NTS/AISSat and SumbandilaSat Satellites – Courtesy of Brochures: SFL/US [11, 12]

The Norwegian AIS transponder is also placed in a Canadian Nano satellite, built by the University of Toronto (UOT), which life span is estimated to three years. The Institute for Aerospace Studies/Space Flight Laboratory (UTIAS/SFL) of UOT has been developed prototype of Generic Nanosatellite Bus (GNB) to fly a verity of payloads, ranging from S-AIS tracking solutions to precision formation flying. With the successful launch of the CanX-2 mission, technological validation is paving the way for the next generation of GNB derived CanX missions.

The COM DEV Ltd launched Nanosatellite Tracking of Ships (NTS) spacecraft at the end of April 2008 following an unprecedented 8-month kick-off to launch cycle, which components are shown in Figure 13 (Left). In Figure 13 (Right) is shown CubeSat SumbandilaSat, microsatellite designed by Stellenbosch University (SU) in Cape Town, South Africa. This satellite launched 17 September 2009 by Russian Soyuz-2 launch vehicle from the Baikonur Cosmodrome will serve for Earth observation with design lifetime of 3 years at an orbit altitude of 500 Km (subject to average sun activity) [2, 5, 11, 12].

8 CONCLUSION

In this article were discussed basic of R-AIS with major class of AIS transponders for installation onboard ships and BS for shore installations. In addition is introduced new S-AIS via Orbcomm Little LEO satellite constellation and via proposed small LEO satellites. Here the question is aroused, namely, whether Little LEO satellites, such as Orbcomm or small satellites can provide full, cost effective coverage and reliable S-AIS service? The total cost of new generation Orbcomm network with 36 satellites in constellation is about 234 million US$ and will be able to provide global coverages except both poles. However, the total cost of the AISSat-1 Nano satellite is approximately 30 million NOK (Norwegian Crown) or about 4 million US$ each.

The another question is, how many Nano satellites will be necessary for total coverage of Earth including polar areas, when in the same time Big LEO Globalstar has 40 satellites in constellation and Iridium with 66 satellites in constellation is providing full coverage of Earth including both poles. Inmarsat GEO satellite operator is only professional network providing near global coverage up to 800 North and South. At present ships are not sailing in Arctic Ocean, but Russian are proposing these routes, so at present only Iridium is able to provide polar coverage.

However, for additional coverage over North Pole Inmarsat needs Hybrid Satellite Orbits (HSO), such as GEO and Medium Earth Orbit (MEO) or GEO with High Elliptical Orbit (HEO) Russian Molniya spacecraft. However, at present Inmarsat is setting advanced new Inmarsat-5 GEO satellite constellation, which will provide broadcasting and broadband satellite office at sea. Perhaps cost of about 50 million US$ to launch one Inmarsat-5 spacecraft is not extremely high for significant professional maritime communications.

REFERENCES

[1] IMO, "AIS Transponders", London, 2015 [http://www.imo.org/OurWork/Safety/Navigation/Pages/AIS.aspx].
[2] Ilcev D.S., "Global Mobile Communications, Navigation and Surveillance (CNS)", Manual, DUT, Durban 2014 [www.dut.ac.za/space_science].

[3] Skoryk I., "Radio and Satellite Tracking and Detecting Systems for Maritime Applications", Doctoral Thesis", Durban University of Technology (DUT), Durban, 2014.

[4] CNS Systems, "VDL 6000 AIS Class-A & Secure Class-A Transponders", Linkoping, Sweden, 2014 [http://www.cns.se/].

[5] Ilcev D. S., "Global Ship Tracking, Automatic Identification System and Determination", Research Group in Space Science, DUT, Presentation, 2011 [www.dut.ac.za/space_science].

[6] Icom, "Class-B AIS Transponder", Osaka, Japan, 2014 [http://www.icom.co.jp/world/].

[7] Orbcomm, "Global Visibility Beyond Coastal Regions", AIS, Rochelle Park, 2015 [http://www.orbcomm.com/networks/ais]

[8] AMSA, "Automatic Identification System (AIS) Class A", Australian Maritime Safety Authority, Canberra, 2015, http://www.amsa.gov.au/forms-and-publications/fact-sheets/aisa_fact.pdf

[9] MRS, Automatic Identification System (AIS), Marine Radio Service, Klang, Malaysia, 2015 http://www.mrs-marine.com.my/index.php/products-services/ais-class-b

[10] Cain J., & Others, "Nanosatellite Tracking of Ships - Review of the First Year of Operations", 7th Responsive Space Conference, Los Angeles, 2009 [http://www.responsivespace.com/Papers/RS7/SESSIONS/Session%20V/6005_Newland/6005P.pdf]

[11] SFL, & Others, "Nanosatellite Tracking Ships: Cost-Effective Responsive Space", Space Flight Laboratory (SFL), Toronto, 2014 [http://utias-sfl.net/wp-content/uploads/IAC2010-S16-1-Nanosatellite-tracking-F-Pranajaya.pdf].

[12] Steyn S., "Satellite Engineering Research at US", Cape Town, 2012 [http://staff.ee.sun.ac.za/whsteyn/Papers2/US_Satellite_Research.pdf].

Satellite Antenna Infrastructure Onboard Inmarsat Spacecraft for Maritime and Other Mobile Applications

D.S. Ilcev

Durban University of Technology (DUT), South Africa

ABSTRACT: In this paper is introduced antenna systems onboard spacecraft for Inmarsat Aeronautical and other Mobile Satellite System (MSS), such as communications, tracking, monitoring and logistics solutions between mobile units and Gateways or Ground Earth Stations (GES) accomplished via Geostationary Earth Orbit (GEO) Satellite Constellation. Inmarsat GEO MSS is employing advanced technology and technique to deliver Voice, Data and Video (VDV) for all mobile applications around the world, excluding Polar Areas. The Inmarsat organization got sufficient funding to implement at first solutions for Maritime applications and in the next stage to develop additional service such as Land (Road and Rail) and finally for Aeronautical applications. Inmarsat team surpassed all problems and challenges getting attribute of only one global mobile satellite operator with professional attribute. With respect to improvements of Space Segment and service in particular, antenna systems onboard Inmarsat spacecraft is discussed. The characteristics of modern Spacecraft antenna and Link Performance with Monobeam and Multibeam Antenna Coverage are introduced. At this point, the possible basic types of antenna installed onboard spacecraft for MSS are presented.

1 INTRODUCTION OF INMARSAT GEO MSS

The first GEO Mobile Satellite Communications (MSC) maritime system was proposed and built by the International Maritime Organization (IMO) from London in 1979, known as the International Maritime Satellite Organization (Inmarsat). At the beginning Inmarsat was not-profit organization for providing maritime satellite communications including distress and safety solutions. In the next stages Inmarsat developed additional services for land (road and rails) and aeronautical applications. For the first decade of operations, Inmarsat leased the Space Segment form Comsat (three Marisat satellites F1, F2 and F3), from ESA (two Marecs satellites A and B2) and from Intelsat (three Intelsat V-MCS A, B and D). These Inmarsat satellites were initially configured in three ocean regions: AOR, IOR and POR, each with an operational satellite and one spare in-orbit. This satellite constellation is known as the first generation of the Inmarsat network. Inmarsat was not responsible for TT&C, but operations were controlled by Inmarsat Network Control Centre (NCC) in London. Then, Inmarsat operated with four 2nd generation Inmarsat-2 birds launched in 1990/92 with a capacity equivalent to about 250 Inmarsat-A voice circuits. These satellites

were built to provide coverage of four ocean regions: AOR-E, AOR-W, IOR and POR by the Space and Communications division of British Aerospace (Matra Marconi Space).

The US Lockheed Martin company built the new spacecraft bus for next Inmarsat-3 generation of spacecraft, based on the GE Astro Space Series 4000, 2.5 m high and with a 3.2 radial envelope centered on a thrust cone. Thus, Matra built the communications payload, antennas, repeater and other electronic equipment. Payload and solar arrays are mounted on N and S-facing panels, while L-band Receiver (Rx) and Transmitter (Tx) reflectors, mounted on E and W panels, are fed by an array of cup-shaped elements. Furthermore, the navigation antenna is located on the Earth-facing panel. The tremendous advantage of the Inmarsat-3 satellites is to concentrate power on particular areas of high traffic within the footprint. Each GEO satellite utilizes a maximum of seven spot beams (A) and one global beam coverage (B) shown in Figure 1.

These satellites also can re-use portions of the L-band for non-adjacent spot beams, effectively doubling the capacity of the satellite. Responding to the growing demands from the corporate mobile satellite users of high-speed Internet access and multimedia connectivities, Inmarsat built fourth generation of Inmarsat-4 satellites as a Gateway for the new mobile and personal satellite broadband network. Inmarsat has awarded European Astrium a 700 million US$ contract to build three Inmarsat I-4 satellites, which will support the new Broadband Global Area Network (BGAN). The BGAN personal and mobile applications have to deliver Internet and Intranet content and solutions, video on demand, videoconferencing, fax, E-mail, phone and LAN access onboard mobiles at speeds up to 432 Kb/s worldwide, compatible with 3G cellular systems.

Finally, Inmarsat has contracted the US Boeing, to build a constellation of new three multipurpose Inmarsat-5 satellites as a part of a global wireless broadband network Inmarsat Global Xpress. The first satellite Inmarsat-5 F1 entered in commercial service on 30 June 2014. The second and third I-5 satellites are on course to launch by the end of 2014 and will provide global coverage during 2015. The spacecraft will provide radio spectrum on L/C and Ka-band for both communication and GNSS (navigation) facilities. The most important parts of Inmarsat spacecraft are antenna for satellite communications and navigation facilities [1, 2, 3, 4].

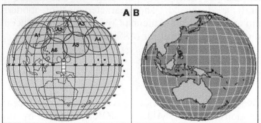

Figure 1. Spacecraft Antennas Spot and Global Beams – Courtesy of Book: by Gallagher [2]

2 BASIC PARTICULARS OF SPACECRAFT ANTENNA

The spacecraft antenna radiates EM energy to the ground stations in both directions, efficiently and in desired path. Satellite antennas act as matching systems between sources of Electro Magnetic (EM) energy and space. The goal in using antennas is to optimize this matching. Here is a list of some of the properties of antennas:

1 Field intensity for various directions (antenna pattern);
2 Total power radiated when the antenna is excited by a current/voltage of intensity (Power Flux Density);
3 Radiation efficiency which is the ratio of power radiated to the total power (Radiation Pattern);
4 The input impedance of the antenna for maximum power transfer (matching); and
5 The antenna bandwidth or range of frequencies over which these properties are nearly constant.

However, spacecraft antennas can also be classified as electrical devices which convert electric currents into radio waves and vice-versa. They are generally used with a radio transmitter and receiver, which are broadly classified in two categories: Transmitting and Receiving antennas.

The difference is in the mode of operation, different functions etc. as the transmitting as well as the receiving antenna, and also difference is mainly in their environmental conditions which lead to their different designs.

Typically an antenna has an array of metallic conductors that are electrically connected. An oscillating current of electrons focused through the antenna by a transmitter creates an oscillating electric field. These fields are time-varying and radiate from the antenna into the space as a moving electromagnetic field wave. Certain properties of antennas such as directional characters result into reciprocity theorem.

The different types of spacecraft antennas are:

1 Wire Antennas (Monopoles and Dipoles) – The dipole is one of the most common used antennas. This spacecraft antenna consists of a straight conductor excited by a voltage from a transmission line or a waveguide and dipoles are easy to make, shown in Figure 2 (A). Wire satellite antennas are used primarily at VHF and UHF-band to provide communications for the Telemetry, Tracking and Command (TT&C) systems. They are positioned with great care on the body of the satellite in an attempt to provide omnidirectional coverage. Most satellites measure only a few wavelengths at VHF frequencies, which make it difficult to get the required antenna patterns, and there tend to be some orientations of the satellite in which the sensitivity of the TT&C system is reduced by nulls in the antenna pattern.

2 Aperture Antennas (Horn Antennas) – A horn is an example of an aperture antenna, which are used in Satellite spacecraft more commonly, shown in Figure 2 (B). Rectangular horn antenna is one of the simplest and most widely used antennas. Horns have been used for more than a hundred years, and today they used in radio astronomy, satellite communications, in communication dishes as feeders, in measurements, etc. Horn antenna is used at MW when for global coverage relatively wide beams are required. A horn is a flared section of waveguide that provides an aperture several wavelengths wide and a good match between the waveguide impedance and free space. It is also used as feeds for reflectors, either singly or in clusters. Horns and reflectors are examples of aperture antennas that launch a wave into free space from a waveguide. It is difficult to obtain gains much greater than 23 dB or beamwidths

narrower than about 10° with horn antennas. For higher gains or narrow beamwidths a reflector antenna or array must be used.

3   Reflector Antennas – The parabolic reflector is a good example of reflectors at microwave frequencies, shown in Figure 2 (C). In the past, parabolic reflectors were used mainly in space applications onboard spacecraft but today they are very popular and are used by almost everyone who wishes to receive the large number of television channels transmitted all over the globe.

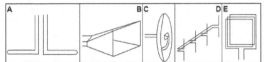

Figure 2. Types of Spacecraft Antenna Systems – Courtesy of Manual: by Ilcev [3]

Reflector antennas are typically used when very high gain or a very narrow main beam is required. Gain is improved and the main beam narrowed with increase in the reflector size. Large reflectors are however difficult to simulate as they become very large in terms of wavelengths. Reflector antennas are usually illuminated by one or more horns and provide a larger aperture than can be achieved with a horn alone. For maximum gain, it is necessary to generate a plane wave in the aperture of the reflector. This is achieved by choosing a reflector profile that has equal path lengths from the feed to the aperture, so that all the energy radiated by the feed and reflected by the reflector reaches the aperture with the same phase angle and creates a uniform phase front. One reflector shape that achieves this with a point source of radiation is the paraboloid, with a feed placed at its focus. The paraboloid, however, is the basic shape for most reflector antennas, and is commonly used for earth station antennas. Satellite antennas often use modified paraboloidal reflector profiles to tailor the beam pattern to a particular coverage zone. Phased array antennas are also used on satellites to create multiple beams from a single aperture, and have been used by Iridium and Globalstar to generate up to 16 beams from a single aperture for their LEO system.

4   Array Antennas – A grouping of several similar or different antennas form a single array antenna, shown in Figure 2 (D). The control of phase shift from element to element is used to scan electronically the direction of radiation. Antenna arrays are able to produce radiation patterns that combined, have characteristics that a single antenna would not. The antenna elements can be arranged to form a 1 or 2 dimensional antenna array. A number of antenna array specific aspects will be outlined using 1dimensional arrays for simplicity reasons. Antennas exhibit a specific radiation pattern, which overall radiation pattern changes when few antenna elements are combined in an array. The array factor quantifies the effect of combining radiating elements in an array without the element specific radiation pattern taken into account. The overall radiation pattern results in a certain directivity and thus gain linked through the efficiency of directivity. Directivity and gain are equal if the efficiency is 100%.

5   Loop Antennas – A loop of wire is used to radiate or receive electromagnetic energy. These antennas can be also used at home to capture signals of Radio or TV channels, shown in Figure 2 (E). An antenna pattern is a plot of the field strength in the far field of the antenna when a transmitter drives the antenna. The gain of an antenna is a measure in dB of the antenna's capability to direct energy in one direction, rather than all around. A useful principle in antenna theory is reciprocity, which means that an antenna has the same gain and pattern at any given frequency whether it transmits or receives. An antenna pattern measured when receiving is identical to the pattern when transmitting.

As stated earlier, the antenna is providing global, spot and multiple beam coverages, but it can provide scanning and orthogonally polarized beams or coverage zones as well. The pattern is frequently specified by its 3-dB beamwidth, the angle between the directions in which the radiated (or received) field falls to half the power in the direction of maximum field strength. However, a satellite antenna is used to provide coverage of a certain area or zone on the Earth's surface, and it is more useful to have contours of antenna gain with maximum strengths of the signal in the middle of the coverage area and with decreasing of signals to the peripheries. When computing the signal power received by an GES from the satellite, it is important to know where the station lies relative to the satellite transmit antenna contour pattern, so that the exact EIRP can be calculated. If the pattern is not known, it may be possible to estimate the antenna gain in a given direction if the antenna boresight or beam axis direction and its beamwidth are known.

All parts of spacecraft antennas, which have to be aligned normally are reflector (main reflector and sub reflector), feed and sometimes also support structures. These substructures allow a pre-assembly of reflectors and feeds subsystem level and an easier integration of the complete antenna onboard spacecraft. The goal of the alignment is to bring all the antenna components in a proper geometric configuration and to get the best or at least the designed RF antenna performance in the test facility and later on in the satellite orbit [1, 2, 3, 5, 6].

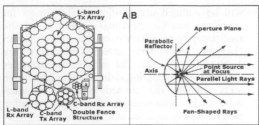

Figure 3. Spacecraft Antenna Systems – Courtesy of Book: by Gallagher [2]

3 SATELLITE ANTENNA SYSTEM ONBOARD SPACECRAFT FOR INMARSAT MSS

The antenna array system of Inmarsat-2 satellite for MSC is illustrated in Figure 3 (A). The satellite antenna system mounted on the spacecraft structure, similar to the transponders, is composed of two main integrated elements: the C/L-band and the L/C-band antenna.

1 Inmarsat-2 C/L-band Arrays – This uplink is actually the feeder link, which operates in the 6 GHz RF range. The signals sent by LES are detected by a C-band receiving array, comprising seven cup-dipole elements in the smallest circle. On the other hand, the L-band transmit antenna is the biggest segment of the whole system, consisting in 43 individual dipole elements, arranged in three rings around a single central element. Thus, this antenna is providing near-global coverage service downlink for MES in the 1.5 GHz RF spectrums.

2 Inmarsat-2 L/C-band Arrays – These arrays is actually the service uplink and operates in the 1.6 GHz RF range. The signals sent by MES in adjacent global coverage region are detected by L-band receiving array, comprising nine cup-dipole elements arranged in a circle. Finally, the C-band transmit antenna consists in seven cup-dipoles for radiation of the feeder downlink to LES in the 3.6 GHz RF spectrum.

3.1 Characteristics of Satellite Antennas

Both transmit antenna array systems are providing a global (wide) footprint on the Earth's surface. However, narrow circular beams from GEO or Non-GEO can be used to provide spot beam coverage. For instance, from GEO the Earth subtends an angle of 17.4o. Antenna beams 5.8o wide can reuse three frequency bands twice in providing Earth disc coverage. The directional properties of antenna arrays can be exploited to permit RF reuse in space communications, which is similar to several radio stations using the same RF being geographically far apart. Earth coverage by seven spot beams (six spots are set out around one spot in the centre) can be arranged by three pairs of beams: 1 and 4, 2 and 5 and 3 and 6, operating on frequencies f2, f3 and f4, respectively. Mutual interference within pairs is avoided by pointing one beam as far away from the other as possible. Coverage of the centre of the disc is provided by a single beam operating on frequency f1.

The main advantage with this spot footprint that is specific Earth areas can be covered more accurately than with wide beams. Furthermore, a greater power density per unit area for a given input power can be achieved very well, when compared with that produced by a global circular beam, leading to the use of much smaller receiving MES antennas. The equation that determines received power (PR) is proportional to the power transmitted (PT) separated by a distance (R), with gain of transmit antenna (GT) and effective area of receiving antenna (AR) and inverse proportional with 4π and square of distance. The relations for PR and GT are presented as follows:

$$P_R = P_T\, G_T\, A_R/4\pi R$$

$$G_T = 4\pi\, A_T/\lambda^2 \qquad (1)$$

where GT = effective area of transmit antenna and λ = wavelength. The product of PT and GT is gain, generally as an increase in signal power, known as an EIRP. Signal or carrier power received in a link is proportional to the gain of transmit and receive antennas (GR) presented as:

$$P_R = P_T\, G_T\, G_R\, \lambda^2/(4\pi R)^2 \; or$$

$$P_R = P_T\, G_T\, G_R/L_P\, L_K \quad [W] \qquad (2)$$

The last relation can be derived with the density of noise power giving:

$$P_R/N = P_T\, G_T\, (G_R/T_R)\, (1/K\, L_P\, L_K) \qquad (3)$$

where LP = coefficient of energy loss in free space, LK = coefficient of EMW energy absorption in satellite channels, TR = temperature noise of receiver, GR/TR is the figure of merit and K = Boltzmann's Constant (1.38 x 10–23 J/K or its alternatively value is –228.6 dBW/K/Hz).

At any rate, PR has a minimum allowable value compared with system noise power (N), i.e., the Carrier and Noise (C/N) or Signal and Noise (S/N) ratio must exceed a certain value. This may be achieved by a trade-off between EIRP (PT GT) and received antenna gain (GR). If the receive antenna on the satellite is very efficient, the demands on the LES/MES are minimized. Similarly, on the satellite-to-Earth link, the higher the gain of the satellite transmit antenna, the greater the EIRP for a given transmitter power. Satellites often have parabolic dish antennas, though there are also other types, such as phased arrays.

The principal property of a parabolic reflector is its ability to turn light from a point source placed at its focus into a parallel beam, mostly as illustrated in Figure 3 (B).

In practice the antenna beam can never be truly parallel, because rays can also be fan-shaped, namely a car headlamp is a typical example. In a microwave antenna the light source is replaced by the antenna feed, which directs waves towards the reflector.

The length of all paths from feed to aperture plane via the reflector is constant, irrespective of their angle of parabolic axis. The phase of the wave in the aperture plane is constant, resulting in maximum efficiency and gain. In such a way, the gain of an aperture (Ga) and parabolic (Gp) type of antennas are:

$$G_a = \eta \ (4\pi \ A/\lambda^2) = 4\pi \ A_E/\lambda^2$$

$$G_p = \eta \ (\pi \ D)^2 / \lambda^2 \qquad (4)$$

where antenna values η = efficiency factor, A = projected area of antenna aperture, $A_E = \eta A$ is effective collecting area and D = parabolic antenna diameter.

Thus, owing to correlation between frequency and wavelength, $f = c/\lambda$ is given the following relations:

$$G_p = \eta \ (\pi \ D \ f/c)^2 = 60,7 \ (D \ f)^2 \qquad (5)$$

where the second relation comes from considering that $\eta \approx 0.55$ of numerical value. If this value is presented in decibels the gain of antenna will be calculated as follows:

$$G_I = 10 \log G_p \qquad (6)$$

For example, a satellite parabolic antenna of 2 m in diameter has a gain of 36 dB for a frequency at 4 GHz and a gain of 38 dB for a frequency at 6 GHz. In such a way, satellite parabolic antennas can have aperture planes that are circular, elliptical or rectangular in shape.

Thus, satellite antenna with circular shape and homogeneous illumination of aperture with a gain of –3 dB has about 47.5% of effective radiation, while the rest of the power is lost. To find out the ideal characteristics it is necessary to determine the function diagram of radiation in the following way:

$$F \ (\delta_0) = s \ (\delta_0)/s \ (\delta_0 = 0) \qquad (7)$$

where parameter s (δo) = flow density of radiation in the hypothetical satellite angle (δo) and s (δo=0) = flow density in the middle of the coverage area. Looking the Geometric Projection of Satellite the relation can be presented by the equation:

$$F \ (\delta_0) = d_0/h = \cos \delta \ \sqrt{(k^2 - \sin^2 \delta_0)}/1 - k \qquad (8)$$

where, as mentioned, $k - R/(R + h)$ sin δ and if δo = δ, the relation is defined by the following equation:

$$F \ (\delta) = k \cos \delta \qquad (9)$$

For GEO satellite the value of ΔL is given as a function of angle δ, which is the distance from the centre of the coverage area, where the function diagram of the radiation is as follows:

$$F \ (\delta) = \Delta L = 20 \log = 20 \log R/(R + h) \cos \delta = 10$$

$$\log R/(1 + 2R/h) \ \text{[dB]} \qquad (10)$$

Therefore, in the case of GEO satellites the losses of antenna propagation are greater around the periphery than in the centre of the coverage area for about 1.32 dB. The free-space propagation loss (LP) and the input level of received signals (LK) are given by the equations:

$$L_P = (4\pi \ d/ \ \lambda)^2$$

$$P_R/S = P_T \ G_T/4\pi d^2 \ L_K \qquad (11)$$

The free-space radio propagation loss is caused by geometrical attenuation during propagation from the satellite transmitter to the receiver [1, 2, 6, 7, 8, 9].

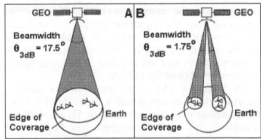

Figure 4. Global Monobeam and Multibeam Antenna Coverage – Courtesy of Book: by Maral [4]

4 LINK PERFORMANCE WITH MONOBEAM AND MULTIBEAM ANTENNA COVERAGE

As stated earlier, the most important parameter of spacecraft transponder and the overall RF link quality depends on the gain of the satellite antenna. From equation (6), it can be seen that the satellite antenna gain is constrained by its beamwidth, whatever the frequency at which the link is operated. So the antenna gain is imposed by the angular width of the antenna beam covering the zone to be served. If the service zone is covered using a single antenna beam, this is referred to as single or monobeam beam satellite coverage, which displays one of these characteristics:

– The satellite may provide coverage of the whole region of the Earth, which is visible from the satellite as a global coverage and thus permit long-distance links to be established, for example from one continent to another with 20 dB bandwidth. In this case, the gain of the satellite antenna is limited by its beamwidth as imposed by the coverage.

- The satellite may provide coverage of only part of the earth (a region or country) by means of a narrow beam (a zone or spot beam), with 3dB beamwidth of the order of 1° to a few degrees.

With single beam antenna coverage, it is therefore necessary to choose between either extended coverage providing service with reduced quality to geographically dispersed GES terminals, or reduced coverage providing service with improved quality to geographically concentrated GES terminals.

Multibeam antenna coverage allows these two alternatives to be reconciled. However, satellite extended coverage may be achieved by means of the juxtaposition of several narrow beam satellite coverages, which each beam providing an antenna gain which increases as the antenna beamwidth decreases (reduced coverage per beam). The link performance improves as the number of beams increases; the limit is determined by the antenna technology, whose complexity increases with the number of beams, and the mass. The complexity originates in the more elaborate satellite antenna technology and the requirement to provide on-board interconnection of the coverage areas, so as to ensure within the satellite payload routing of the various carriers that are unlinked in different beams to any wanted destination beam.

In Figure 4 (A) is presented that a satellite provides global coverage with a single satellite beam (monobeam) of beamwidth and in Figure 4 (B) is illustrates that satellite supports spot beams with beamwidth of a consequently reduced coverage, known as multibeam satellite coverage. In both cases, all GES terminals in the satellite network are within the correspondent satellite coverage or in LOS with satellite. Multibeam coverage is providing the following advantages:

1. Impact on the Earth Segment – The satellite communication link performance is evaluated as the ratio of the received carrier power C to the noise power special density N0 and is quoted as the C/N0 ratio, expressed in Hz. The expression for $(C/N0)_U$ for the uplink (U) is given by the following equation:

$$(C/N_0)_U = (EIRP)_{station} (1/L_U) (G/T)_{satellite} (1/k) \text{ [Hz]} \tag{12}$$

Assuming that the noise temperature at the satellite receiver input is $T_{satellite}$ = 800 K = 29 dBK and is independent of the beam coverage (this is not rigorously true but satisfies a first approximation), let LU = 200 dB and neglect the implementation losses. This equation becomes (all terms in dB) and can be presented as:

$$(C/N_0)_U = (EIRP)_{station} - 200 + (G_R)_{satellite} - 29 +$$
$$228.6 = (EIRP)_{station} + (G_R)_{satellite} - 0.4 \text{ [dBHz]} \tag{13}$$

where value $(G_R)_{satellite}$ is the gain of the receiving satellite antenna in the direction of the GES transmitting terminals. This relation is represented by the two cases considered receiver:

- Global coverage (θ3 dB = 17.5°), which implies $(G_R)_{satellite}$ = 29 000/(θ3 dB)2 ≈ 20 dBi.
- Spot beam coverage (θ3 dB = 1.75°), which implies $(G_R)_{satellite}$ = 29 000/(θ3 dB)2 ≈ 40 dBi.

The expression for $(C/N0)_D$ for the downlink (D) is given by:

$$C/N_0)_D = (EIRP)_{satellite} (1/L_D) (G/T)_{station} (1/k) \text{ [Hz]} \tag{14}$$

Assume that the power of the carrier transmitted by the satellite is PT = 10 W = 10 dBW. Let LU = 200 dB and neglect the implementation losses. Thus, this equation becomes (all terms in dB):

$$(C/N_0)_D = 10 - 200 + (G_T)_{satellite} + (G/T)_{station} + 228.6$$

$$= (G_T)_{satellite} + (G/T)_{station} + 38.6 \text{ [dBHz]} \tag{15}$$

This relation is represented for the two cases considered transmitter:

- Global coverage (θ3 dB = 17.5°), which implies $(G_T)_{satellite}$ = 29 000/(θ3 dB)2 ≈ 20 dBi.
- Spot beam coverage (θ3 dB = 1.75°), which implies $(G_T)_{satellite}$ = 29 000/(θ3 dB)2 ≈ 40 dBi.

In case those values indicate the reduction in $(EIRP)_{station}$ and $(G/T)_{station}$ the transmission system is changing from a satellite with global coverage to a multibeam satellite with coverage by several spot beams. In this case, the multibeam satellite permits an economy of size, and hence cost, of the earth segment. For instance, a 20 dB reduction of $(EIRP)_{station}$ and $(G/T)_{station}$ may result in a tenfold reduction of the antenna size (perhaps from 30 m to 3 m) with a cost reduction for the GES terminal for more than 100 times. If an identical GES is retained (a vertical displacement towards the top), an increase of C/N0 is achieved which can be transferred to an increase of capacity, if sufficient bandwidth is available, at constant signal quality in terms of Bit Error Rate (BER) [1, 4, 10, 11].

2. Frequency Reuse – Frequency reuse consists of using the same frequency band several times in such a way as to increase the total capacity of the satellite network without increasing the allocated bandwidth (B). In the case of a multibeam satellite the isolation resulting from antenna directivity can be exploited to reuse the same frequency band in separate beam coverages.

The frequency reuse factor is defined as the number of times that the bandwidth is used. In theory, a multibeam satellite system with M single-polarization antenna beams, each being allocated the bandwidth, combines reuse by angular separation and reuse by orthogonal polarization may have a frequency reuse factor equal to 2M. This signifies that it can claim the capacity which would be offered by a single beam satellite with single polarization using a bandwidth of M x B. In practice, the frequency re-use factor depends on the configuration of the service area which determines the coverage before it is provided by the satellite. If the service area consists of several widely separated regions (for example, urban areas separated by extensive rural areas), it is possible to reuse the same band in all beams. The frequency reuse factor can then attain the theoretical value of M. Figure 4 (B) shows an example of multibeam coverage.

On the other hand, multibeam coverage is providing the following disadvantages:

1 Interference Between Beams – In practical reality the interference generation within a multibeam satellite system is called self-interference. Thus, the effect of self-interference appears as an increase in thermal noise under the same conditions as interference noise between systems. At this point, it must be included the term (C/N0)1, which expresses the signal power in relation to the spectral density interference.

Taking account of the multiplicity of sources of interference, which become more numerous as the number of beams increases, relatively low values of (C/N0)1 may be achieved and the contribution of this term impairs the performance in terms of (C/N0)T of the total link. As modern satellite systems tend to re-use frequency as much as possible to increase capacity, self-interference noise in a multibeam satellite link may contribute up to 50% of the total noise.

2 Interference Between Coverage Areas – A satellite payload using multibeam coverage must be in a position to interconnect all network Earth stations and consequently must provide adequate interconnection of the entire coverage areas. The complexity of the payload is added to that of the multibeam satellite antenna subsystem, which is already much more complex than that of a single beam satellite. Different techniques, depending on the onboard satellite processing capability (no processing, transparent processing, regenerative processing, etc.) and on the network layer, are considered for interconnection of coverage:
  – Interconnection by transponder hopping (no on-board processing);
  – Interconnection by onboard spacecraft switching (transparent and regenerative processing); and – Interconnection by beam scanning.

Multibeam satellite systems make it possible to reduce the size of GES terminal and hence the cost of the Earth segment infrastructure. Frequency reuse

from one satellite beams to other permits an increase in capacity without increasing the bandwidth allocated to the system.

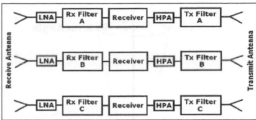

Figure 5. Multibeam Antenna Coverage Transponder – Courtesy of Book: by Swan [6]

However, interference between adjacent satellite channels, which occurs between beams using the same frequencies, limits the potential capacity increase, particularly as interference is greater with earth stations equipped with small antennas.

The simplest form of a payload with multibeam antenna radiation is illustrated in Figure 5. A three-transponder payload uses one transponder per coverage circle. At this point, there is not connectivity between satellite coverage areas in this simple transponder.

However, this payload could be designed so each transponder antenna illuminates three coverage circles and provide connectivity, but would cover three times area with just one-third the gain.

The relatively simple changes to multiple small beams have significant consequences as:

1 The area covered by each beam is much smaller, increasing the satellite antenna gain and allowing smaller and less expensive ground and mobile terminals;

2 The same total RF power can be used to carry more traffic, and/or reduce the RF power;

3 The same transponder bandwidth can be used multiple beam antennas, greatly increasing the available bandwidth of the satellite:
  – This allow the same bandwidth to be reused, increasing the amount that can be accommodated within the bandwidth; and
  – A terminal has to be tuned to the correct RF to function and retuned if it is moved.

4 Connectivity between beams, if required for the mission, must be provided by additional hardware on the satellite since a single uplink does not encompass the entire coverage area. The satellite beams may be formed by individual feeds (circles) and by mechanical or electronic satellite beam former. However, mechanical beam formers use fixed wave-guide components to control the RF phase and amplitude. Usually there is one amplifier for each transponder in each composite beam. Electronic beam formers use electronic RF phase shifters, and sometimes

51

provide electronic amplitude control, to produce the multibeam. There are many radiating elements and each usually has its own amplifier.

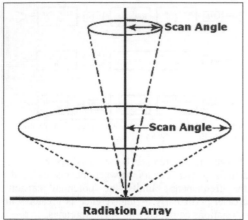

Figure 6. Antenna Scan Angles – Courtesy of Book: by Swan [6]

Satellite antenna radiators fall into two categories, reflector and direct radiating antennas. Reflector antenna uses a feed that indirectly radiates the energy towards the illuminated area of users, while direct radiating antenna radiates the energy direct to the coverage area. The satellite spot beam antenna can be pointed in various directions within a cone characterized by the scan angle, which illustrated in Figure 6.

In this way, a direct radiating array can radiate at large scan as illustrated in Figure 6. Such arrays are attractive for LEO communication satellites because they operate over large scan angle than reflectors antenna, so they requires a scan angle of 63° to cover its field of view. In contrary, a GEO communication satellite requires a scan angle of 7° to cover the entire visible coverage circle on the Earth surface.

In addition to the fact that satellite antenna gain decreases as much the scan angle increases, and including polarization purity decreases as well. In such a way, the LEO satellite constellations require a direct radiating antenna. No commercial GEO satellite currently uses this antenna, while reflector antennas. At MEO the situation is less clear and both direct radiating and reflector antennas have been proposed for this orbit [1, 6, 12, 13, 14].

5 CONCLUSION

The design and configuration of spacecraft antenna systems for MSS needs to be compact and robust especially for global coverage beam. Spacecraft mounted payloads usually require very accurate tracking performance capabilities for all mobile applications. For example, a spacecraft mounted flexible antenna applies high accuracy and precision control to perform its mission serving ships, land vehicles and aircraft. The precision and accuracy is necessary to achieve the desired performance typically requires the use of a high gain feedback control system and accurate plant knowledge.

On the other hand, the physical characteristics of antennas for ships and aircraft applications may be quite different, but both have to be designed compact for harsh environments and very extreme operating temperatures. These requirements will be difficult to achieve because the compact antenna has two major electrical disadvantages such as low gain and wide beam coverage, and because directional antenna has very heavy components for satellite tracking and getting satellite in the focus. However, a new generation of powerful satellite constellations with values of high EIRP and G/T performances should permit the design of compact and lightweight mobile satellite antennas.

The current Inmarsat-4 Spacecraft user link or mobile antenna system consists of a 9 meters deployable reflector and a feed array with 120 helical elements. Over 220 simultaneous RF beams are created by applying vector weights to the feed elements under the control of the onboard Digital Signal Processor (DSP). The simultaneous transmission and reception requires the achievement of very low levels of Passive Inter-Modulation (PIM). The mission requires continuous coverage over fixed ground cells for orbital inclinations of up to 3°. This is achieved by uploading modified beam weights on a daily basis. This requires a large number of beam weights to be pre-synthesized.

In addition to the pervious, the Inmarsat team has placed a contract for the procurement of a fifth generation spacecraft system to fulfill the modern communications requirements of mobile terminal users worldwide into the 21st century. The system requirements for the Inmarsat-5 antenna systems onboard spacecraft is to service the projected communications capacity needs, focusing upon the L, C and Ka-band antenna system requirements which are one of the key technology development areas of the program. The critical technology aspects of the antenna design needed to provide efficient implementation of the system requirements such as global and spot coverages for Voice, Data and Vide (VDV) and VDVoIP transmissions over the globe up to 750 of Elevation angles on North and South latitudes.

REFERENCES

[1] Ilcev D. S., "Global Mobile Satellite Communications for Maritime, Land and Aeronautical Applications", Springer, Boston, 2005.

[2] Gallagher B. "Never Beyond Reach", Book, Inmarsat, London, 1989.

[3] Ilcev D. S., "Global Mobile CNS", Manual, DUT, 2011.

[4] Maral G. & Other, "Satellite Communications Systems", Wiley, Chichester, 2009.

[5] Ilcev D. S., "Global Aeronautical Communications, Navigation and Surveillance (CNS)", Volume 1 & 2, AIAA, Reston, 2013.

[6] Swan P.A. & Other, "Global Mobile Satellite Systems: A Systems Overview", Kluwer AP, Boston, 2003.

[7] Rudge A.W. & Others, "The Handbook of Antenna Design", Volume 1 & 2, IEE, London, 1986.

[8] Kantor L.Y. & Others, "Sputnikovaya svyaz i problema geostacionarnoy orbiti", Radio i svyaz, Moskva, 1988.

[9] Maini A.K. & Agrawal V., "Satellite Technology – Principles and Applications", Wiley, Chichester, 2007.

[10] Stacey D., "Aeronautical Radio Communication Systems and Networks", Wiley, Chichester, 2008.

[11] Kadish J.E. & Other, "Satellite Communications Fundamentals", Artech House, Boston-London, 2000.

[12] Richharia M., "Mobile Satellite Communications - Principles and Trends", Addison-Wesley, Harlow, 2001.

[13] Ohmory S., Wakana. H & Kawase S. "Mobile Satellite Communications", Artech House, Boston, 1998.

[14] Zhilin V.A., "Mezhdunarodnaya sputnikova sistema morskoy svyazi - Inmarsat", Sudostroenie, Leningrad, 1988.

Synthesis of Composite Biphasic Signals for Continuous Wave Radar

V.M. Koshevyy, I.V. Koshevyy & D.O. Dolzhenko
Odessa National Maritime Academy, Odessa, Ukraine

ABSTRACT: Among the signals, which can be used in vessel's continuous wave (CW) radar, there is a class of discrete biphasic signals, which allows providing the ideal properties of the periodical autocorrelation function. Suggested signals have a special structure due to amplitude modulation, which is necessary to provide capability of radar with Search and Rescue Transponder (SART). The basic modes of operation of radar are also reserved.

As amplitude modulation has its drawback – not efficient use of signal energy, the consideration of the signals with uniform amplitude in the class of proposed biphasic signals was provided. The Mismatched filtering were be used by means special weighting functions for obtaining the necessary correlation properties. On the base of these signals the composite signals of any length could be constructed with proper correlation properties and low sensitivity enough to Doppler shifts.

Nowadays marine radar with continuous radiation attract attention of radar designers due to their possibility to reduce significantly the peak power of radiation and thus to improve environmental ecology and electromagnetic compatibility with other radio electronic devices on a vessel. Concerning to technical realization it should be noted that modern electronic technologies and digital techniques allow the processing of long duration signals. It is very important for this kind of radar the correct choice of the radiation signals and methods of their processing in the receivers.

One of the possible approaches to the choice of such signals has been considered in [1]. Since the search and rescue transponder (SART) has no compression filter, for its work in the class of continuous signals it is required amplitude modulation and peak factor other than one, in particular, a class of binary signals has been presented in [2] and has the form:

$$\vec{s} = [s_0 s_1 s_2 \ldots s_{N-1}] = [a \; b \; b \ldots \; b]$$

On the basis of this structure it was built a class of signals, which ensures to co-operate CW radar and SART.

For the work of marine radar the signals with ideal correlation properties are of interest. This helps improve immunity of signals against interference such as clutter. Therefore, within the continuous discrete signals with non-uniform structure there were considered signals with zero side lobe level of periodic autocorrelation function (PAF) which are defined as follows [2]:

$$a = |a| \cdot e^{i\varphi_a} = -\frac{N-2}{2}, \text{ where } N \geq 2 \qquad (1)$$

$$b = |b| \cdot e^{i\varphi_b} = 1$$

Such signals have a peak factor:

$$H = \frac{S_{n\max}^2}{\sum_{n=0}^{N-1} \frac{S_n^2}{N}} = \frac{N \cdot (N-2)^2}{(N-2)^2 + 4 \cdot (N-1)} \qquad (2)$$

In order to reduce the peak factor there was proposed a method in [3, 4] based on the known property of element-wise multiplication of signals with mutually prime periods [5]. This increases the coherence of the resulting signal. In particular, an example was presented, where the resultant signal is obtained due to the product of two signals (1) and $N = N_1 \cdot N_2 = 21 \cdot 4 = 84$; the peak factor has a value $H = 17$ (according to the left part of (2)).

Expression for the calculation of Periodic cross-ambiguity function has the form

$$\chi(k,l)=\sum_{n=0}^{N-1}w_n^* s_{(n+k)}e^{i\frac{2\pi nl}{4N}},\qquad(3)$$

where k is the discrete interval of time and l is discrete of frequency with step $\Delta f = \dfrac{1}{4NT_0}$; T_0 is elementary pulse duration; w_n is the filter coefficient.

Below in Fig. 1 plots of three sections of the periodic ambiguity function (AF) of the signal with $N = N_1 \cdot N_2 = 21 \cdot 4 = 84$ are shown.

factor $H = 9.5$ which leads to a noticeable sensitivity to Doppler frequency shift, which is undesirable. Therefore we consider the option, in which we can further reduce the peak factor. To do this, we can choose the smaller parts for product $N = N_1 \cdot N_2 \cdot N_3 = 3 \cdot 4 \cdot 7 = 84$. This will decrease the peak factor (according to (2)) of the signal to a value $H = 4.7$. At the same time, it is expected to reduce the side lobes of the cross sections of AF with a Doppler frequency shift.

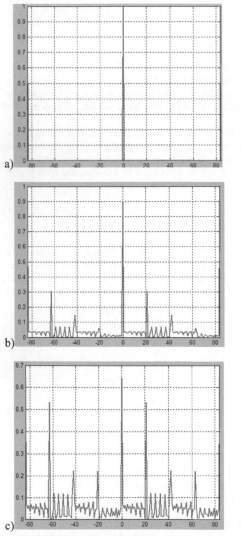

Fig. 1. Sections of AF of signal $N = N_1 \cdot N_2 = 21 \cdot 4 = 84$: a) l=0; b) l=1; c) l=2.

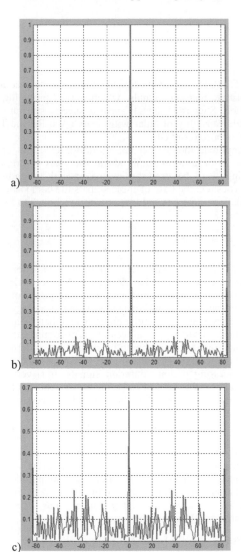

Figure 2. Sections of AF of signal
$N - N_1 \cdot N_2 \cdot N_3 - 3 \cdot 4 \cdot 7 - 84$: a) l=0; b) l=1; c) l=2.

As it can be seen from the figure, a periodic autocorrelation function has ideal correlation properties. But the product of two signals has a drawback – not enough decreasing the value of peak

As it's expected and seen in Fig. 2 above, sensitivity to the Doppler frequency shift was reduced.

56

The presented approach is effective and can be used for obtaining signals with long periods. Consider, for example, a signal in the form of a product of four signals: $N = N_1 \cdot N_2 \cdot N_3 \cdot N_4 = 4 \cdot 5 \cdot 7 \cdot 9 = 1260$ (AF is on Figure 3) and peak factor is equal to $H = 5$.

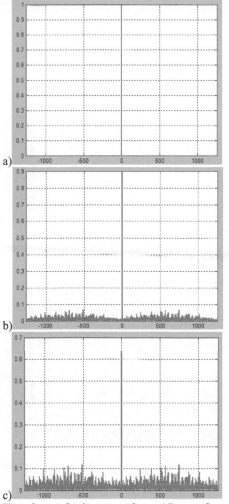

a)

b)

c)

Figure 3. Sections of AF of signal $N = N_1 \cdot N_2 \cdot N_3 \cdot N_4 = 4 \cdot 5 \cdot 7 \cdot 9 = 1260$: a) l=0; b) l=1; c) l=2.

It was confirmed that we have received periodic signal of long duration with a small peak factor and low sensitivity to the Doppler frequency shifts.

Thus, we can widely modify the peak factor of signals (1), obtaining an ideal correlation properties in the AF zero section, and with an increase in the coherent part of the signal we can reduce the side lobes of AF.

Further increasing of segments of resulting signal will lead to an increase in the peak factor. For the individual modes of radar let's consider the case

when the peak factor is one: $H = 1$. In this case, the form of the signal is determined according to:

$$[abb...b], \text{ where } a = -1, \quad b = 1 . \tag{4}$$

Thus to provide ideal properties of periodic autocorrelation function it can be used mismatched processing that is development of an approach described in [4]. The filter coefficients can be calculated according to the expression found:

$$w_0 = \frac{a(a+N-1)-2a-(N-2)}{1-a} ; \quad w_n = 1$$

($n = 1 \div N - 1$). \hfill (5)

Arising some losses in signal-to-noise ratio can be found from the following expression:

$$\rho = \frac{\left[a\left[a(a+(N-1))-2a-(N-2)\right]+(N-1)(1-a)\right]^2}{\left[\left[a(a+(N-1))-2a-(N-2)\right]^2+(N-1)(1-a)^2\right] \cdot \left(a^2+(N-1)\right)} \tag{6}$$

If $a = -\dfrac{N-2}{2}$, then there has no losses in signal-to-noise ratio $\rho = 1$.

If $a = -1$; then the losses in signal-to-noise ratio can be calculated from the relation:

$$\rho = \frac{4 \cdot (N-2)^2}{\left[(N-3)^2+(N-1)\right] \cdot N} \tag{7}$$

As an example there is a signal (4) when N=9, and on the Figure 4 there is cross- ambiguity function (CAF) of it.

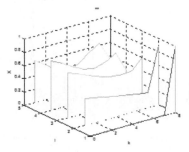

Figure 4. CAF of signal (2) N=9

Here is an example where the resulting product signal is composed of signals (4) with relatively prime periods $N = N_1 \cdot N_2 \cdot N_3 \cdot N_4 = 4 \cdot 5 \cdot 7 \cdot 9 = 1260$.

The value of the loss in signal-to-noise ratio for the resulting signal will be a product of the values ρ of its component signals [6]: $\rho = \rho_1 \cdot \rho_2 \cdot \rho_3 \cdot \rho_4 = 1 \cdot 0.9 \cdot 0.65 \cdot 0.5 = 0.29$.

A cross- ambiguity function of product of the four signals $N = 4 \cdot 5 \cdot 7 \cdot 9$ in the case when the signals have the form $[a11...1] a = -1$ is illustrated below in Figure 5.

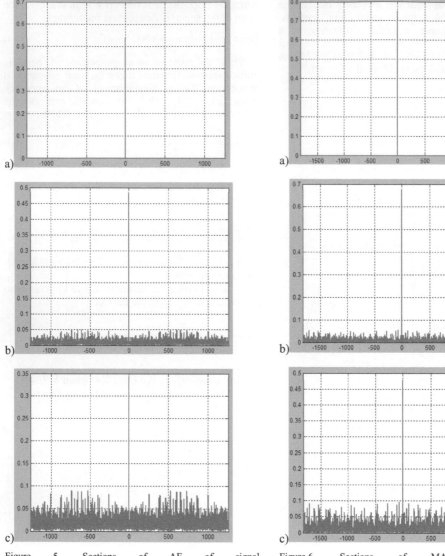

Figure 5. Sections of AF of signal $N = N_1 \cdot N_2 \cdot N_3 \cdot N_4 = 4 \cdot 5 \cdot 7 \cdot 9$: a) *l*=0; b) *l*=1; c) *l*=2

Figure 6. Sections of MAF of signal $N = N_1 \cdot N_2 \cdot N_3 \cdot N_4 = 4 \cdot 5 \cdot 7 \cdot 13$ with Barker code: a) *l*=0; б) *l*=1; в) *l*=2.

However, increasing N increases the losses in signal-to-noise ratio. Thus, when N> 11, the effectiveness of such signals will fall, because of significant losses in the signal-to-noise ratio. As possible solutions is using the additional classes of signals, including Barker codes. For example, consider a signal $N = N_1 \cdot N_2 \cdot N_3 \cdot N_4 = 4 \cdot 5 \cdot 7 \cdot 13$ which includes a Barker code $N_4 = 13$.

The presence of positive side lobes of the correlation function at the Barker code gives low losses in signal to noise ratio $\rho = 0.96$ [7]. Resulting $\rho = 0,57$.

Below in Fig. 6 CAF of such signal is presented:

However, the continuous mode has some difficulty in the practical implementation, since it requires the availability of two antennas on the vessel.

As a solution, we can use a quasi-continuous mode [8], when the entire signal is superimposed another signal with intervals of zeros. A disadvantage of such signals is that they are not suitable for all ranges of elements.

Another way is to use periodic signals of a certain duration [1], but they have a disadvantage which is a significant side-lobe level of AF equal $\frac{1}{N}$.

The third approach is the following: to radiate a certain signal in each period and due to complementarity zero side lobes of resulting aperiodic cross-corrnelation function will be provided.

Thus, based on the considered signals are constructed composite signals of arbitrary length with suitable correlation properties and different peak factors, and in particular equal to unity with low sensitivity to the Doppler shift.

REFERENCES

[1] V. M. Koshevyy, D. O. Dolzhenko. The Selection of Signals which Enable Continuous Wave Radar in Conjunction with Existing SART [in Russian], Navigation (Судовождение), Odessa: ONMA, №21, 2012, pp124-129.

[2] V. M. Koshevyy, D. O. Dolzhenko. The Synthesis of Periodic Sequences with Given Correlation Properties // Proc. of IEEE East–West Design & Test Symposium (EWDTS'11), Sevastopol, Ukraine, September 9–12, 2011. – P. 341 – 344.

[3] V. M. Koshevyy, D. O. Dolzhenko. Selection of Signals for Continuous Wave Radar [in Russian], Automation of ship technical aids, Odessa: ONMA, №18, 2012, pp. 63-73.

[4] V. M. Koshevyy, D. O. Dolzhenko, The Signals of Marine Continuous Radar for Operation with SART, 10th International Navigational Symposium on "Marine Navigation and Safety of Sea Transportation" TRANS-NAV 2013, Gdynia.

[5] R. C. Titsworth, Correlation properties of cyclic sequences, Ph. D. Thesis, Calif. Inst. Tech., 1962.

[6] V. P. Ipatov, Periodic Discrete Signals with Optimal Correlation Properties, Radio I Svyaz, Moscow,1992.

[7] V. M. Koshevyy, A. O. Shapovalova,Optimization of filters for Periodic Cross-Ambiguity function side lobes suppression under additional constraints [in Russian], Navigation (Судовождение), Odessa: ONMA, №20, 2011, pp. 99-108.

[8] Pat. USA № 3727222.

Zero Levels Formation of Radiation Pattern Linear Antennas Array with Minimum Quiantity of Controlling Coefficients Weights

V.M. Koshevyy & A.A. Shershnova
Odessa National Maritime Academy, Ukraine

ABSTRACT: The algorithm of radiation pattern linear antennas array formation for vessel's radar is suggested. Only two controlling elements of array are needed for obtaining practically full rejection of side lobe level of pattern linear antenna array at any azimuth angle outside the main lobe area. This gives the possibility separation of signals from vessels with big and small Target Cross Sections, equals range and close azimuth angels. The results of Radiation Pattern calculations for different situations are given. The possibility of twinning of rejection points are shown.

The modern radars are the high-powered technical mean of navigation and take important part to ensure maritime safety. But they aren't free from lots of drawbacks. The sufficient side lobes level of ship's radar antenna array diagram is the one of such drawback. It's lay to impossibility of separate signal observation from the big vessel (which has big effective reflecting surface (ERS)) and from small vessel, which has small ERS, and which are situated on the same distance and have azimuth angle's close value. The solution of this problem can be received in ship radar antenna pattern formation with using linear antenna array with controlled elements. The optimal methods of formation such diagrams are known and associated with the need of all array elements tuning by difficult algorithms [5]. At the same time the reaching of antenna pattern side lobes required suppression level may be with the simple methods by using the arrays with limited number of tunable weight coefficients of spatial filter (antenna array elements), for example, when there are only two of such tunable weight coefficients.

In this case all of the receiving antenna array of spatial filter's weights coefficients W_i of the processing, except two (first and last : W_1, W_N), are fixed (selected under the condition of providing the required antenna pattern side lobe's level) ($W_2; W_3; ...; W_{N-1}$). Value of the two tunable weights coefficients are selected for carried out the condition of providing zero values in two points (θ_1, θ_2) of the reception pattern. The expressions, which are describing the reception pattern of linear array

antenna $G(\theta)$ for this case, may be written in the following form.

$$G(\theta) = G_{N-2}(\theta) - \gamma_1(\theta) \cdot G_{N-2}(\theta_1) - \gamma_2(\theta) \cdot G_{N-2}(\theta_2) = \sum_{i=1}^{N} W_i \cdot e^{-j2\pi(N-1)\frac{d}{\lambda}\sin\theta} \quad (1)$$

where GN-2(θ) – partial diagram,

$$G_{N-2}(\theta) = \sum_{i=2}^{N-1} W_i \cdot e^{-j2\pi(N-1)\frac{d}{\lambda}\sin\theta} \quad (2)$$

$$\gamma_1(\theta) = \frac{e^{j2\pi(N-1)\frac{d}{\lambda}\sin\theta_2} - e^{-j2\pi(N-1)\frac{d}{\lambda}\sin\theta}}{e^{j2\pi(N-1)\frac{d}{\lambda}\sin\theta_2} - e^{j2\pi(N-1)\frac{d}{\lambda}\sin\theta_1}};$$

$$\gamma_2(\theta) = \frac{e^{-j2\pi(N-1)\frac{d}{\lambda}\sin\theta} - e^{-j2\pi(N-1)\frac{d}{\lambda}\sin\theta_1}}{e^{j2\pi(N-1)\frac{d}{\lambda}\sin\theta_2} - e^{j2\pi(N-1)\frac{d}{\lambda}\sin\theta_1}} \quad (3)$$

$\varphi = 2\pi\frac{d\sin\theta}{\lambda}$ =signal phase; λ = wave's length; d = distance between antenna's array elements, θ = angle between the normal to the axis of the array antenna and direction of coming signal.

(N-2)fixed weight coefficients may selected under condition of additional suppression average level of reception diagram's side lobes (2) with possible widening the main lobe of antenna array (1).

So, the full number of the coefficients, which creates reception array diagram, is equal N.

The expression for fixed weight coefficients has the next form[4]:

$$W_f = D^{-1} \cdot 1 \qquad (4)$$

where D^{-1} – inverse matrix; 1 – identity column-vector, which consists of all unit numbers;

Matrix D is formed as follows:

$$D = \sum_{L_1}^{L_2} 1 \cdot Q_l \cdot 1^t \cdot Q_l^* \qquad (5)$$

where Q_l – diagonal matrix:

$$Q_l = \begin{bmatrix} e^{j2/2\pi\frac{d}{\lambda}\sin\Delta\theta} & 0 & 0\ldots & 0 \\ 0 & e^{j3/2\pi\frac{d}{\lambda}\sin\Delta\theta} & 0 & \ldots & 0 \\ 0 & 0 & 0 & \ldots & e^{j(N-1)/2\cdot 2\pi\frac{d}{\lambda}\sin\Delta\theta} \end{bmatrix}$$

$\Delta\theta_f$ = an interval, between suppressing points of the spatial diagram $G_{N-2}(\theta)$; 1^t = transposed matrix; L_1, L_2 = upper and lower bounds for suppressed points.

Choosing in Q_l (5) step $\Delta\theta_f$, and also values L_1 and L_2 we can provide different values of average level of side lobes $G_{N-2}(\theta)$, but with varying degrees of its main lobe widening.

Thus, if

$$\Delta\theta_f = arcsin\frac{1}{2(d/\lambda)(N-2)}$$

(ratio $d/\lambda = 0,5$ was set everywhere), $L_1 = 0$ and $L_2 = N-2$ from (4), (5) we are getting the case where all $W_i = 1$ ($i = 2 \div N-2$).

That is the case of fully coherent to reception diagram $G_{N-2}(\theta)$ (equable correction), which corresponds the condition with absence of main lobe widening (Fig.1, a). Such approach was considered in [1] and [2].

If

$$\Delta\theta_f = arcsin\frac{1}{4(d/\lambda)(N-2)}\ , \text{ and } L_1 = 3\ ,\ L_2 = 2N-3,$$

the unequal values of weight coefficients can be obtained (uneven correction), which provides compromise between average value of side lobes suppression (more than -30 dB) and relatively small main peak's extension (about 10%) (Fig.1, b).

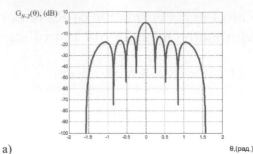

a)

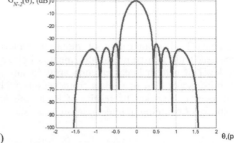

b)

Figure 1. Partial reception diagram: a – with equal weight coefficients values; b - with unequal weight coefficients values

This diagrams are showing, that average level of suppression with different W_i less, than in case, when $W_2 = W_3 = \ldots = W_{N-2}$.

The expression for tunable weight coefficients has the following form:

$$W_l = \frac{G_{N-2}(\theta_2)\ e^{j2\pi 2\pi-1)\frac{d}{\lambda}sin\theta_1} - G_{N-2}(\theta_1)\ e^{j2\pi 2\pi-1)\frac{d}{\lambda}sin\theta_2}}{e^{j2\pi 2\pi-1)\frac{d}{\lambda}sin\theta_2} - e^{j2\pi 2\pi-1)\frac{d}{\lambda}sin\theta_1}}; \qquad (6)$$

$$W_N = \frac{G_{N-2}(\theta_1) - G_{N-2}(\theta_2)}{e^{j2\pi(N-1)\frac{d}{\lambda}sin\theta_2} - e^{j2\pi(N-1)\frac{d}{\lambda}sin\theta_1}}. \qquad (7)$$

Fig.2,a-c shows the calculation by expressions (1), (2), (3) for case of common-mode processing in partial antenna (all values $W_2 = W_3 = \ldots = W_{N-2} = 1$). In this case:

$$\Delta\theta_f = arcsin\frac{1}{2(d/\lambda)(N-2)}\ ,\ L_1 = 0\ \text{и}\ L_2 = (N-2).$$

Losses in antenna's directivity to fully common-mode reception diagram:

$$\rho = \frac{|G(0)|^2}{N\sum_{n=1}^{N}|W_n|^2}. \qquad (8)$$

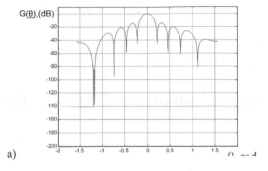

a)

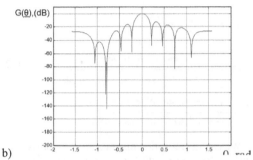

b)

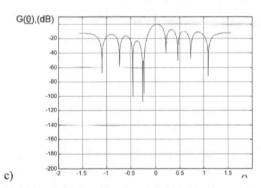

c)

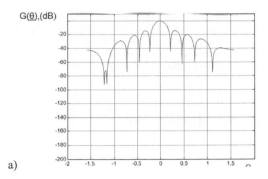

a)

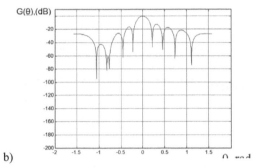

b)

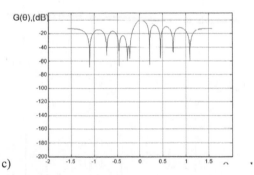

c)

Figure 2. Reception diagram (partial diagram equable correction):

a - $\theta_1 = -1.2001, \theta_2 = -1.1718, \rho = -0.2238, dB$; $б$ -

$\theta_1 = -0.8137, \theta_2 = -0.7854, \rho = -0.2784, dB$;

c - $\theta_1 = -0.2576, \theta_2 = -0.2293, \rho = -1.4107, dB$

The similar type of expressions, as (6), (7) were obtained for the signal time processing in the case of frequency selection [1], [6] and the time-frequency task selection signal [7].

As we can see, the suppression of this points is high enough, and the side lobes level between the suppressed points not high (about – 80 dB). Losses in antenna's directivity (8) are increase with drawing closer to main lobe, and in considered case isn't more than -1,4 dB.

Figure 3. The equable correction of reception diagram (of partial diagram) with increased in twice distance between suppressed points:

a – $\theta_1 = -1.2143, \theta_2 = -1.1577, \rho = -0.2228, dB$; b –

$\theta_1 = -0.8278, \theta_2 = -0.7712, \rho = -0.2749, dB$;

c – $\theta_1 = -0.2717, \theta_2 = -0.2152, \rho = -1.3647, dB$

For comparison, on fig.3, a-c was considered the case, when the distance between suppressed points was increasing in twice. The side lobes level between suppressed points is increase, but wasn't more than -40 – -60 dB.

Now let's consider the case, when:

$$\Delta\theta_f = arcsin \frac{1}{4(d/\lambda)(N-2)} \ , \ L_1 = 3 \ and \ L_2 = (2N-3).$$

63

That is the case, when we get unequal values of weight coefficients, which are obtained by (4), (5), which provide the compromise between level of side lobes suppression (about 30 dB) and not high main lobe widening (about 10%) (fig.4, *a-c*).

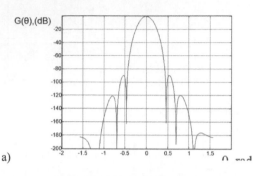

a)

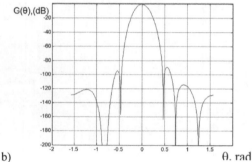

b)

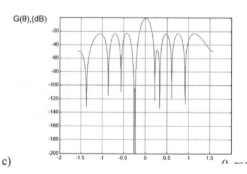

c)

Figure 4. Reception diagram with equable correction (of partial diagram):

$a - \quad \theta_1 = -1.2001, \theta_2 = -1.1718, \rho = -0.4202,$ dB ; $\quad b -$

$\theta_1 = -0.8137, \theta_2 = -0.7854, \rho = -0.4620,$ dB ;

$c - \quad \theta_1 = -0.2576, \theta_2 = -0.2293, \rho = -2.7293,$ dB

With the same suppressed points θ_1 и θ_2 we received less average side lobes level within the given area at the expense of some main lobe widening.

It interesting to note, than if instead of reception diagram, which described by (1), we use the expression for symmetrical form with real coefficients (11), (12) in following form:

$$G^{(1)}(\theta) = G^{(1)}{}_{N-2}(\theta) - \gamma_1^{(1)}(\theta) \cdot G^{(1)}{}_{N-2}(\theta_1) - \gamma_2^{(1)}(\theta) \cdot G_{N-2}^{(1)}(\theta_2)$$

where $\quad G_{N-2}^{(1)}(\theta) = G_{N-2}(\theta) \cdot e^{j\pi(N-1)\frac{d}{\lambda}\sin\theta}$, (10)

$$\gamma_1^{(1)}(\theta) = \frac{\cos\left[-j2\pi(N-1)\frac{d}{\lambda}\sin\theta_1\right]}{\cos\left[-j2\pi(N-1)\frac{d}{\lambda}\sin\theta_2\right] - \cos\left[-j2\pi(N-1)\frac{d}{\lambda}\sin\theta_1\right]} - \frac{\cos\left[-j2\pi(N-1)\frac{d}{\lambda}\sin\theta\right]}{\cos\left[-j2\pi(N-1)\frac{d}{\lambda}\sin\theta_2\right] - \cos\left[-j2\pi(N-1)\frac{d}{\lambda}\sin\theta_1\right]}$$ (11)

$$\gamma_2^{(1)}(\theta) = \frac{\cos\left[-j2\pi(N-1)\frac{d}{\lambda}\sin\theta\right]}{\cos\left[-j2\pi(N-1)\frac{d}{\lambda}\sin\theta_2\right] - \cos\left[-j2\pi(N-1)\frac{d}{\lambda}\sin\theta_1\right]} - \frac{\cos\left[-j2\pi(N-1)\frac{d}{\lambda}\sin\theta_1\right]}{\cos\left[-j2\pi(N-1)\frac{d}{\lambda}\sin\theta_2\right] - \cos\left[-j2\pi(N-1)\frac{d}{\lambda}\sin\theta_1\right]}$$ (12)

We can get reception diagram with suppressing in four points with symmetrical positioning relative to main lobe. As an example, on figures (12) – (14) are showed the next diagrams:

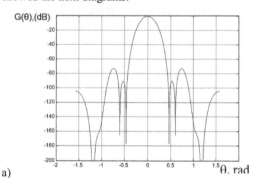

a)

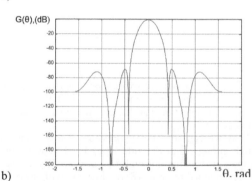

b)

64

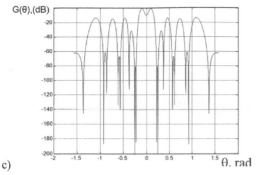

G(θ),(dB)

c)

Figure 5. Reception diagram with real coefficients (symmetrical):

a – $\theta_1 = -1.2001, \theta_2 = -1.1718, \rho = -1.0848$, dB ; b

– $\theta_1 = -0.8137, \theta_2 = -0.7854, \rho = -1.5527$, dB ;

c – $\theta_1 = -0.2576, \theta_2 = -0.2293, \rho = -2.4359$, dB

Such away, the algorithm of radiation pattern linear antennas array formation for vessel's radar is suggested, which provide high enough side lobes level suppression of reception diagram with only two controlling elements, this gives the possibility separation of signals from vessels with big and small Target Cross Sections, equals range and close azimuth angels. It's impossible to reach by using standart vessel's antennas. As in the case of equal (N-2)- weight coefficients, which forms the partial diagram, as in the case of uneven correction, it's provided zero level of side lobes within the given area with reduced average side lobe's level in case of equable correction. Whcrein, the algorithm of tunable weight coefficients formation (6) and (7) are easy for practical implementation. The possibility of twinning of rejection points is shown.

CONCLUSIONS

Nowadays the ship's radar antennas are antenna arrays, which consists N-elements as line slot antennas or loud-hailer antennas, which are situated in horizontal plane and aren't tuned.

In this paper the simplified variant of controlling antenna array is offered, in which only two ultra elements are tuned – first and last, which provides

the suppression of any antenna array diagram side lobe in azimuth plane.

The obtained expressions for module and phase of this tuning elements are implemented by appropriate attenuators and phase shifters. It gives the control principle by way of moving the reception diagram laydowns.

By removing on the control panel the appropriate controller, which connected to two tunable antenna array elements, it's possible to move the suppression zone to different azimuth angles.

The attenuators and phase shifters control principle is received and can be easy realized on base of the simplest microprocessor.

In cases, when the probability of fogging the small ship by the bigger one on radar screen, by the way of turning the additional controller, which is on the radar panel, the suppression of needed side lobe of reception diagram can be received.

It gives the possibility of practical use of the antenna reception diagram proposed algorithm in the azimuth plane of ship radar for voiding the small targets from the big ones.

REFERENCES

1. V. Koshevyy, V. Lavrinenko, 1981, The target's selection on based on the discrete structurc with a minimum quantity of controlled elements. «Izvestia VUZ. Radioelectronika», t. 24, №4, pp. 105 – 107.
2. V. Koshevyy, A. Shershnova, 2013, The formation of zero levels of Radiation Pattern linear Antennas Array with minimum quantity of controlling elements, Proc. 9 Int. Conf. on Antenna Theory and Techniques (ICATT-13), Odessa, Ukraine, pp.264-265.
3. V. Koshevyy , M. Sverdlik, 1974, About the possibility of full side lobes level suppression of ambiguity function in the given area. – « Radio Eng. Electron. Phys.», t. 19, № 9, pp. 1839 – 1846.
4. V. Koshevyy, V. Lavrinenko , S. Chuprov, 1975, The efficiency of quasi-filter analysis. «RIPORT », VIMI. №2, – p. 7.
5. Y. Shirman, V. Mandjos, 1981, Theory and technics of radar information processing under interferences. M. Radio I Svyaz, – 416c.
6. V. Koshevyy, 1982, Moving target systems indication synthesis with the inverse matrix size restrictions. - «Izvestia VUZ. Radioelectronika», т.25, № 3, С. 84-86.
7. V. Koshevoy, M. Sverdlik, 1973, About influence of memory and pass-band of generalized V-filter to efficiency of interference suppression. « Radio Eng. Electron. Phys.», t.18, №8, pp. 1618-1627.

Radio Refractivity and Rain-Rate Estimations over Northwest Aegean Archipelagos for Electromagnetic Wave Attenuation Modelling

E.A. Karagianni & A.P. Mitropoulos
Electronics Laboratory, Department of Naval Sciences, Hellenic Naval Academy, Hellas

N.G. Drolias & A.D. Sarantopoulos
Division of Climatology - Applications, Hellenic National Meteorological Service, Hellas

A.A. Charantonis
CEDRIC Laboratory, Conservatoire National des Arts et Metiers, Paris, France

ABSTRACT: By utilizing meteorological data such as relative humidity, temperature, pressure, precipitation height and precipitation duration at Skyros and Naxos Islands, Hellas (24,32°E, 38,50°N - Skyros) from four recent years (2010 – 2013), the effect of the weather on Electromagnetic wave propagation is studied. The EM wave propagation characteristics depend on atmospheric refractivity and consequently on Rain-Rate which vary in time and space randomly. Therefore the statistics of radio refractivity, Rain-Rate and related propagation effects are of main interest. This work investigates the differences in monthly, seasonal and daily amounts of rainfall, for every synoptic 3 and 12-hour interval and the change of refractivity as observed in Northeastern Aegean Archipelagos.

1 INTRODUCTION

1.1 *Humidity of the atmosphere*

Water is the only substance that can be found naturally in all three states (gas, liquid, solid) in the atmosphere. Humidity is the general term to show the existence of the "water" in the atmosphere. A number of parameters are used to expresses the humidity reveals its importance in many scientific domains. Such parameters are water vapor pressure (e), saturation vapor pressure (e_s), relative humidity (H), dry temperature (θ), precipitation height (h), precipitation time interval (t), pressure (P), water vapor density (ρ).

Water in any state affects the transmission of the electromagnetic wave. When the wave passes through a humid atmospheric layer, part of its energy is reflected or scattered, another is absorbed and the rest is passing without any change. Therefore the electromagnetic wave is attenuated. The attenuation is caused by the scattering and absorption of electromagnetic waves by drops of liquid water. The scattering diffuses the signal, while absorption involves the resonance of the waves with individual molecules of water. Absorption increases the molecular energy, corresponding to a slight increase in temperature, and results in an equivalent loss of signal energy.

In strong precipitation phenomena (due to the movements of the "drops") in addition to the signal being attenuated, the system noise temperature is increased and the polarization is changed. These three effects cause degradation in the received signal quality. At C-band the effects are minor whilst at higher frequencies, such as Ka-band, the attenuation increases. Prediction of the influence of these factors is very important in telecommunications systems design.

Attenuation due to rain can lead to the perturbations of the wireless, mobile, satellite and other communications. Rain events produce unavailability of microwave links, which may lead to economical losses or even license loosing.

1.2 *The raindrop*

Water molecules are dipoles (hydrogen bonding). The raindrop's dipoles act as an antenna with low directivity, which re-radiates the electromagnetic wave energy. Water is a loss-making dielectric medium. The relative dielectric constant of water is high, compared to the dielectric constant of the surrounding air. It depends on the medium temperature and the operating frequency of the telecommunications system. One of the problems in the prediction of electromagnetic wave power losses is the description of shape of the raindrop, strongly depended on the size. It is known, that only very small raindrops are like spheres. Such drops exist in clouds. The shape of raindrops, that are larger than 1 mm in diameter, is no more spherical. They are not tear-shaped, as it commonly presented. The shape of falling large raindrops is more like a hamburger shape. Therefore, horizontally polarized waves suffer greater attenuation than vertically polarized waves.

The raindrop size distribution and the drop shape relation have great variation in different precipitation conditions. In heavy precipitation phenomena the raindrop size is composed of lots of median and small raindrops rather than giant raindrops [1], [7]. Both shape and size are depending from the dynamics principles where a raindrop is deforming as it falls in air, breaking into smaller fragments. The topological change from a big drop into smaller stable fragments is accomplished within a time scale much shorter than the typical collision time between the drops [2], [3], [8]. Regarding strong precipitation phenomena, the size of a typical raindrop, is about 3 millimeters.

When the wavelength approaches this length, attenuation is significant. Wavelength in free space and frequency are related by the well-known equation c=λ·f, where λ is the wavelength, f is the carrier frequency, and c is the speed of the electromagnetic wave in free space

$$c = \frac{1}{\sqrt{\varepsilon_0 \cdot \mu_0}} \approx 3 \cdot 10^8 \text{ m/s} \tag{1}$$

where ε_0 and μ_0 are the free space permittivity and permeability constants in F/m and H/m respectively.

For example, at the S-band of 2,4 GHz, the wavelength is 125 millimeters in free space and thus it is 40 times larger than a raindrop and the signal passes through the rain with relatively small attenuation [4], [9]. At X-band with a frequency of 10 GHz, the wavelength is 30 millimeters and at Ka-band with a carrier frequency of 20 GHz, the wavelength is 15 millimeters. At these frequencies, the wavelength and raindrop size are comparable and the attenuation is quite large.

1.3 *The relative refractive index*

When the propagation medium is a material different than air, the speed of an electromagnetic wave depends on the relative dielectric constant known as relative permittivity ε_r and to the relative permeability μ_r, with the following formula:

$$v = \frac{1}{\sqrt{\varepsilon \cdot \mu}} = \frac{c}{\sqrt{\varepsilon_r \cdot \mu_r}} \tag{2}$$

where $\varepsilon = \varepsilon_0 \cdot \varepsilon_r$ and $\mu = \mu_0 \cdot \mu_r$, are the permittivity and permeability in F/m and H/m respectively, and ε_r and μ_r are the relative permittivity and permeability of the medium [6], [11].

The relative refractive index, n is defined as

$$n = \sqrt{\varepsilon_r \cdot \mu_r} \tag{3}$$

Values for the relative refractive index, are presented in Figure 3 for four years observations in Skyros Island at Northwest Aegean Sea (Figure 2).

2 RAIN MODELS FOR WAVE ATTENUATION

2.1 *Rain attenuation*

The electromagnetic wave attenuation due to rain (the rain attenuation) is one of the most noticeable components of excess losses, especially at frequencies of 10 GHz and above. Considerable research has been carried out to model rain attenuation mathematically and to characterize rainfall throughout the world. The standard method of representing rain attenuation is through the power-law relationship [10], [14], [15].

$$L_r = k \cdot R^\alpha \cdot L \tag{4}$$

where L_r is the rain attenuation (in dB), R is the rain rate (in mm/h), L is an equivalent path length (in km), and k and α are empirical coefficients, functions of the operating frequency (f), polarization (k_h or α_h - h for horizontal polarized waves and k_v or α_v - v for vertical polarized waves) and temperature (T).

For horizontally polarized waves, k_H and α_H parameters are given in Table 1 for selected S and X band frequencies [12], [16].

Table 1. Parameters for horizontally polarized waves (h) in S and X-band.

Frequency (GHz)	$k_H(\times 10^{-3})$	α_H
2,5	0,1321	1,1209
5	0,2162	1,6969
10	12,17	1,2571

The equivalent path length depends on the angle of elevation of the communication link, the height of the rain layer (h), and the latitude of the earth station. (Figure 1)

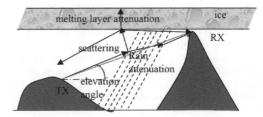

Figure 1. Path length through rain

The rain rate enters into equation 4 because it is a measure of the average size of the raindrops. When the rain rate increases that means it rains harder, the rain drops are larger and thus there is more attenuation. Also, the conversion of the radar reflectivity factor to rain rate is a crucial step in weather radar measurements. It has been common practice for over 60 years now to take for this conversion a simple power law relationship between them, using the classical exponential raindrop size distribution [17], [18].

2.2 *Rain Models*

Rain models differ principally in the way the effective path length L is calculated. Two rain models are widely used, the Crane model and the ITU-R model (International Telecommunication Union's Radio-communication sector). The two-component Crane model takes into account both the dense center and the fringe area of a rain cell [11], [19]. In the design of a telecommunications' link a margin is included to compensate for the effects of rain at a given level of availability. The statistical characterization of rain begins by dividing the world into rain climate zones. Within each zone, the maximum rain rate for a given probability is determined from actual meteorological data accumulated over many years.

The methods of prediction of the rain attenuation can be categorized into two groups: the physical models and the empirical models. The physical models attempt to reproduce the physical behavior involved in the attenuation processes while the empirical methodologies are based on measurement databases from stations in different climatic zones within a given region. The empirical methods are used widely.

The factor $\gamma_r = L_r/L$ (in dB/Km), where L_r and L are defined in equation 4, is called the specific rain attenuation. One of the most widely used rain attenuation prediction methods is the empirical relationship between this specific rain attenuation γ_r (in dB/km) and the rain rate R(in mm/h) [12], [13]

$$\gamma_r = k \cdot R^\alpha \tag{5}$$

2.3 *Data Collection*

When it is impossible to gather data for calculations of the specific attenuation due to rain and atmospheric relative refractive index, the values recommended by the ITU-R can be used. But the recommended values are not always exact. The easiest telecommunication system design is for operating frequencies below 10 GHz because for these frequencies atmospheric absorption and rainfall loss generally are neglected. Due to the fact that these bands are highly congested, in telecommunication systems design, frequencies above 10 GHz are often used. But, the higher the

operating frequency, the greater attenuation due to hydrometeors (rain, cloud, fog, snow) is observed.

In this paper, our attention would be concentrated on the attenuation due to rain using collecting data for Northwest Aegean for a period of four years.

3 RAIN RATE

For the determination of the rain attenuation, the main parameter used is the rain rate R, which is expressed in mm/h. The rain rate can be described as the thickness of the precipitation layer, which felled down over the time period of one hour in the case when the precipitation is not evaporated, not soaked into the soil, and is not blown away by the wind. The evaluation of R-value is of crucial importance in the rain attenuation prediction. The rain attenuation depends on the meteorological conditions in the considered place.

Consider that the maximum rain rate is 20 mm/h with a probability of 99.99%. If the rain attenuation for this rain rate is compensated by adding sufficient margin to the link estimation, there will be a 99.99% probability that the signal will be received with the specified performance. Thus there is a probability of 0.01% where the anticipated attenuation will be exceeded. The translation is that there is a possibility of unavailability for 53 minutes over the entire year, assuming that the year has 8766 hours.

3.1 *Integration Time*

As mentioned above, the R-values are expressed in mm/h. Time intervals between the readings of rainfall amount in many cases are unrealistic. The period of time between the readings of the rainfall amount values is called integration time τ and it is a very important parameter, because it can significantly change the R-value. High R-values are hidden when τ is long.

Consider an example where the duration of the rain was 30 minutes. The total amount of the precipitation was 20 mm. It did not rain during remaining 30 minutes of the hour neither during the remaining 2,5 hours for the interval of 3 hours. Thereby, if we count the average R-value for that hour it will be 20 mm/h. If we count the average R for the measuring duration of 3 hours, it will be 6,6 mm/h. But if we count R for every minute of that hour, we will find that R=40 mm/h, because in every of those 30 rainy minutes the amount of the precipitation was 0,66 mm/min. That is why the average R-values are unreliable.

3.2 *The "one-minute" rain rate*

Almost all rain attenuation methods require "one–minute" rain rate value. The "one-minute" rain rate value R is expressed in mm/h. This value can be defined as the R–value for 0.01% of time of the year, obtained using the rainfall amount value, which was measured in τ=1 min and multiplied by 60 [10], [11].

However, in our case, 12-hourly instances data are used. There are various models for conversion of R (τ-min) into R (1-min). [5], [6], [9]. One of such conversion models [14] is presented in the following equation

$$R_{1\,min}=(R_{\tau\,min})^d \quad (6)$$

$$d=0.987\,\tau^{0.061} \quad (7)$$

where $R_{\tau\,min}$ is rain rate value measured in τ minutes ($\tau\geq1min$) and $R_{1\,min}$ the "one-minute" rain rate value.

However, the results presented in this study are referred to the precipitation duration in minutes and then the result is divided by 60. It is valuable to be reported that both results - method used and model of equations 6 and 7 -coincide with each other at low R-values but at high R-values there is an enormous deviation upwards with the worst result at the highest R-value which is duplicated.

Figure 2. Skyros Island in the Northwest Aegean Archipelagos

On the other hand, the main advantage of the "Worst-month" model which was proposed by ITU-R [12] is that only the worst-month statistics must be collected, although it is appropriate in cases when the required reliability of the radio system is other than 99.99%. This month is not necessarily the same month in different year. The fraction of time when the threshold value of rain rate (so, and rain attenuation value) was exceeded is identical to probability that the threshold value of rain rate would be exceeded [15], [19].

4 DATA ANALYSIS

Our survey is focused on those measurements referred as Northwest Aegean Archipelagos, namely, Skyros Island as well as in Naxos Island which is

approximately 110 nautical miles south. The differences in monthly, seasonally and daily amounts of rainfall are observed in Skyros island of Hellas (for every synoptic 3 or 12-hour interval).

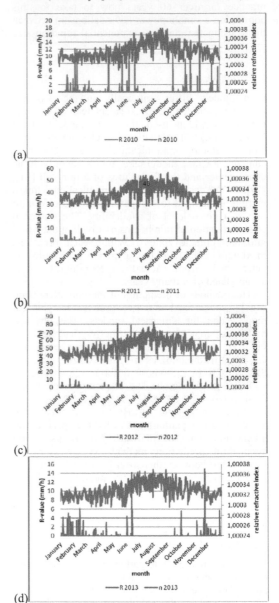

(a)

(b)

(c)

(d)

Figure 3. Rain rate and relative refractive indexes for the period January 2010 – December 2013 for Skyros Island.

The precipitation amount as a meteorological index on Skyros has a maximum value of 74 mm for the period January 2010 to December 2013. This value appeared in May 2012 for 55 minutes, giving a R-value of 81 mm/h. The day following this storm, an R-value of 43 mm/h (36 mm for 50 minutes) was observed. The next maximum values for precipitation height appeared in June 2011 (64 mm) for 1 hour and 20 minutes and in February 2012 (53 mm) for 6 hours, giving R-values at 48 mm/h and 8,8 mm/h respectively.

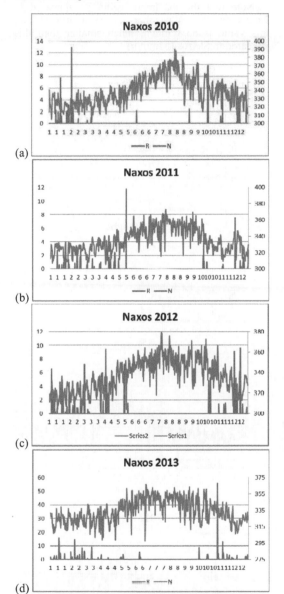

(a)
(b)
(c)
(d)

Figure 4. Rain rate and refractivity values for the period January 2010 – December 2013 for Naxos Island.

Regarding maximum R-values as presented in Table 4, the next maximum appeared in December 2011 (24,7 mm/h) with a 45 mm rain height within 1 hour and 49 min, giving R=25 mm/h. R-values in between 10 mm/h and 24 mm/h observed only for

the 0,1% per year (34 hours total) having an average precipitation height at 18 mm and a maximum at 48 mm. All other values are bellow 10 mm/h (1520 hours in 4 years with an average at 4,9 mm/h.

Following observed parameters values are presented for precipitation time interval (τ), precipitation height (h), pressure (P), dry temperature (θ), relative humidity (H), refractivity (N) and Rain-Rate (R) at a specific parameter at its minimum (Table 2) and maximum value (Table 3) respectively.

Table 2. Observed parameters values at a specific parameter at its minimum value (in bold).

τ minutes	H mm	P mbars	θ °C	H %	N	R mm/h
80	3,4	**981,6**	11,4	81	318	3
0	0	1021,8	**-0,6**	65	310	0
10	0	1006,9	32,4	**19**	293	0
0	0	1006	27,4	19	**288**	0

Table 3. Observed parameters values at a specific parameter at its maximum value (in bold).

τ minutes	H mm	P mbars	θ °C	H %	N	R mm/h
720	20,8	1016,9	20,8	88	336	2
0	0	**1035**	2,4	58	212	0
0	0	1000,9	**35,2**	31	321	0
0	0	1020,7	22	**100**	380	0
0	0	1013,3	29	78	**388**	0
55	**74**	1009,9	15,4	88	341	**81**

Table 4. Observed parameters values for rain-rate (R), relative humidity (H), relative refractive index (n) and water vapor density (ρ) for a specific maximum R-value per month for 4 years.

Month	R mm/h	H %	n-1 (x10^{-6})	ρ g/m^3
January	11	84	317	5,7
February	12	88	331	10,1
March	4	93	338	10,3
April	12	86	334	10
May	81	89	340	11,6
June	48	57	329	10,6
July	1	68	340	12,5
August	2	76	354	15,3
September	24	87	363	16,9
October	12	82	341	12,2
November	19	91	346	12,4
December	25	84	326	7,8

5 REFRACTIVITY

Another parameter to consider is the relative refractive index of the atmosphere given in equation 3, which affects the curvature of the electromagnetic wave path and graces the fading phenomenon. For example, the anomalous electromagnetic wave propagation can cause disturbances to radar work, because variation of the refractive index of the atmosphere can induce loss of radar coverage.

Accurate prediction of losses due to these factors can ensure a reliability of the telecommunication system, decrease the equipment cost and protect people.

5.1 *Refractive Index*

The relative refractive index, n, for the troposphere, is computed by [20]

$$n = 1 + N \cdot 10^{-6} \qquad (8)$$

where N is the radio refractivity expressed by:

$$N = N_{dry} + N_{wet} = 77,6 \cdot \frac{P}{T} + 3,732 \cdot 10^5 \cdot \frac{e}{T^2} \qquad (9)$$

with P being the atmospheric pressure (in mbars), e the water vapour pressure (in mbars), and T the absolute temperature (in ^{0}K).

This expression may be used for all radio frequencies. The relationship between water vapour pressure e and relative humidity is given by:

$$e = \frac{H \cdot e_s}{100} \qquad (10)$$

where

$$e_s = EF \cdot a \cdot \exp\left[\frac{\left(b - \frac{t}{d}\right) \cdot t}{t + c} \right] \qquad (11)$$

where a=6.1121mb is the vapour pressure at the triple point and it has the same unites with e_s and

$$EF = 1 + 10^{-4} \cdot \left[7,2 + P \cdot \left(0,0032 + 5,9 \cdot 10^{-7} \cdot t^2\right) \right] \qquad (12)$$

where t is the temperature (in $^\circ$C), P is the pressure (in mbars), H is the relative humidity (in %) e_s is the saturation vapour pressure (in mbars) at the temperature t (in $^\circ$C) and the coefficients b, c and d for water for the measured temperature, are: b=18.678, c=257.14, d=234.5.

Vapour pressure e in mbars is obtained from the water vapour density ρ (in g/m^3) using the equation:

$$e = \frac{\rho \cdot T}{216,7} \qquad (13)$$

where ρ is given in g/m^3.

Skyros annual averages for the period 2010 to 2013 are N_{dry}=272, EF_{water}=1,001, e_s=20,3, e=13,02, N_{wet}=57,07, N=328,95, n=1,0003 and ρ=9,67 while Naxos averages which is located south from Skyros for the same period, are θ=18,8^{0}C dry temperature, P=1013,77 mbars, H=68,44% relative humidity, N=336.18 and ρ=11,3.

It is noticeable that our statistics based at 12-hour measurements 6:00 and 18:00, and this for the following reason: During the 12-hour measurements, the station, recorded summed with the last 3 hour and all measurements were preceded during that 12 hour and were given separately in each 3-hour observation.

Note that the maximum value R, appeared in Naxos, on 5th of November 2013 at 06:00 UTC and this value is 55,8 mm/h, much smaller than this appeared in Skyros (81mm/h).

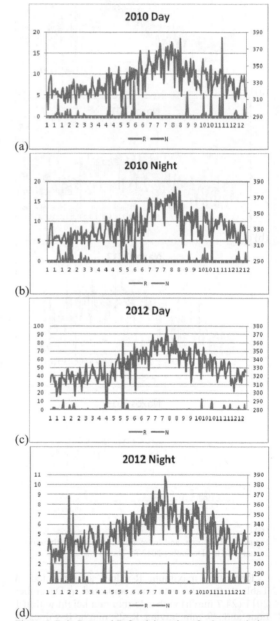

Figure 5. Rain Rate and Refractivity values for 2 years during day and night.

72

6 DISCUSSION

The main models for calculation of electromagnetic wave attenuation due to atmospheric humidity and heavy precipitation phenomena, were revised. In Northwest Aegean Sea, Hellas, when the reliability of the radio system of 99,99% is required, R(1 min) value equals to 81 mm/h. The attenuation of horizontally polarized electromagnetic waves is greater than the attenuation of vertically polarized electromagnetic waves. The dependency of the average specific electromagnetic wave attenuation due to rain on the operating frequency was determined.

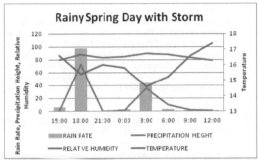

Figure 6. Hourly parameters presentation on a rainy day with storms in May 2012

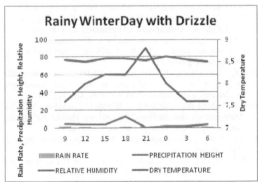

Figure 7. Hourly paremeters presentation on a drizzle day at the end of January 2011.

The variations of the atmospheric humidity, temperature and pressure cause the fluctuations of the atmospheric refractive index. In Skyros, the atmosphere refractivity fluctuates between 288 and 388 for nights. This interval is shortening by 14% for day periods. In heavy precipitation phenomena (R>10mm/h) refractivity values are between 317 and 363 with N=340 at the maximum R-value. There have been observed 4 cases with very strong precipitation phenomena (R-values at 24, 25, 48 and 81mm/h for τ-values 50, 109, 80 and 55 minutes respectively). For these cases the initial interval for refractivity is shortening by 85% and 66%

respectively for days and nights, having a minimum of 329.

Except the fact values of refractivity are increased at nights as shown in Figures 5, the same phenomenon exists for summer periods as it is observed in diagrams in Figure 5. The explanation of this phenomenon is based on the fact that refractivity is inversely proportional to the temperature.

Moreover, as it is shown in Figure 6, in strong precipitation phenomena, the temperature decreases and the relative humidity increases. The relative refractive index will be increased as it is proportional to the water vapor pressure (which in turn is proportional to the relative humidity) and inversely proportional to temperature. The same phenomenon is ob-served when the precipitation has very small rain-rates although the temperature, pressure relative humidity changes - and consequently refractivity and relative refractive index changes - occur a couple of hours later as it is presented in the diagram of Figure 7.

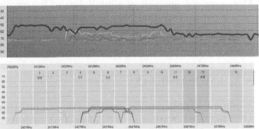

Figure 8. Measurements in moderate precipitation phenomena showed a power loss of about 10 to 12 dBm at 2,4 GHz

Measurements performed in a rainy day, for 40 minutes, with a wireless system at 2,4 GHz with minimum connection requirements approximately at -80 dBm. The first measurement was performed without rain, at free space and the signal's power was varied from -60 to -53 dBm. With the use of a directional antenna the receiving power was between -37 to -29 dBm. In the case of rain with R-value at 12 mm/h the signal's power varied in between -71 to -64 dBm. This 10 to 12 dBm attenuation was expected due to the path loss equation (4), [21]. and measurements are in accordance with theoretical results.

7 CONCLUSION

The EM wave propagation characteristics depend on atmospheric refractivity. Nevertheless, atmospheric refractivity varies in time and space randomly. Therefore the statistics of atmospheric refractivity and related propagation effects are of main interest. This work investigated the major differences between radio refractivity changes for Northwest

Aegean Archipelagos. Radio refractivity values were calculated from measured meteorological parameters (relative humidity, temperature and pressure) during a recent period of 4 years. The results showed that radio refractivity fluctuates between 288 and 388 but in strong precipitation phenomena where $R \geq 24mm/h$ it fluctuates in between 326 to 363 (N-units).

The rainfall rate exceeded for a probability of 0,01% of the average year and the location (24,32°E, 38,50°N) is 48 mm/h and it is in accordance with previous published studies and ITU recommendations [22], [23]. Rainfall events with rain rates of 81, 48, 25 mm/h were observed in Skyros. The durations of these precipitations were 55 minutes, 80 minutes and 109 minutes respectively. The percentages of the time were 0.01 %, 0.015 % and 0,02 % respectively.

Incorporating data on extreme weather events like heavy precipitation impacts, into GIS maps would be of tactical advantage for military operations [24]. More ore measurements have to be performed for various conditions of precipitation. Moreover, the analysis of rainfall data of the longer period (of several decades) and several points must be carried out to determine the parameters involved in rain attenuation prediction.

ACKNOWLEGMENTS

Authors would like to acknowledge the contribution of Hellenic National Meteorological Service, Division of Climatology-Applications for providing meteorological data used in this study as well as the contribution of Dr. Th. Charantonis and Mr. Ch. Petropoulos for their valuable advice. Also, they would like to thank Mr. Konstantinos Alvertos, MSc candidate, for his contribution in experimental results, performing signal's power measurements. Last but not least, authors would like to acknowledge the contribution of Commodore (ENG) E. Stelioudakis and generally the Hellenic Navy for supporting this research.

REFERENCES

[1] W.-Y. Chang, T.-C. C. Wang, P.-L. Lin, The Characteristics of Raindrop Size Distribution and Drop Shape Relation in Typhoon Systems from 2D-Video Disdrometer and NCU C-Band Polarimetric Radar, Journal of Atmospheric and Oceanic Technology, vol. 26, 2009

[2] E. Villermaux, B. Bossa, Single-drop fragmentation determines size distribution of raindrops, Nature Physics, 2009.

[3] J.S. Marshall, W.M. Palmer, The distribution of raindrops with size, Journal of Meteorology, 5, 1948.

[4] R.A. Nelson, Rain How it Affects the Communications Link, Applied Technology Institute, 2000

[5] Y. Karasawa, T. Matsudo, One-minute rain rate distributions in Japan derived from A Me DAS one-hour rain rate data, IEEE Transactions on Geoscience and Remote Sensing, 1991

[6] O. N. Okoro, G. A. Agbo, J. E. Ekpe and T. N. Obiekezie, Comparison of hourly variations of radio refractivity for quiet and disturbed days during dry and rainy seasons at Minna, International Journal of Basic and Applied Sciences, vol. 2, 2013

[7] Recommendation ITU-R P.1815-1, 2009, Differential rain attenuation.

[8] M. Tamošiūnaitė, S. Tamošiūnas V. Daukšas M. Tamošiūnienė M. Žilinskas, Prediction of Electromagnetic Waves Attenuation due to Rain in the Localities of Lithuania, Electronics and Electrical Engeneering, No. 9, 2010.

[9] S. Tamošiunasa, M. Tamošiunien, M. Žilinskasa, Calculation of Electromagnetic Wave attenuation due to rain using Rainfall Data of Long and Short Duration, Lithuanian Journal of Physics, Vol.47, No. 2, 2007.

[10] P. A. Owolawi, Rainfall Rate Probability Density Evaluation and Mapping for the Estimation of Rain Attenuation in South Africa and Surrounding Islands, Progress In Electromagnetics Research, Vol. 112, 2011

[11] J. M. Gomez, Satellite Broadcast Systems Engineering, Artech House, 2002

[12] Recommendation ITU-R P.838-3, Specific attenuation model for rain for use in prediction methods, 2005

[13] R. L. Freeman, Radio systems design for telecommunication, Wiley, 2007

[14] F. Moupfouma, L. Martin, Modelling of the rainfall rate cumulative distribution for the design of satellite and terrestrial communication systems, International Journal of Satellite Communications and Networking, v.13, 1995.

[15] M. Tamošiūnaitė, M.Žilinskas, M. Tamošiūnienė, S. Tamošiūnas, Atmospheric Attenuation due to Humidity, Elecromagnetic Waves, Edited by V. Zhurbenko, 2011

[16] Recommendation ITU-R P.676-10, Attenuation by atmospheric gases, 2013

[17] R. Uijlenhoet, Raindrop size distributions and radar reflectivity–rain rate relationships for radar hydrology, Hydrology and Earth System Sciences, vol. 5, 2001.

[18] J. Bosy W. Rohm J. Sierny J. Kaplon, GNSS Meteorology, International Journal on Marine Navigation and Safety of Sea Transportation, Vol. 5, 2011.

[19] R. Crane, Electromagnetic Wave Propagation through Rain, Wiley, 1996

[20] Recommendation ITU-R P.453-10, The Radio refractive index: its formula and refractivity data, 2012

[21] E.A. Karagianni, A.P. Mitropoulos, A.G. Kavousanos-Kavousanakis, J.A. Koukos and M.E. Fafalios, Atmospheric Effects on EM Propagation and Weather Effects on the Performance of a Dual Band Antenna for WLAN Communications, Nausivios Chora, Volume 5, 2015, in press.

[22] A. D. Papatsoris, K. Polimeris, I. Sklari, A. A. Lazou, Rainfall Characteristics for Radiowave Propagation Studies in Greece, IEEE Antennas and Propagation Society International Symposium, 2008.

[23] Recommendation ITU-R P.837-6, Characteristics of precipitation for propagation modeling, 2012.

[24] J. W. Weatherly and D. R. Hill, The Impact of Climate and Extreme Weather Events on Military Operations, ADA432260, 2004

Concepts of the GMDSS Modernization

K. Korcz
Gdynia Maritime University, Gdynia, Poland

ABSTRACT: The current status of the Global Maritime Distress and Safety System (GMDSS) has been described. On the base of a current status of the GMDSS, the different concepts of the GMDSS modernization has been presented. The future of the maritime communication has been discussed as well.

1 INTRODUCTION

The Maritime Safety Committee (MSC) at its 81st session, in 2006, decided to include, in the work programmes of the Safety of Navigation (NAV) and Radiocommunications and Search and Rescue (COMSAR) Sub-Committees, a high priority item on "Development of an e-navigation strategy".

It needs to be noted that the development of e-navigation is an ongoing process and that the "Development of an e-navigation strategy implementation plan", as a next step of the project, was included in the work programmes of the COMSAR, NAV and Standards of Training and Watchkeeping (STW) Sub-Committees by MSC 85, in 2008.

Without a doubt, one of the fundamental elements of e-navigation will be a data communication network based on the maritime radiocommunication infrastructure. Taking into account the above, in 2009 at MSC 86, the Committee agreed to include in the COMSAR Sub-Committee work programme, a sub-item on "Scoping exercise to establish the need for a review of the elements and procedures of the GMDSS" under the work programme item on "Global Maritime Distress and Safety Systems (GMDSS)".

In the aftermath of this work in 2012 MSC 90 agreed to include in agenda of the COMSAR, NAV and STW Sub-Committees a high priority item on "Review and modernization of the Global Maritime Distress and Safety System (GMDSS)", with a target completion year of 2017, assigning the COMSAR Sub-Committee as the coordinating organ.

After the changes in the organization of the work of IMO subcommittees at the end of 2013, the issue of "Review and modernization of the Global Maritime Distress and Safety System" is, like the issue of "Development of an e-navigation strategy implementation plan", the competence of the new Sub-Committee on Safety of Navigation, Communication and Search and Rescue (NCSR), resulting from connections of the NAV and COMSAR Sub-Committees. In this context, during the discussion on the modernization of the GMDSS there are the different approaches to carry it out.

2 GMDSS FUNDAMENTALS

The original concept of the GMDSS is that search and rescue authorities ashore, as well as shipping in the immediate vicinity of the ship in distress, will be rapidly alerted to a distress incident so they can assist in a coordinated search and rescue operation with the minimum delay. The system also provides for urgency and safety communications and the promulgation of maritime safety information (MSI).

2.1 Functional requirements

The GMDSS lays down nine principal communications functions which all ships, while at sea, need to be able to perform (SOLAS, 2014):

1 transmitting ship-to-shore distress alerts by at least two separate and independent means, each using a different radiocommunication service;
2 receiving shore-to-ship distress alerts;
3 transmitting and receiving ship-to-ship distress alerts;
4 transmitting and receiving search and rescue co-ordinating communication;

5 transmitting and receiving on-scene communication;
6 transmitting and receiving signals for locating;
7 transmitting and receiving maritime safety information;
8 transmitting and receiving general radiocommuni-cation from shorebased radio systems or networks;
9 transmitting and receiving bridge-to-bridge communication.

Distress alerting is the rapid and successful reporting of a distress incident to a unit which can provide or co-ordinate assistance. This would be a rescue co-ordination centre (RCC) or another ship in the vicinity. When an alert is received by an RCC, normally via a coast station or a land earth station, the RCC will relay the alert to SAR units and to ships in the vicinity of the distress incident.

Search and rescue (SAR) co-ordinating communications are the communications necessary for the co-ordination of ships and aircraft participating in a search and rescue operation following a distress alert, and include communications between RCCs and any on-scene co-ordinator (OSC) in the area of the distress incident.

On-scene communications are the communications between the ship in distress and assisting units relate to the provision of assistance to the ship or the rescue of survivors.

Locating is the finding of a ship/aircraft in distress or its survival craft or survivors.

Ships need to be provided with up-to-date navigational warnings and meteorological warnings and forecasts and other urgent maritime safety information (MSI). MSI is made available by *Promulgation of MSI* by the responsible administration means.

General radiocommunications in the GMDSS are those communications between ship stations and shore-based communication networks which concern the management and operation of the ship and may have an impact on its safety.

Bridge-to-bridge communications are inter-ship safety communications conducted from the position from which the ship is normally navigated.

2.2 *GMDSS Sea areas*

Radiocommunication services incorporated in the GMDSS system have individual limitations with respect to the geographical coverage and services provided. The range of communication equipment carried on board the ship is determined not by the size of the ship but by the area in which it operates.

Four sea areas for communications within the GMDSS have been specified by the IMO. These areas are designated as follows (SOLAS, 2014):

– Sea area A1 – an area within the radiotelephone coverage of at least one VHF coast station in which continuous DSC alerting is available.
– Sea area A2 – an area, excluding sea area A1, within the radiotelephone coverage of at least one MF coast station in which continuous DSC alerting is available.
– Sea area A3 – an area, excluding sea areas A1 and A2, within the coverage of an Inmarsat geostationary satellite in which continuous alerting is available.
– Sea area A4 – an area outside sea areas A1, A2 and A3 (the polar regions north and south of 70° latitude, outside the Inmarsat satellite coverage area).

2.3 *Ship requirements*

In accordance with SOLAS Convention every ship shall be provided with radio installations capable of complying with the functional requirements throughout its intended voyage (see Sec. 2.1). The type of radio equipment required to be carried by a ship is determined by the sea areas through which a ship travels on its voyage (see Sec. 2.2).

Every ship shall be provided with:
1 VHF radio installation capable of transmitting and receiving:
   – DSC on the channel 70;
   – radiotelephony on the channels 6, 13 and 16;
2 Search and Rescue Locating Device (SRLD);
3 receiver capable of receiving international NAVTEX service broadcasts if the ship is engaged on voyages in any area in which an international NAVTEX service is provided;
4 radio facility for reception of maritime safety information (MSI) by the Inmarsat enhanced group calling (EGC) system if the ship is engaged on voyages in any area of Inmarsat coverage but in which an international NAVTEX service is not provided; however, ships engaged exclusively on voyages in areas where an HF direct-printing telegraphy maritime safety information service is provided and fitted with equipment capable of receiving such service, may be exempt from this requirement;
5 satellite emergency position-indicating radio beacon (EPIRB).

In addition to the above requirements every ship engaged on voyages in sea area A1 or A1 and A2 or A1, A2 and A3 or A1, A2, A3 and A4 shall be provided with an additional radio installation.

2.4 *Order of priority of communications*

In accordance with Radio Regulations all stations in the maritime mobile service and the maritime mobile-satellite service shall be capable of offering four levels of priority in the following order (RR. 2012):

1 distress communications;
2 urgency communications;
3 safety communications;
4 other communications.

The transmission of *a distress call* indicates that a mobile unit or person is threatened by grave and imminent danger and requires immediate assistance.

The urgency call indicates that the calling station has a very urgent message to transmit concerning the safety of a mobile unit or a person (for example medical advice or medical assistance).

The safety call indicates that the calling station has an important navigational or meteorological warning to transmit.

The other communication means any communication beyond distress, urgency and safety communications, for example public correspondence (any telecommunication which the offices and stations must, by reason of their being at the disposal of the public, accept for transmission).

3 CURRENT STATUS OF THE GMDSS

Since implementation of the GMDSS some changes both of technical and regulatory nature have occurred.

One of the most important GMDSS changes has concerned the Inmarsat. In 1999, Inmarsat became the first intergovernmental organization to be transformed into a private company. It caused that at present Inmarsat is recognised as a leader in global mobile satellite communication field. Inmarsat B, launched in 1993, was first maritime fully digital service. Inmarsat C, introduced in 1994, is one of the most flexible mobile satellite message communication systems. Inmarsat Fleet service provides both ocean-going and coastal vessels with comprehensive voice, fax and data communications. Fleet 77 (introduced in 2002) fully supports the GMDSS and includes advanced features such as emergency call prioritization, as stipulated by IMO Resolution A.1001 (25). Fleet F77 also helps meet the requirements of the International Ship and Port Facility Security (ISPS) code, which enables the cost-effective transfer of electronic notices of arrival, crew lists, certificates and records. Because Fleet 77 is IP compatible, it supports an extensive range of commercially available off-the-shelf software, as well as specialized maritime and business applications. Fleet 77 also ensures cost-effective communications by offering the choice of Mobile ISDN or MPDS channels at speeds of up to 128kbps.

FleetBroadband is Inmarsat's next generation of maritime services delivered via the Inmarsat-4 satellites. It is commercially available since the second half of 2007.

Inmarsat had over 175,000 registered GMDSS-capable mobile terminals at the end of October 2014, of which more than 154,000 were Inmarsat C.

It should be also noted that Inmarsat E service (L-band EPIRB) ceased to be supporting GMDSS in 2006 and Inmarsat A service – in 2007.

Instead of the Inmarsat E service, the new Cospas-Sarsat Geostationary Search and Rescue System (GEOSAR) has been introduced by Cospas-Sarsat as completion of the Low-altitude Earth Orbit System (LEOSAR). These two Cospas-Sarsat systems (GEOSAR and LEOSAR) create the complementary system assisting search and rescue operations (SAR operations).

At the same time, the Cospas-Sarsat ceased satellite processing of 121.5/243 MHz beacons on 1 February 2009.

It is also worth to note that in 2010 AIS-Search and Rescue Transmitter (AIS-SART) was introduced. So, shipboard GMDSS installations include one or more Search and Rescue Locating Devices (SARLD). These devices may be either an AIS-SART or a SART (Search and Rescue Transponder).

And at the end, as the result of the hard work of International Maritime Organization (IMO) and other bodies, two new systems have been introduced:

– Ship Security Alert System – SSAS (in 2004),
– Long-Range Identification and Tracking of ships – LRIT (in 2009).

Although the SSAS and LRIT systems are not a part of the GMDSS, in the direct way they use its communication means.

Besides the above-mentioned main changes the following should be noted as well (Korcz, 2013):

– the cessation of aural 2 182 kHz listening watch;
– the requirement for passenger ships to have aviation VHF;
– the non-introduction of VHF DSC EPIRBs despite their inclusion in SOLAS;
– the introduction of the SafetyNET service together with Inmarsat-C and mini-C terminals;
– the introduction of a range of non-SOLAS Inmarsat services;
– the closure of much of the worlds "public correspondence" radiotelephony, Morse, radio-telex and VHF coastal radio networks;
– the advent of 406 MHz digital distress beacons, with or without GNSS;
– the development of the Galileo GNSS;
– the civil availability of Glonass GNSS;
– the development of Man Overboard devices using DSC and AIS;
– proposals for EPIRBs equipped with AIS;
– the major decline in the use of radio-telex at sea;

- the introduction of HF e-mail data (non-SOLAS) services for routine communications;
- the increase in availability of terrestrial and satellite communications networks;
- the increase in the bandwidth supported by many communications platforms, both terrestrial and satellite;
- the further development of the internet, and associated communications software, including "social media", and mobile phone "apps";
- the increased use of e-mail by crews, and for ship's business;
- the widespread take-up of data-intensive technology for ship's business (i.e. e-commerce, digital photography for cargo and ship-related business, maintenance, medical applications, stability calculations, cargo calculations, engineering analysis, numerous reporting requirements for customs, immigration, quarantine, etc.);
- increasing broadband wireless access along coastlines;
- the further use of non-GMDSS earth stations aboard vessels;
- increased satellite bandwidth available in C-band and Ku-band via geostationary satellites;
- greater access to radiocommunications equipment by crews outside the GMDSS installation (i.e. cellular phones, and broadband wireless);
- the massive deployment of cellular telephony and broadband wireless along populated coastlines;
- the advent of much cheaper and smaller radio-electronics equipment;
- radio-navigation equipment/computing equipment product choice increasing and more equipment is being combined, particularly in the non-SOLAS sector;
- the commencement of the non-GMDSS "505" Inmarsat service;
- the new Arctic NAVAREAs/METAREAs;
- the advent of new satellite providers seeking participation in the GMDSS;
- the increased availability of affordable non-SOLAS satellite providers;
- the advent of software-defined radio (SDR);
- the development of Voyage Data Recorders (VDR);
- the use of wireless local area networks aboard ships;
- the advent of short-range wireless device interconnectivity (PCs, etc.);
- the use of marine pilots carrying their own PC with a range of marine software products;
- the advent of AIS Application-Specific Messages;
- AIS-equipment aids to navigation, including virtual and synthetic AtoN;
- the advent of Integrated Bridge Systems.

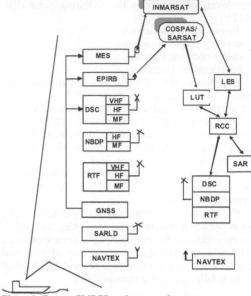

Figure 1. Current GMDSS equipment and systems

Taking into account the above changes, up to date equipment and systems used in the GMDSS is showed in Fig. 1 (Korcz, 2013). Used in Figure 1 devices and systems abbreviations mean (Fig. 1):
- MES - Mobile Earth Station;
- LES - Land Earth Station;
- Inmarsat - Satellite System;
- EPIRB - Emergency Position Indicating Radio-Beacon;
- LUT - COSPAS/SARSAT Local User Terminal;
- DSC - Digital Selective Calling;
- NBDP - Narrow Band Direct Printing;
- RTF - Radiotelephony;
- GNSS - Global Navigation Satellite System - for support (for example GPS- Global Positioning System);
- SARLD - Search and Rescue Locating Device;
- NAVTEX System;
- RCC - Rescue Coordination Centre.
- SAR – Search and Rescue Service.

At a regulatory level a modification of the GMDSS is coordinated by two international organizations: International Maritime Organization (IMO) and International Telecommunication Union (ITU).

IMO modifications are mainly concerning the amendments to Chapter IV of the International Convention for the Safety of Life At Sea (SOLAS) and the proper IMO resolutions. From the Radiocommunication point of view, the most important modification was adoption by IMO of Resolution A.1001(25) dated 29.11.2007 on Criteria for the Provision of Mobile Satellite Communication Systems in the GMDSS and revision of Chapter IV of IMO SOLAS Convention extends the International Mobile Satellite Organization (IMSO) oversight to GMDSS Services provided by any satellite operator which fits criteria [Korcz, 2011].

ITU modifications are mainly concerning the amendments to Radio Regulations. Since 1999, these amendments were adopted by six World Radiocommunication Conferences (WRCs), which are held every three to four years. These World Radiocommunication Conferences were held in 1995, 1997, 2000, 2003, 2007 and 2012 (Korcz, 2013).

From the point of view of the GMDSS modernization, among many important outputs of the WRC-12, the two documents are very important: Resolution 359 (WRC-12) "Consideration of regulatory provisions for modernization of the Global Maritime Distress and Safety System and studies related to e-navigation" and Resolution 360 (WRC-12) "Consideration of regulatory provisions and spectrum allocations for enhanced Automatic Identification System technology applications and for enhanced maritime Radiocommunication".

4 GMDSS MODERNIZATION

As a result of the work on "Scoping Exercise to establish the need for a review of the elements and procedures of the GMDSS", the Maritime Safety Committee (MSC) at its 90 session decided to include, in the work programme of the Radiocommunications and Search and Rescue (COMSAR) Sub-Committee, a high priority item on "Review and modernization of the Global Maritime Distress and Safety System (GMDSS)".

This new work item is to review the GMDSS, and then to develop a modernization programme. The modernization programme would implement findings of the review, include more modern and efficient communications technologies in the GMDSS, and support the communications needs of the e-navigation.

At the beginning of the GMDSS modernization process a lot of issues were considered but the most important considerations concern the fundamental issues presented in Sec. 2.

4.1 Functional requirements

In considering the nine functional requirements (see Sec. 2.1), the following over-arching issues is considered (NCSR, 2014):
- the possible need for inclusion of security-related communications in the GMDSS; and
- the possible need to develop a clearer definition of "General Communications".

The Ship Security Alert System (SSAS) does not involve communication with other ships or with coast radio stations. Therefore, those communications are neither ship-to-ship nor ship-to-shore communications. Communications are addressed to a designated competent authority. Therefore, *security-related communications* should not be a functional requirement of the GMDSS but chapter IV should include a requirement for ships to be capable of security related communications, and a definition of "security-related communications" is also required. A definition of "security-related communications" is proposed as follows (NCSR 1):

"Security-related communications means communications associated with the update of security levels, security incidents or threast thereof and security-related information prior to the entry of a ship into a port."

The existing definition in SOLAS regulation IV/2.1.5, defines *general radio communications* as "operational and public correspondence traffic, other than distress, urgency and safety messages conducted by radio". Coast radio stations (Government owned) which provided public correspondence facilities when the GMDSS was first designed have now all largely been closed down. However, facilities for public correspondence are still required. These communications are now being achieved using commercial services which are not normally associated with coast radio stations and the term public correspondence is no longer widely used. For the Modernized GMDSS it is therefore proposed to change the term Public correspondence to "Other communications" and include a new capability for Other communications but not as part of the GMDSS functional requirements.

The definition of urgency and safety communications (Radio Regulations) includes the following communications:
1 navigational and meteorological warnings and urgent information;
2 ship-to-ship safety of navigation communications;
3 ship reporting communications;
4 support communications for search and rescue operations;
5 other urgency and safety messages; and
6 communications relating to the navigation, movements and needs of ships and weather observation messages destined for an official meteorological service.

Operational communications are now, therefore, covered under the definition of urgency and safety communications. It is proposed to redefine the term "General communications" by aligning it with the Radio Regulations. The new definition proposed is:

"General communications means operational communications, other than distress conducted by radio".

So, the new text of functional requirements for the Modernized GMDSS is proposed as follows:

1 performing the GMDSS functions as follows:
   - transmitting ship-to-shore distress alerts by at least two separate and independent means, each using a different radiocommunication service;
   - receiving shore-to-ship distress alert relays;
   - transmitting and receiving ship-to-ship distress alerts;
   - transmitting and receiving search and rescue coordinating communications;
   - transmitting and receiving on-scene communications;
   - transmitting and receiving signals for locating;
   - transmitting and receiving safety-related information;
   - receiving Maritime Safety Information (MSI);
   - transmitting and receiving general communications; and
   - transmitting and receiving bridge-to-bridge communications,
2 transmitting and receiving security-related communications, in accordance with the requirements of the International Ship and Port Facility Security Code; and
3 transmitting and receiving other communications to and from shore-based systems or networks.

4.2 *GMDSS Sea areas*

Four sea areas have been defined according to the coverage of VHF, HF and MF Coast Radio Services and Inmarsat Services as given in Sec. 2.2. During the review it was noted that extensive use was made of VHF communications and, therefore, sea area A1 should be retained. Because of considerable use of MF voice communications, it was finally concluded that sea area A2 should be retained as a separate sea area as well.

The definition of the boundary between sea area A3 and A4 is currently defined by Inmarsat coverage, but Inmarsat might not always be the only GMDSS satellite provider. In future, the Organization might recognize regional or global satellite systems to provide GMDSS services in an A3 sea area, each of them providing coverage different to the current A3 sea area. It was considered that HF should remain a requirement for sea area A4 and an option for sea area A3. Recognizing that other options for the definition of sea areas A3 and A4 could be developed, three different options for the definition of sea areas A3 and A4 were identified as follows:
OPTION 1
Sea area A3 means an area, excluding sea areas A1 and A2, within the coverage of a recognized mobile satellite communication service using geostationary satellites in which continuous alerting is available.

Sea area A4 means an area outside sea areas A1, A2 and A3.

Above Option is the most similar to the current definitions, except that the reference to Inmarsat has been deleted. Option 1 does not facilitate the introduction of non-geostationary satellite systems. The boundary between sea areas A3 and A4 would depend upon the satellite system used and could be different for different ships.
OPTION 2
Sea area A3 means an area, excluding sea areas A1 and A2, within the coverage of a recognized mobile satellite communication service in which continuous alerting is available between [70 or 76] degrees North and South.

Sea area A3- Sub means a sub-area within sea area A3, within the regional coverage of a recognized mobile satellite communication service in which continuous alerting is available.

Sea area A4 means an area outside sea areas A1, A2 and A3.

Sea area A4-R means a sub-area within sea area A4, within the regional coverage of a recognized mobile satellite communication service in which continuous alerting is available.

Above Option defines a clear boundary for the A3 sea area and, as such, might be helpful to an Administration in issuing safety radio certificates to ships.
OPTION 3
Sea area A3 means an area, excluding sea areas A1 and A2, within the coverage of a recognized mobile satellite communication service in which continuous alerting is available as may be defined by the Organization.

Sea area A4 means an area outside sea areas A1, A2 and A3.

Above Option defines the sea area A3 as somewhere where satellite coverage is available. The boundary between sea areas A3 and A4 would depend upon the satellite system used and could be different for different ships. The safety radio certificate would require details of the geographical area in which the ship is permitted to sail. Availability of a global satellite system would result in not having a sea area A4 for ships that are certificated to use a global system.

4.3 *Ship requirements*

In future, if other satellite service providers are recognized by the Organization, the safety radio certificates of the ship should be required to define the geographic area in which the ship is permitted to operate. The detail of the geographical areas covered by all the different satellite service providers will be given in the GMDSS Master Plan. Taking into account the changes of the functional requirements (see Sec. 4.1) and GMDSS Sea areas (see Sec. 4.2) changes in the basic equipment of ships will also have to be introduced (see Sec. 2.3).

4.4 *Order of priority of communications*

The Radio Regulations provide the existing order of four levels of priority (see Sec. 2.4). The four priorities are needed for communications and operational use in general, including voice, maritime safety information, as well as other text and data messages. Priorities for text and data messages can be used to sort message displays in order of importance or the way in which they are displayed. However, two priorities are sufficient for controlling the radiocommunication link, for example by using pre-emption.

It is concluded, therefore, that the four levels of priority should be retained, and applied to voice, text, and data messages and that there is no need to revise Radio Regulations. Automated systems should give priority to category distress. Automated systems should also give priority to categories urgency and safety (ahead of other communications category).

5 CONCLUSIONS

Discussion on the modernization of the GMDSS (marine radiocommunication) is still in very early stage. The main problems to be solved at this stage of work concern, discussed in the article, the concepts of modernization of the GMDSS, connected with the appropriate changes in the SOLAS Convention.

More than twenty years have passed since the time when the Global Maritime Distress and Safety System (GMDSS) became introduced. Planning for the GMDSS started more than thirty years ago, so some elements of it have been in place for many years. There have been numerous advances in the use of maritime radiocommunication to maritime safety, security and environmental protection during this period. But now there are some obsolete GMDSS equipment and systems or the ones that have seldom or never been used in practice.

The GMDSS is a system of systems, which like a "bus" going along the road, and certain technologies get off the bus and certain new technologies get on board the bus. The idea that the GMDSS and related systems are static is rather incorrect. Reference to the section 3 proves that the GMDSS continues to be a dynamic system, whilst still supporting the essential functions of the GMDSS (Korcz, 2011).

On the other hand there are a lot of the new digital and information technologies outside the GMDSS.

In the Author's opinion, the future of the GMDSS is closely connected both with the development of the e-navigation project and the role of the radiocommunication in this process. It is envisaged that a data communication network will be one of the most important parts of the e-navigation strategy plan (Korcz, 2009).

In order to realize efficient and effective process of data communication for e-navigation system, the existing radio communication equipment on board (GMDSS), as well as new radio communication systems should be recognized.

Apart from the satellite systems, the GMDSS MF, HF and VHF equipment and systems (Fig. 1) should be also used as a way of data communication for the e-navigation system, provided that this equipment and systems will be technically improved by means of (Korcz, 2013):
– digitization of the analogue communication MF, HF and VHF channels;
– application of high-speed channel to GMDSS;
– utilization of SDR (Software Defined Radio) technology;
– adaptation of IP (Internet Protocol) technology to GMDSS;
– integration of user interface of GMDSS equipment; and
– any other proper technology for GMDSS improvement.

Analyzing the existing communication systems for inclusion in the modernized GMDSS, there are the opinions that mobile internet services, mobile telephone services, Broadband wireless access (BWA), eg. Wimax/mesh networks wireless Local Area Networks and non-regulated Satellite Emergency Notification Devices (SENDs), although more and more used by the public, including non-SOLAS ships for alerting, are not the appropriate means and, therefore, should not form part of the international system. More consideration is needed to decide which systems, relying on older or inefficient technologies, might be considered for replacement by more modern systems.

The following new equipment, systems and technologies, currently not included in GMDSS, might be included in the modernized GMDSS (COMSAR, 2013 and NCSR, 2014):
– AIS, including Satellite monitoring of AIS and additional AIS channels for identification but not alerting;
– HF e-mail and data systems;
– VHF data systems;
– Application Specific Messages over AIS or VHF data systems;
– NAVDAT;
– Modern satellite communication technologies;
– Additional GMDSS satellite service providers;
– Hand-held satellite telephones in survival craft;
– Hand-held VHF with DSC and GNSS for survival craft;
– Man Overboard Devices; and

- AIS and GNSS-equipped EPIRBs.

In considering the GMDSS modernization the following developments that are already under way should be taken into account as well:
- proposed FleetBroadband FB500 terminal for GMDSS;
- Inmarsat MSDS (Maritime Safety and Data Service) over BGAN FleetBroadband;
- proposed Cospas-Sarsat MEOSAR network (will be probably fully operational in 2018) and retirement of the Cospas-Sarsat LEOSAR network;
- new GNSS projects (eg. Galileo);
- introduction of new broadcasts to ships using systems in the 495-505 kHz band; and

the following changes in the GMDSS which might be expected to occur in the near to mid-term:
- the provision of hand-held satellite telephones in survival craft;
- the provision of handheld VHF with DSC and GNSS for survival craft;
- all SOLAS EPIRBs to be fitted with GNSS;
- EPIRB-AIS;
- EPIRB-AIS to be used with/in voyage data recorders;
- MOB-AIS;
- next generation AIS;
- development of an electronic equipment;
- using AIS in a new "ranging mode";
- additional AIS channels for channel management and data transfer;
- the use of Channels 75 and 76 for enhanced detection via satellite;
- harmonized digital bands in VHF RR;
- the further evolution of Maritime Safety Information broadcast systems, taking into account the ongoing work in IHO and WMO;
- an eventual replacement of MF NAVTEX with a new higher data-rate system in the 495-505 kHz band;
- a new channel plan for Appendix 17 HF bands to permit new digital emissions, whilst still allowing Morse and NBDP, and protection for GMDSS MSI;
- the introduction of an e-navigation communications suite independent of the GMDSS, but with interconnections with it.

It should be also noted that Inmarsat Global Limited has informed of its intention to close the Inmarsat-B Service from 31 December 2016.

During work on review and modernization of the GMDSS it is necessary first to identify real user needs and secondly to realize that the modernization of the maritime radiocommunication should not be driven only by technical requirements. In addition, it is necessary to ensure that man-machine-interface and the human element will be taken into account including the training of the personnel.

The lessons learnt from the original development and operation of GMDSS should be taken into consideration in the modification of GMDSS as well.

Furthermore, the continuous and open process is needed to ensure it remains modern and fully responsive to changes in requirements and evolutions of technology and it will meet the expected e-navigation requirements. In order to ensure it, a mechanism for continuous evolution of the GMDSS in a systematic way should be created as well.

In this approach to development of the GMDSS it is very important that the integrity of GMDSS must not be jeopardized.

It should be noted that in this context, for the process of the GMDSS modernization, regulatory decisions taken on the World Radiocommunication Conference being held in 2015 (WRC-15) and in 2018 (WRC-18) will be of great importance.

And finally it should be noted that a key to the success of the review and modernization process of the GMDSS is not only that the work is completed on time, but also that it has the flexibility to implement changes ahead of schedule. In this approach to development of maritime radiocommunication it is also essential that the integrity of the GMDSS must not be threatened.

REFERENCES

International Maritime Organization (IMO). 2014. International Convention for the Safety of Life At Sea (SOLAS), London
International Telecommunication Union (ITU). Radio Regulations (RR), Geneva 2012
Sub-Committee on Radiocommunications, Search and Rescue - COMSAR 17. 2013. Report to the MSC, International Maritime Organization (IMO), London
Sub-Committee on Radiocommunications, Search and Rescue - NCSR 1. 2014. Report to the MSC, International Maritime Organization (IMO), London
Korcz K. 2009. Some Radiocommunication aspects of e-Navigation. 8th International Navigational Symposium on Marine Navigation and Safety of Sea Transportation, TRANS-NAV 2009, Gdynia
Korcz K. 2011. Yesterday, today and tomorrow of the GMDSS. 9th International Navigational Symposium on Marine Navigation and Safety of Sea Transportation, TRANS-NAV 2011, Gdynia
Korcz K., 2013. Modernization of the GMDSS, 10th International Symposium on Marine Navigation and Safety of Sea Transportation, TRANS-NAV 2013, Gdynia.

Decision Support System

Information, Communication and Environment – Marine Navigation and Safety of Sea Transportation – A. Weintrit & T. Neumann (eds.)

Decision Support System

Supporting Situation Awareness on the Bridge: Testing Route Exchange in a Practical e-Navigation Study

T. Porathe
Norwegian University of Science and Technology, Norway

A. Brodje & R. Weber
Chalmers University of Technology, Sweden

D. Camre & O. Borup
Danish Maritime Authority, Denmark

ABSTRACT: In a simulator study parts of the ACCSEAS project's e-Navigation route exchange concept termed "intended routes" has been tested in a full mission bridge simulator using experienced bridge officers in port approach scenarios. By "intended routes" we mean a service where ships underway send a number of waypoints ahead of their present position, from their voyage plan; thus sharing their intentions with ships within radio range. Other ships "intended routes" become visible on the ECDIS screen on request and can be queried for where my own ship is when the other ship is at the cursor indicated point on the displayed "intended route". Observation, focus group interviews and questionnaires were used to capture qualitative data on professional acceptance, the concept, procedural changes, functions and interface. The tested service was very well received with high acceptance ratings.

1 INTRODUCTION

Sometimes people misunderstand each other's intentions. That happens at sea as well as in all walks of life. To prevent accidents at sea the International Maritime Organization (IMO) has established the International Regulations for Preventing Collisions at Sea (COLREGS). They are supposed to unambiguously determine which ship is to stand on and which is to give way in a collision avoidance situation. However misunderstandings occur, a collision in the English Channel in 1979 can serve as an example:

The Liberian bulk carrier *Artadi* was proceeding NE in the Traffic Separation Scheme (TSS) in the Dover Strait in restricted visibility (see Figure 1). The French ferry *St-Germain* was approaching from the east. She was spotted in good time on the radar of the *Artadi*. Coming from starboard, *St-Germain* was the stand-on ship according to rule 15 of the COLREGS, however, according to rule 19 both ships should give way in this case of restricted visibility. The pilot and master of the *Artadi* expected *St-Germain* to keep speed and course and started to make a starboard turn to give way. However, on-board the *St-Germain* the intention was not at all to cross the traffic separation scheme diagonally in front of *Artadi*, but instead to turn port and follow outside the boarder of the NE going traffic lane until the traffic cleared and she could

make the crossing at a right angle (according to rule 10c). In the subsequent collision two persons were killed (Kwik, 1984; Office of the Commissioner for Marine Affairs, 1979).

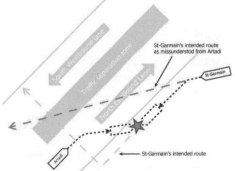

Figure 1. Misunderstanding intentions: The collision of French train ferry *St-Germain* with Liberian bulk carrier *Artadi* in the English Channel, 1979.

1.1 *ACCSEAS*

One of the surprising findings in the recently concluded EU project ACCSEAS (Accessibility for Shipping, Efficiency Advantages and Sustainability) was how the development of off-shore wind turbines would restrict shipping in the southern part of the North Sea in the future. Looking at the plans for future wind farms the project came up with the map in Figure 2 (ACCSEAS, 2013). The polygons are planned wind mill parks in some stage of concession. We here see a clear trend: the shipping industry must in the future be prepared to share ocean space with a lot of new actors, not only wind energy, but different sorts of off-shore farming as well.

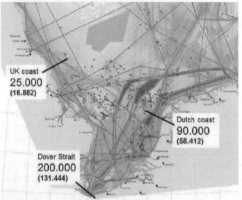

Figure 2. The south part of the North Sea. Number of ship predictions for 2020+ (2012 numbers in parenthesis). The polygons are planned areas for wind turbines. The fussy lines are 2012 traffic density plot summarized in the solid lines, darker polygons areas are TSS separation zones. (ACCSEAS, 2013)

The problems of navigation caused by one of these new installations might be exemplified with the newly constructed Thornton Bank wind mill park outside Zeebrugge on the Belgian coast (see Figure 3).

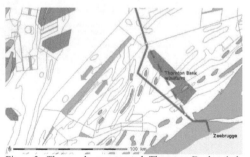

Figure 3. The newly constructed Thornton Bank wind mill park has forced the P&O ferry to change its route causing possible risks of misunderstandings.

The P&O ferry between Zeebrugge and Hull in the U.K. needed to change its route due to the newly constructed park. While the old track (dashed) gave some indications of the ferry's intentions, the new track (filled line) headed straight towards the main English Channel TSS at a right angle as if the intentions were to cross straight over. This TSS is the most trafficked route in the world with 133 000

passing's in 2012 (ACCSEAS, 2013). Ships coming up the TSS towards the North Sea could potentially misunderstand the intentions of the big ferry approaching form starboard. And as more wind mill parks appear we are getting closer to the street-like situation we are used to in road traffic in cities. Only, cars have direction indicators which are something we do not have on ships. The only indication of the intent of the approach P&O ferry would be to see the destination through the AIS static message and try and deduce the intentions from there.

That is why one of the suggested solutions from the ACCSEAS project is a service aimed at showing ships intentions to other vessels in the vicinity.

1.2 *Tactical and strategic route exchange*

Situation awareness is a fundamental property for humans driving any kind of vehicle. Endsley defined it in 1988 as ""the perception of elements in the environment within a volume of time and space, the comprehension of their meaning, and the projection of their status in the near future." Not only knowing what is going on, but also being able to predict what will happen in the future is crucial for navigation ships with large inertia.

With regards to the future whereabouts we will use the following taxonomy when talking about ships future positions (see Figure 4).

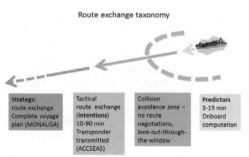

Figure 4. Taxonomy of route exchange based on how long in advance the service aims to predict vessels future positions. The levels are explained in the text.

1.3 *Predictors*

On the very close scale ships use *predictors*. Predictors may be able to look some 3 to 15 minutes into the future and are typically very reliable on short time ranges. They simply extrapolate present speed and heading a number of minutes into the future. This type of simple prediction is typically used by AIS and ARPA radar where targets show course and speed by the direction and length of a vector, sometimes augmented by symbols for rate of turn (ROT) prediction extrapolating ships present turn speed when conducting a turn.

Because ships are heavy and cannot change course or speed very rapidly, predictors are reliable up to some 3 minutes. Longer predictions rely on the ships not making any course speed changes.

The predictor is often used to investigate possible close quarter situations by changing the time setting for own and other vessels course speed vector on the ARPA or the ECDIS onboard.

1.4 Strategic route exchange

In the other end of the scale we have long term strategic route exchange which is part of the Ship Traffic Management (STM) concept which is being investigated by the MONALISA project. Every ship is mandated to make a berth-to-berth voyage plan before leaving port. In the old days the voyage plan was a pencil line on the paper chart; today the voyage plan resides in the electronic chart system, the ECDIS. The main objectives for sharing voyage plans are safety and efficiency. By coordinating voyage plans collisions might possibly be avoided by awareness of upcoming congestions. By coordinating voyage plans with availability of port facilities, fuel and emissions might be saved and efficiency in the transportation chain increased.

Although it is easy to find the present position and destination of any ship though AIS data in the Internet, the voyage plan is considered to be of business interest and not to be shared with anybody. Strategic route exchange therefore involves a coordination center doing the route coordination.

1.5 Tactical route exchange

On a level between the short range predictors and ships entire voyage plans, we have the *Intended routes*. The idea here is to transmit a number of waypoints ahead of the ships present position with the AIS message (or some future system) and so show any ships intentions some 60-90 minutes ahead. Presently we have tested sending out 8 waypoints. The shown intentions will then differ in length depending on the density of waypoints.

The intended route should be integrated in the ECDIS and shown on demand not to clutter the screen with all ships intended routes.

1.6 Collision avoidance zone

In earlier tests with intended routes users intuitively started using the intended routes to negotiate behavior for collision avoidance (see Porathe, Lutzhoft, & Praetorius, 2013) it became clear that there needed to be a psychological cut-off distance where navigators stop using computer systems to negotiate evasive maneuvers and start using basic COLREGS based on visual observation and ARPA. The radius of that zone would be dependent on several factors like traffic density, vessel type, speed and weather. The range might typically be 6-10 miles.

However, having said that, it was found in the earlier study that negotiating by clicking and dragging waypoints in the intended route might be a way of avoiding to enter into a close quarter situation. Provided it was done in good time.

2 METHOD

Human (or User)-Centered Design is a design philosophy that aims to involves the user throughout the design process from early context enquiries on through prototype design and different level user tests (Norman, 1988).

In research projects like the ACCSEAS new solutions are tested in very early phases of the development process with the goal to investigate *professional acceptance* of a new service. Mainly qualitative data is collected with methods like Usability testing. Usability is defined as "the extent to which a product can be used by specified users to achieve specified goals with effectiveness, efficiency, and satisfaction in a specified context of use." (ISO 9241-11) Also learnability and safety are important aspects to investigate.

In this test users (ship officers and VTS operators) were asked to use the Intended route service during different port approach scenarios. The testes had a high level of ecological validity through the use of full mission bridge simulators and VTS simulators. During the test observers on the bridges and in the VTS filmed and asked questions to the participants who were encouraged to think aloud. After the scenarios a debriefing session was held. The participants were also asked to fill in a survey rating their *professional acceptance* of the Intended route service.

2.1 The e-Navigation Prototype Display

To be able to test the Intended route service an ECDIS-like test platform had been developed by the Danish Maritime Authority: The EPD (E-navigation Prototype Display). The platform contained enough ECDIS features to be able to replace the ordinary ECDIS in the full mission bridge simulators used.

An EPD Shore system had also been developed for use in the VTS center. All systems had the ability to exchange route information such as the Intended route service (but also Suggested routes from shore to ship and strategic route exchange as explained above).

By right-clicking on an AIS target and selecting "Show intended route" in the EPD the intended route of the vessel was shown (if the vessel indeed had the Intended route service, which was not always the case as was though realistic). There was also a choice of "Show all ships intended routes". In Figure 5 the portrayal of the Intended route service is shown.

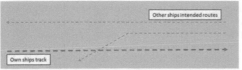

Figure 5. This is how the own ships route (below) and two ships intended routes (top) was portrayed in the EPD.

There was also a feature that allowed the navigator to query other ships intended routes based on the planned speed that had been entered into the voyage plan. By moving you cursor over the other ships intended route a *CPA Guidance Line* would appear connecting the cursor with the point on your own ships track where you would be when the other ship was on the position of the cursor (given planned route and speed was kept). The portrayal of this feature is shown in Figure 6.

Figure 6. The CPA Guidance Line connecting the point the cursor is at on another vessels intended track with the position the own vessel is calculated to be at the same time.

The same feature is used for the *CPA Alert* feature. If the calculate CPA of any vessel transmitting Intended routes becomes less that a predefined distance, e.g. 0.5 mile, CPA Guidance Lines calculated for each minute becomes visible highlighted in yellow together with the risk vessels intended track and an audible alarm(see Figure 7).

2.2 *Simulator study*

The study took place during four days in the end of September 2014 at the Simulator Centre at the Department of Shipping and Marine Technology at Chalmers University of Technology in Sweden.

Two Transas 5000 bridges and one VTS station were used, using parallel worlds so that the two ships could not see each other but interacted instead with target ships controlled by the simulator instructor station. The VTS could see both ships on different screens.

Two other e-Navigation services (the Suggested route, and the NoGo area services) were also tested. These results are presented separately.

Figure 7. The CPA Alert feature warning for a possible future close quarter situation. Note here the difference between ARPA CPA (based on present route and speed) and Route CPA based on the Intended route.

2.3 *Scenarios*

Five scenarios in the river and approach to the Humber Estuary were suggested by experts from ABP Humber. The area was chosen because high ship density and changing tidal situations. The area had also a VTS service.

Own ship in all scenarios was a180 m long ro-pax ferry. An overview of the geographical limits of the 5 scenarios can be seen in in Figure 8.

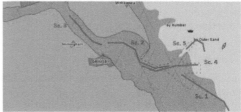

Figure 8. The image shows an overview of the five scenarios in the Humber River areas in eastern U.K.

In the first scenario the ferry was bound for Hull approaching the Sea Reach TSS in northerly gale. In order to avoid heavy rolling with beam seas the vessel was to re-route using the Rosse Reach TSS.

The second scenario involved a special transport (wind turbine propeller) with extensive width. Thus requiring an exclusion zone according to port regulations. Own vessel was re-routed by the VTS using the Sunk Dredged Channel (see Figure 9).

The third scenario was traffic congestion outside Immingham Oil and BulkTerminals. Several large ships was maneuvering in the area and the ferry was re-routed by the VTS using the Foul Holme Channel.

The fourth scenario was a contravention scenario in the TSS off Spurn Head. A deep draught vessel needed to use the inbound TSS and the ferry was asked to use "the southern part of the inbound TSS" when passing.

Figure 9. Screen shot from the EPD during scenario 2: own ship in Sunk Dredged Channel outbound ship passing through the main channel.

The final scenario involved a vessel leaving Anchorage A for Immingham with a strong south bound current. This situation has several times caused ships to hit the North New Sand N-cardinal light buoy.

Most of the scenarios involved the VTS at Spurn Head sending out route suggestions (which is disseminated in a separate conference paper), but all scenarios involved several other ships. Thereby reflecting the Humber area being one of the busiest ports in the UK.

The five scenarios took two days with familiarization, briefings and debriefings between all scenarios and the final discussion at the end.

2.4 Participants

11 professional British, Swedish and Danish bridge officers, harbor masters, pilots and VTS operators with experience from traffic in the Humber area were used for the test. All the participants were male from age 32 to 58, with a mean age of 47 years. They all had a sea time ranging from 12 to 30 years, mean 22 years.

Each bridge was manned with two bridge officers, which would be realistic considering that the situation was approach to port with restricted waters and heavy traffic.

The Spurn Head simulated VTS was manned with two VTS operators from the actual VTS center. The VTS operators were available from the whole test except on the last day when a Gothenburg VTS operator took over the chair (after having worked together with the Humber operators the day before).

3 RESULTS

The comments from the videos and discussions were analyzed and the results are below presented in four levels: conceptual, procedural, functional and HMI (Human-Machine Interface). Quotes are from the video recordings.

3.1 Conceptual level

All the participants agreed that this service was valuable. "I might not have said so three days ago, but now having used it: Yes, the concept is very good. Provided the data that is displayed is correct." (said by Humber pilot with 12 years' experience). On a question if someone in the group was against the concept, there were head shakes and silence. Several of the participants soon got used to the service where they could see ships intentions; one said "after having used the system for six hours I find it annoying not being able to see ships intentions." (He was referring to the fact that some of the target vessels did intentionally not send out Intended routes.)

3.1.1 Training

Several participants talked about the importance of "correct data": that the voyage plan was updated and correct and from berth-to-berth. It is necessary that the bridge personal are trained and can handle the system. Generally today, the Humber participants explained, the tankers coming into river have very good passage plans because they are heavily vetted. The general cargo and bulk carries, however, generally tend to have a voyage plan that either stops at the pilot station, or – if it goes all the way to the berth do so by a couple of haphazard waypoints. If the displayed data is not correct it could be a dangerous concept: you think you know where someone is going, but instead they are going a completely different route. E.g. there might be a change in the voyage plan and because the 2nd officer responsible for voyage planning is not on watch, the new intentions are not displayed.

One of the pilots said: On a big ship like the P&O ferry they have the time and people to do the voyage plan prudently with the right speed on all legs, etc. But on a small coaster they will just click out the waypoints, they don't have the time or the people to do anything else. "So my concern is not so much the quality of the proposed system, much more so, the quality of the people onboard that must be able to use the system."

3.1.2 Cluttering

Going into details, the participants felt that it is important that rather than displaying all ships Intended routes all the time (which would clutter the display) you can (as indeed was the case) "interrogate" the display for intentions of vessels of interest.

3.1.3 Turning off transmission intended routes

A discussion took place of what to do if a ship for some reason had to deviate from its route. One suggestion from the developers was that there should be an easy (or maybe even automatic) way of turning off the transmission of the Intended route if the vessel for some reason deviated too far from its intended voyage plan. One of the pilots answered "Yes, having no data is better than having the wrong data." There was an agreement that it could be a good thing if the system stopped sending route intentions if the ship was some predefined distance from its intended route for a predefined amount of time. But for minor deviations from the intended route, like overtaking, or giving extra space in a meeting situation, no one in the group felt it was necessary to stop sending, or changing the intended route. It would be obvious why the deviation was made.

There was also an agreement that the Intended route service should not be used as a collision avoidance tool in close quarters situations.

3.1.4 *Use in approaches and open sea*

It was felt that the Intended route service was probably being more important in open seas than in port approaches like the Humber River, because there is already a risk mitigation service like pilot onboard and VTS that keeps an eye on things. But for ships coming to the pilot station it is good, but there are also uses on the river. Approaches to junction points is an example where the Intended route service can be very valuable, for example a small ship leaving the Baltic Sea destined for Rotterdam may equally well take a route via The Sound, the Great Belt or the Kiel Canal. Being able to see the intended route makes it possible for an overtaking ship to place itself on the proper side of the other ship.

3.1.5 *Trust*

One of the participants said on a question if he would trust an Intended route, that he would trust it in the same way that he today trusts the AIS information. "I will not trust 100 percent, but it is helpful."

3.1.6 *Planned speed vs. current speed*

There was a major discussion on whether planned speed or current speed should be used when calculating a ships future position. The Intended route service as it was implemented in the prototype system was using the planned ETA in all waypoints to calculate where own and other ships would be at a certain time. The planned speed was based on the notion that ships should be at their final destination precisely in the planned arrival time. However, one of the pilots commented that in reality ships will not be following their planned speed exactly why the ETAs in different waypoints (at least the closest ones) instead should reflect the actual, current, speed of a vessel. "You always want to go a little bit faster to make sure that you can make your ETA Rush to

wait. You will burn a little bit more fuel, but it cost more to let the stevedores, the lorries, etcetera wait."

3.1.7 *Pre-checked Alternative routes*

An interesting issue brought up was use of *alternative routes*. When you are doing you berth-to-berth voyage plan you may make e.g. two alternative routes on either side of e.g. an island or a bank. Both of them will be checked for UKC etc. One of them would be the preferred one (visible as the Intended route) but the officer could easily change to the alternative route if the weather or traffic situation so demands

3.2 *Procedural level*

3.2.1 *Workload*

It was discussed if the Intended route service would increase workload compared with today to a point where you would need to have an extra person on bridge just to run the system. Observation during the test scenarios showed that the usability of the system was not optimal yet and the participants were given help when they did not know how to activate a feature. Several participants commented however that they would expect the handling of the service to be smooth once they mastered the system. The test scenarios took place close to port or in the approach and this is where you would normally be two persons on the bridge. In a deep sea passage there would be only one officer on the bridge, but then the situations would normally be a lot calmer. "The workload remains the same, but the system will increase the quality of decision making," was one comment.

The Intended route service might lessen workload for the pilot as the rest of the bridge team can see the intentions and future whereabouts of other vessels. One of the pilots mentioned that he spent a lot of time explaining to the captain or watch officer what was the intentions of other ships in the area leaving berth or entering into the approach channel.

One of the VTS operator said that, given the VTS had Traffic Organization Service (TOS) authority, the Intended route service would greatly increase the opportunity and possibility to organize the traffic. This would be of great value but would also increase the workload in the VTS.

3.3 *Functional level*

Normally you have your ECDIS off-centered with most of the space in front of your ship and very little space behind you. But sometimes you are overtaken by a much faster ship. If you use route CPA as a filter for turning on Intended routes automatically you might get too many intended routes visible cluttering the screen, but it would be nice if you could have a "guard zone" astern which would turn on Intended routes only from overtaking ships. It would probably be necessary to have a "harbor" and a "sea" mode with different route CPA filter settings.

An issue could be that you are making an approach. You investigate the other vessels intended routes and you make a strategy for how you want to deal with upcoming meetings. Then one of the vessels changes his intended route. The chance is that you will not notice that. It might be useful with some form of highlighting of changed intentions.

3.4 *HMI level*

3.4.1 *Intuitive use*

The user friendliness of the system was discussed. It was pointed out that it was important that all watch officers onboard could use the system so that updates of Intended routes did not have to wait for that the responsible navigation officer (normally 2nd mate) was on watch. "But I think if we were here for another week we would be a lot quicker and comfortable with it. It is not a difficult system to use. It is more a question of familiarity, rather than the system being complicated."

3.4.2 *Cluttering*

During the first round of tests users commented on the HMI that it was difficult to distinguish intended routes from each other as they all had the same light green color, and also to know which track belonged to which ship (the label with ship information was only shown on mouse-over on the vessel AIS target triangle). Because we had the programmer present during the tests the interface was updated for the next set of trials starting the day after. In the new HMI an Intended track could be queried by pointing at it with the cursor. The track would then become highlighted in a darker green color, the vessels icon would become highlighted with a circle and the position on the intended track line where the cursor pointed would be connected to the own ship's position at the same time by a CPA Guidance Line (see Figure 6). These lines could be used to query another ships track about the closest point of approach (CPA). The second round of participants found theses new features useful and de-cluttered the interface somewhat.

Overtaking another vessel on a similar route is still difficult because the intended route of the other vessel may be hidden by your own route.

It was also mentioned that routes needed to be transparent so that they did not hide e.g. depth figures.

"The green color of the intended routes makes them difficult to see; especially if you got more than one. Maybe you could use different colours; you need to be able to separate one vessel from another."

The text and symbols are too small in the EPD. "When you get to our age you cannot see such small print"

3.5 *Survey*

The participants were asked to summarize their impressions about the service in a survey with three questions. Only 9 of the 11 participants answered the survey as 2 had to leave early.

1 **What is your opinion about the tested Intended routes concept?** All the 9 answering participants answered Good or Very good. No-one answered I don't know, Bad or Very Bad.

2 **Do you think a similar Intended routes concept will become reality in the future?** On this question all 9 participants answered Probably or Most probably. No one answered I don't know, Probably not or Most probably not.

3 **What is your professional opinion about the system tested?** On this question the participants were asked to rank their acceptance on a scale between 0 and 5 where 0 was "Totally unacceptable", 1 was "Not very acceptable", 2 was "Neither for, nor against", 3 was "Acceptable", 4 was "Very acceptable" and 5 was "Extremely acceptable". The mean acceptance score from the 9 answering participants was 3.7, somewhere between "Acceptable" and "Very acceptable".

4 DISCUSSION

It is of course of outmost importance that "intended routes" are understood as just "intentions" and not as a deterministic future. This was discussed very much during the 4 day test, but the concept of intended routes (as indicated by the very name of the service) seemed to be fully understood by the participants. There was an agreement that there should be a function that allowed ships to stop sending Intended routes if they for some reason had to change their intentions and did not have time to change the voyage plan on the chart display, but for minor offsets, like overtaking another vessel, or giving extra room in a close quarter situation, they did not feel it was necessary to turn off the Intended route. What was going on would be obvious to everyone.

The scenarios chosen were normal everyday situations and they were based on real scenarios that were either described to us by the Humber pilots and VTS operators on a previous focus group meeting held in Hull several months before. In one case (The number 4 scenario, contravention in the TSS) the scenario was based on a AIS video provided by the Spurn Head VTS. In no case did we see what we considered any dangerous behavior by the participants onboard or ashore.

4.1 Planned versus current speed

The discussion on planned versus current speed was interesting. One major idea with route exchange on the STM level is to make a new energy and emission saving paradigm with slow steaming and just-in-time-arrival replace the old wasteful full-speed-ahead, then anchor and wait paradigm. To make such a system work ships would be expected to follow their planned voyage plan exactly. This will be necessary in order to calculate safety feature using *dynamic separation* where no two ships would be allowed to set out on a voyage plan where they would be at the same place at the same time. So both from a fuel and emission saving, as well as a safety perspective, it would be essential that current and planned speed was the same. This is not the case today, where the present paradigm is "rush to wait", as was mentioned by one of the participants. Keeping a very exact speed down to a tenth of a knot, according to the voyage plan, would be difficult manually and would require a *speed pilot*. (Which like an autopilot automatically keeps the set speed.) The advanced speed pilots needed are today only used by some ferry lines, and are not common in the merchant fleet. A ships speed is also depending on being within a limited window of propeller revolutions and the speed resulting from the number of revolutions will depend on wind, sea state and depth. So while waiting for engines and speed pilots that will allow an exact voyage plan to kept, the use of current speed to calculate ETA in all waypoints except those designated as "critical" (e.g. final destination, arrival at a lock or passing a congested area where traffic management is essential) might be a solution.

4.2 Alternative routes

The alternative route suggestions brought up during the test is maybe more relevant to the strategic route planning of the MONALISA project than the tactical Intended route of the ACCSEAS, but never the less very interesting. If the Ship Traffic Coordination Centre suggested in MONALISA would be aware of both the preferred and the alternative routes, they could, if need be, use the Alternative as a new strategic route suggestion. That way the shipping companies would retain more control of the suggestions made by the STCC (which was mentioned as imperative by a cruise ship captain during another simulation in 2013).

4.3 Workload

It was unclear whether the Intended route service would increase or lessen the workload on the bridge. In the scenarios tested the situation was port

approach with two officers on the bridge. At times one of them would be occupied handling the chart system. To a large extent this could be because they were not proficient with the system (and several also said that the system felt easy to learn given ample time to practice). The system lends itself to making intentions clear and could be used for meeting and overtaking situations, but the system must not be used for collision avoidance in close quarters situations.

The user interface seemed to be intuitive and relatively easy to work with: A participant added a new WP and dragged it to starboard to indicate to a stand-on vessel from starboard that he had the intention to go astern of him. First time user, 50 sec.

5 CONCLUSIONS

The Intended route service was considered a valuable concept.

Intended routes should be displayed on a need to know basis, being able to customize and not to clutter the screen

The green, dashed representation was considered OK if the route of a particular vessel was highlighted on rollover to make its track more salient. The routes should also be transparent not to hide important information.

Use current speed to calculate the next 8 waypoints used for the intended route service (unless one of the waypoints is the final destination or otherwise designated as "critical", e.g. arrival at a lock).

REFERENCES

ACCSEAS. 2013. *Baseline and Priorities Report*, v. 2. http://www.accseas.eu/content/downloadstream/2743/2572 0/file/ACCSEAS+Baseline+and+Priorities+Report+v2.pdf [acc. 2015-01-31]

Endsley, M. R. 1988. Design and evaluation for situation awareness enhancement. In *Proceedings of The Human Factors Society 32nd Annual Meeting* (pp. 9 –101) Santa Monica, C.A.: Human Factors Society.

Kwik, K. H. (1984). Sailing Problems Within and Near Traffic Separation Schemes. *Journal of Navigation*, 37: 398-406.

Norman, D. 1988. *The Design of Everyday Things*. Basic Books.

Office of the Commissioner for Marine Affairs. 1979. *Decision of the Commissioner of Marine Affairs, R. L. in the Matter of the Collision of the Liberian Bulk Cargo Vessel ARTARDI (O.N. 3592) and the French RoRo/Train Ferry SAINT GERMAIN in the Dover Strait on 21 February 1979*. Monrovia, Liberia: Ministry of Finance.

Porathe, T., Lutzhoft, L. & Praetorius, G. 2013. Communicating intended routes in ECDIS: Evaluating technological change. Accident Analysis and Prevention, Elsevier 2013 Nov; 60:366-70

PARK Model and Decision Support System based on Ship Operator's Consciousness

S.W. Park, Y.S. Park & J.S. Park
Korea Maritime and Ocean University, Busan, S.Korea

N.X. Thanh
Ho Chi Minh City University of Transport, Ho Chi Minh, Vietnam

ABSTRACT: The recent situation at sea shows that, as the number of the vessels increases, the port traffic and the encounter situations also increase. Hence, minor and major marine accidents occur consistently. However, in the guidelines of the ship officers, COLREG regulation (International Convention for Regulations for preventing collision at sea) only gives the subjective expressions to avoid the collisions between the vessels, which is quite ambiguous in executing its actions to avoid the collisions. This study used the collision risk model (PARK model) that reflects the ship operator's consciousness which introduces the alarm systems and shows the degree of the risk on the screen. Moreover, it was tested in the vessels and ports with high traffics to confirm the effectiveness of the model.

1 INTRODUCTION

Ships make a voyage by using all available means such as RADAR, ECDIS and other navigation equipment that consistently develop in terms of their level of performance and convenience. Nevertheless, collision accidents continue to occur around the world, bringing forth the need to develop a system to prevent and quantify such risk of collision.

Therefore, we developed a tool to quantitatively evaluate the degree of danger from the perspective of the ship operator. They also validated system that would provide the ship operator with information required in avoiding collision.

The research to quantify the risk of collision has been implemented mostly based on the 'fuzzy theory' that could help evaluate marine traffic situations. However, the theory has its limitation for utilization as a model to support the decision-making process of a ship operator since the level of risk felt by the operator is not reflected (Kim, J. S et al 2011)

In order to reflect the psychological perception of the degree of danger, we divided the consideration factors into internal and external factors. The internal factors are the factors that affect degree of danger at sea and external factors affect the ship during navigation such as kinds of ship the own ship encounters and the angle of encounter. Using such factors, we conducted a survey on ship operators and ran simulations to quantify and evaluate a numerical

value on ships that navigate the coastal waters of South Korea and developed a tool based on such information.

Based on the evaluation model, the risk of collision between own ship and other ship as well as among other ships were assessed by using AIS information. A system was also developed to distinguish and display such data by each ship and waters. Finally, the system was validated in actual waters.

2 BACKGROUND

2.1 Needs and Purpose

For the last five years (2009-2013), among the marine accidents happened in Korean waters, a total of 423 collision cases took place and 411 cases of them (95.5% of the total) occurred due to the negligence of the ship operators. It can be seen that the most of the collision accidents were caused by the negligence of the ship officers. In order to reduce these marine accidents, any factors related to the ship operators must be corrected.

There are many parameters in the marine traffic. The negligence of the ship operators can be seen as an important factor among these parameters and is the main cause of the collision accidents. However, such ship operators have their own subjective minds which then show different degree of risk on the same situations occurred. This means that the ship operators voyage by the subjective judgments while VTS officers (VTSO) too, control the ships by subjective judgments. The purpose of this investigation is to voyage or control the ship by the objective judgment through the quantification of risk.

2.2 *Questionnaire survey*

65 VTSO of the Busan Port, with the high traffic, Busan New Port, Ulsan Port and Masan Port were surveyed about the need of the tool which displays the risk of collision and its effectiveness.

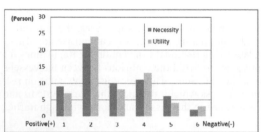

Figure 1. VTSO's response about a tool which displays the risk of collision

Figure 1 shows an increasingly positive response as an answer goes to the left of the X-axis controller, and one gets a negative answer to the right. The Y-axis is the number of VTSO whom answered. 31 of the VTSO (51%) answered positively to the need of the tool, while 31 of them (52%) answered positively to the usefulness of the tool. As shown above, more than 50% of the VTSO answered that they need this tool, which displays the risk of collision and that this tool will also be useful when used at work.

Table 1 shows a survey of the need for a tool which displays the risk of collision depending on the age of the VTSO. On the other hand, table 2 shows a survey on the effectiveness of the tool depending on the age of the VTSO. To conclude, it can be seen as the young VTSO about 20-30 years old tend to answer more positively when compared with the other VTSO of the age around 40-60. This seems to reflect the characteristics of the young generation familiar with the new changes.

Table 1. VTSO's response about need for a tool which displays the risk of collision

	1	2	3	4	5	6	Total
	(Positive)				(Negative)		
20-30's	2	7	2	4	0	0	15
40-60's	7	15	8	7	6	2	45

Table 2. VTSO's response about effectiveness for a tool which displays the risk of collision

	1	2	3	4	5	6	Total
	(Positive)				(Negative)		
20-30's	2	9	1	3	0	1	16
40-60's	7	15	7	10	4	2	43

3 USED MODEL & TOOL

3.1 *PARK model*

Due the increased volume of trade around the world including South Korea, the number of arrival and departure has been on the rise. As a result, encounters with other vessels in coastal waters have become more frequent, escalating the potential of marine accidents. Although navigation equipment have been evolving, collision accidents continue to occur due to the limitation of ship maneuver by experience and information provided to avoid collision.

In order to ensure safe navigation by providing appropriate information to ship operators, this paper used the Potential Assessment of Risk Model (PARK Model) made in 2011 (Nguyen, X. T. 2014). The PARK model is an evaluation model that assesses the degree of collision risk between ships. This model has been developed through the verification of ship handling simulator experiment and statistical analysis of questionnaire survey on 183 ship operators. It is based on ship operator's consciousness to gather and calculate directly from data of navigation instrument as AIS.

The development basis of the PARK Model is a survey that was conducted on internal and external factors that affect the ship operators' recognition of degree of danger. Internal factors are tonnage, LOA, breadth, career, license factor and position. External factors are encounter situation, direction, inner harbour, outside harbour, speed and distance. The survey used a seven–point-criteria with which respondents assessed the level of danger for different situations such as encounter angle of 045, 090, 135, overtaking and head-on situation. It was implemented on the analysis of variance and multiple comparisons on survey results to identify the difference of factors. Using the regression analysis, this model had been made.

In order to validate the model derived from the survey, we used a simulated ship steering device and performed experiments on 35 different cases. We were able to confirm the validity of the model's factors and came up with an equation as follows:

$$\text{Risk value} = 5.081905 + T_p + T_f + L_f + W_f + C_f + L_f + + P_f + 0.002517L + C_f + S_f + Hi/o + S_p - 0.004930 \times \times S_d - 0.430710 \times D \tag{1}$$

T_p: Own ship type factor

T_f: Own ship Ton factor (Ton)
W_f: Own ship width factor (m)
C_f: Ship operator's career factor
L_f: license factor
P_f: position factor
L :   Own ship LOA (m)
C_f: crossing factor for other ship
S_f: side factor for other ship
$H_{i/o}$: in/out harbor factor of own ship
S_p: Own ship's speed factor (kt)
S_d: speed difference between ships (kt)
D :   distance (NM)

In the equation stated above, the degree of danger is expressed through a value ranging from one to seven. The range between one and three is recognized as a safe situation, while the value between three and five is unsafe, and that above five is dangerous.

3.2 *COllision Risk Alarm and course adviser System (CORAS)*

The CORAS was developed to verify the appropriateness of the PARK Model for use as a technological factor of e-Navigation. The CORAS utilizes AIS information of ships and evaluates the risk of collision between the own ship and other ship as well as that among other ships. It results of the evaluation, which is information on the degree of collision risk, is categorized by ship and waters. The flow chart of the CORAS is described in Figure 2. Based on AIS data, this can calculate own ship's collision risk against other ship. With the degree of risk, it makes an alarm under constant value. It also shows risk circle, so operator can select safe course (Park, Y. S. 2013).

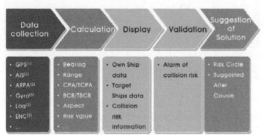

Figure 2. Flow chart of CORAS

The composition of the CORAS is seen in Figure 3. The AIS Pilot plug is used to receive AIS information of other ships, which is passed on to the CORAS using a Wi-Fi transmitter.

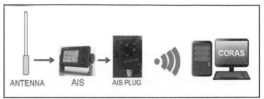

Figure 3. Composition of the CORAS

In an environment where Wi-Fi is available, the ship operator will be able to use easily the CORAS in a remote setting.

The display system of the CORAS is shown in Figure 4. The basic display system is similar to the RADAR and it can verify the degree of danger by vessels and waters through a colour code

On this figure 4, ① is Own ship and displayed at the center. ② is Other ship. Information of other ships is displayed using the AIS information. ③ is Own ship information. Information of own ship including GPS position, HDG, COG, SOG is displayed. ④ is Other ship information. Information of the other ship including name, GPS position, relative HDG/DISTANCE, COG, SOG, CPA, TCPA, BCR, BCT is displayed ⑤ is Degree of risk. The degree of risk of collision is displayed by the PARK Model. ⑥ is Degree of risk around the area. Degree of risk around own ship is displayed through colour code (1 ~ 4: green, 4 ~ 5: yellow, 5 ~ 7: red)

⑦ is Cursor information. Displays relative HDG/distance of the other ship from the mouse cursor's position and own ship. ⑧ is Monitor calibration. Scale of monitor, direction of heading, rearrangement of other ship's information.

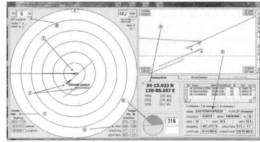

Figure 4. Display of the CORAS

The CORAS displays ships with a collision risk of 5.0 and above, which indicates a dangerous situation, with a red line for clear identification. It also calculates the degree of risk of collison with every other vessels 360 degrees around the own ship. For the lowest level of collision risk for every one degree is displayed on the screen with its corresponding colour code to indicate the level of risk by waters.

4 APPLICATION & VERIFICATION

4.1 *Verification with real ship*

We conducted the verification in the coastal water in order to validate the level of risk calculated by the model. To this end, the degree of danger subjectively felt by the ship operator and that indicated by the PARK Model were compared.

Verification took place on the training ship, HANBADA of KMOU (Korea Maritime and Ocean University) during a coastal voyage on 15 April 2014. In the experiment, 2 near collision situations occurred. It wasn't easy to encounter such dangerous situations in a real situation. So, the verification requested other vessel to create 2 near collision situations. Overall, the weather conditions and visibility were very good.

Figure 5 shows the scene of the two vessels encountered by head-on situation, range with other ship is around 2 cable.

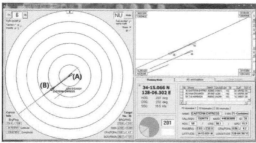

Figure 5. Experimental environment on South sea of Korea (Source : Yonhap news)

4.1.1 *Verification of situation 1*

On April 15, 2014 15:50LT (+9), the Eastern Express (B) and the test vessel (A) (T/S HANBADA) came into a head-on encounter situation. Figure 6 shows the head-on situation between ships before collision 2NM.

Figure 6. Display of encounter with the Eastern Express

Figure 7 shows the average degree of danger felt by ship operators and that indicated by the PARK Model in the case of encounter between the Eastern Express and the test vessel. The ship operator

recognized the situation as being dangerous when the distance with the other ship was 3NM.

When the distance came up to 1NM, the OOW (Officer Of the Watch) altered the course of the ship to avoid a collision. Despite the slight gap in human perception and the result of the PARK Model, the result shows that the tendency is similar.

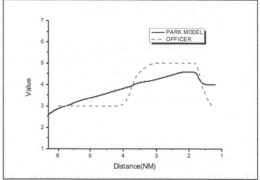

Figure 7. Graph showing degree of risk by PARK model and officers (Eastern express)

4.1.2 *Verification of situation 2*

On April 15, 2014 18:10LT(+9), M/V Saenuri (D) and the test vessel (C) came into a crossing encounter situation. Figure 8 shows the situation 1.7 NM before encounter

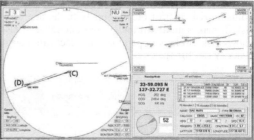

Figure 8. Display of encounter with Saenuri

Figure 9 shows the average degree of danger felt by ship operators and that indicated by the PARK Model in the case of encounter between the Saenuri and the test vessel. It can be seen on the graph that the ship operator judged the situation to be dangerous when the distance with the other ship was 3NM. When the distance was shortened to 0.2NM, the OOW changed the course to prevent a collision.

There was a slight difference between the ship operator's degree of risk and Park Model's degree of risk, however their tendency were similar. Difference between the size of the risk could be from the limits of the model, which its accuracy should be improved in the future.

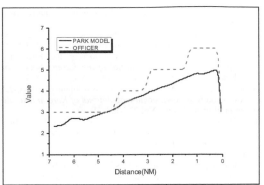

Figure 9. Graph showing degree of risk by PARK model and officers (Saenuri)

4.2 *Verification in Busan port*

The verification was implemented to confirm and verify ability of the CORAS to identify multiple risk factors. Figure 10 shows display of congested waterway in CORAS. It took place in Busan port, one of the busiest ports in South Korea. We installed the CORAS on the training ship, T/S HANBADA that was moored on KMOU's wharf, from which we received AIS information. The narrow entrance of Busan port gave rise to various situations, from which Figure 10 was derived.

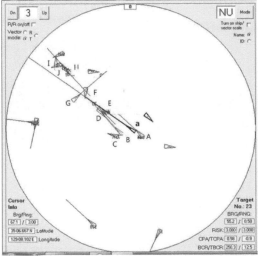

Figure 10. CORAS Display of congested situation in Busan port

With the CORAS, we could observe multiple encounter situations, check the degree of danger from the perspective of the ownship (a) and provide a less dangerous course from the ship's position. Furthermore, it displayed in red dangerous situations from which the course shall be altered immediately (degree of danger of 5.0 and above).

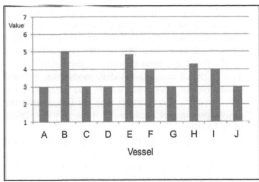

Figure 11. Risk of every ship on screen

Figure 11 indicates the risk value of every vessel, the number of which in the case was ten, related to the ownship. Vessel B and Vessel E shows the highest degree of danger. Vessel B is the overtaken vessel and Vessel E was in a head-on situation with the own ship. The ship operator is able to refer to red-coloured information in a congested water to figure out other ships that require immediate response. Moreover, the tool allows users to confirm the degree of danger for each waters and when to change the course.

5 CONCLUSIONS

For the last five years (2009-2013), among the marine accidents happened in the Korean waters, 95.5% of them occurred due to the negligence of the ship operators. Any factors related to the ship operators should be corrected to reduce marine accidents. To focus on ship's operator, we used PARK model, which is an evaluation model that assesses the degree of collision risk between ships. And this model reflects ship operator's recognition. This study developed a CORAS that can be confirmed by the development of ship collision risk assessment model which applied ship operator's point of view. Not only that, this tool was used in the actual ship and at the Busan Port in order to be verified. Conclusions of this study are as follows:

1 In the survey of the VTSO around the Busan Port, 51% of them gave the positive answers to the tool which predicts the risk of collision.

2 This study developed the CORAS to verify the appropriateness of the PARK model. With the CORAS, operator can check easily degree of danger by vessel and waters through a colour code.

3 The CORAS was applied to real ship. On situation 1, two vessels encountered by head-on situation, range with other ship is around 1 NM. On situation 2, two vessels encountered by crossing situation, range with other ship is around

0.2NM. Even though their tendencies were similar, there was a slight difference between the ship operator's degree of risk and PARK Model's degree of risk. So, this is the part where the model must continue to adjust in the future.

4 The CORAS using the PARK model was applied at the Busan Port and was able to determine the risk for each vessel and it was confirmed that the dangerous vessel could be prominently displayed in red colour.

In next step, this CORAS is ought to be applied in the field of Vessel traffic control Center. Then compare the degree of risk from the VTSO and the tool, and enhance the accuracy of the tool.

REFERENCES

COAA 2012.ShipPlotter,*www.coaa.co.uk/shipplotter.htm*

E-Navigation tests in Southsea (S.Korea) *www.yonhapnews.co.kr/bulletin/*

Heo, T. Y., Park, Y. S. and Kim, J. S. 2012. A Study on the developmnet of Marine Traffic Risk Model for Mariners, *J. Korean Soc. Transp.* Vol.30, No.5 : 91-100.

Kim, J. S. 2014. A Study on the Development of Marine Traffic Assessment Model based on Vessel Operator's Risk Consciousness, Ph D. thesis, *Mokpo maritime university* : 7-8.

Kim, J. S., Park, Y. S., Heo, T. Y., Jeong, J. Y., Park, J. S. 2011. A Study on the Development of Basic Model for Marine Traffic Assessment Considering the Encounter Type Between Vessels, *Journal of the Korean Society of Marine Environment & Safety.* Vol.17 No.3 : 227-233

Kim, K. I. 2013. A Study on the Development of Vessel Collision Risk Assessment tool, *Journal of Korean Institute of Intelligent System,* Vol.23 No.3 :29-30.

Lee, H. K. 2011. Development of Marine Traffic Supporting System through ES Model for VTS, Ph D. thesis, *Pukyong national university* : 37

Nguyen, X. T. 2014. A Study on the Development of Real Time Supporting System (RTSS) for VTS Officers, Ph D. thesis, *Korea maritime and ocean university* : 60-65.

Park, J. S., Park, Y. S. and Nha, S. J. 2014.*Marine Traffic Engineering &Policy* :291-292

Park, Y. S. 2013. Estimation and Alarm Technic of Collision Risk for SOLAS Vessels including Non-SOLAS Vessel, *Journal of the Korean Society of Marine Environment & Safety,* Vol.2103 No.11 : 145-147.

Park, Y. S., Kim, J. S., Kim, C. S., Jeong, J. Y., Lee, H. K. and Jeong, E. S. 2013. A Study on the Marine Traffic Risk Assessment by using Ship Handling Simulator, *Journal of Korean Society of Marine Environment,* Vol.19 No.2 : 138-144.

Statistics of Collision near Korea port, *http://www.kmst.go.kr /statistics/*

Multi-objective Route Optimization for Onboard Decision Support System

R. Vettor & C. Guedes Soares

Centre for Marine Technology and Ocean Engineering (CENTEC), Instituto Superior Técnico, Universidade de Lisboa, Portugal

ABSTRACT: An onboard decision support system for the route selection is being developed. The intrinsic multi-objective nature of the problem is kept by adopting a widely tested and robust genetic algorithm allowing the optimization of conflicting objective such as the expected time of arrival (ETA), the fuel oil consumption (FOC) and the safety, resulting in a set of routes approximating the optimum in the Pareto meaning. The complete modelling of the ship responses in waves and engine performances for different sea-states, ship speeds and headings will also be described. Finally a test case will allow to assess the potentialities and lacks of the tool and to discuss about the further improvements of the system.

1 INTRODUCTION

One of the most crucial aspects in the ship management is the correct choice of the route. Once a ship has been built, consumptions, emissions and operability level do not only depend on the over all quality of the vessel. How it is managed in terms of cargo distribution and route selection can strongly affect the safety and the efficiency of the journey (IMO, 2009). A large amount of information regarding land distribution, dangerous or restricted areas and weather forecast in the operating area, ship's behavior in sub-ideal environment conditions in terms of sea-keeping responses and propulsion system have to be collected and jointly analyzed. Traditionally all this burden was entrusted to the masters' experience (Motte, 1985) cause to the impossibility to perform reliable weather forecast and to solve complicated optimization algorithms. Nowadays, besides acting as an emergency measure to keep or bring ships out of weather hazards (Mackie and Houghton, 1992), sophisticated routing software are increasingly recognized as an important contribution to safe, reliable and economic ship operation, that will become more and more an irreplaceable aid to the human experience for more awareness decisions.

Although commercial marine weather services already exist (Ocean Prediction Center website), they are still not considered completely reliable by the ships' masters, as revealed by a series of interview (Prpić-Oršić et al., 2015). The most

common claim is that weather routing software would push the vessel into too heavy storms, compromising the safety. This might be due to a too strong emphasis to the economical factor when (often) the optimization method is not intrinsically multi-objective, typically Dijkstra's algorithm (Cormen et al., 2001) is adopted (Chen, 2011) or to a poor modelling of the ship behaviour in real sea conditions, which neglects important factors (Guedes Soares, 1995; Orlandi & Bruzzone, 2012) or dangerous situations (IMO, 1995, Krata & Szłapczynska, 2012). Nevertheless there is room and need for a deeper investigation in this field in order to create more reliable software to be tailored on the requirements of the ship where they are equipped.

Since the accurate prediction of voyage duration is one of the main tasks in ship navigation (Guedes Soares et al., 1998), the first generation of weather routing methods were thought only to optimize time, as for instance James (1957), Zoppoli (1972), Spaans (1986). The more recent concerns on energy efficiency and reduction of emissions (IMO, 2009; Prpić-Oršić et al., 2015) and the increasingly competitiveness in maritime trades pressured researches into multi-objective optimization. The Dijkstra's algorithm was successfully applied in the Indian Sea (Padhy et al., 2008), but it requires summarizing the objectives in a single goal function. More recently (Skoglund, 2012) the same algorithm has been slightly modified to account for multi-objectives, simply saving the Pareto-optimal results encountered during the iteration process. It showed noteworthy improvement in reducing the fuel consumption with respect to the original one. The increasing computational capabilities allowed to develop more sophisticated methods like the GIS grid based multi-objective methodology (Pacheco and Guedes Soares, 2007) and the more and more frequent employment of evolutionary programming and genetic algorithms: Hinnenthal & Saetra (2005), Szłapczynska & Smierzchalski (2007), Marie and Courteille (2009) and Maki et al. (2011). The importance of accounting for uncertainties for a conscious evaluation of the hazards is stressed in Papatzanakis et al. (2012).

An advanced onboard decision support system for the selection of the preferable route is being prepared. The guideline that drives the development is a strong attention to the needs of the master, in order to be able to provide, together with the indication of the routes that better fits the requirements, all the necessary information for an aware final decision. The first vital requirement is a reliable and detailed modelling of the ship behaviour in seaways. Both the motions induced by rough weather and the combined effect of waves and wind in the propulsion performances are considered. To deal with all the complex interacting sub-systems of the weather routing optimization, a class will be dedicated to each component. To design the code with this framework since the beginning is fundamental since it will allow separately treating each problem (such as sea-keeping, propulsion system, weather, etc.), to easily replace methods and algorithms if needed and to tailor the code to the specific requirements in the real implementation.

The main objective of this paper is to describe the detailed assessment of the sea-keeping responses and the propulsion system performances for many conditions of sea-states, ship speeds and headings. Section 2 is dedicated to this issue. In section 3 a brief description of the code, which adopt a multi-objective optimization algorithm for the search of the Pareto-optimal solutions is given. Finally an example will show the results of a run of the code in a specific case.

2 SHIP PERFORMANCE

When the optimal routes have to be identified, first of all it is essential to be able to describe all the significant aspects of the ship in a seaway, which allows to compute the objectives. Under adverse weather conditions, the ship behaviour differs significantly from the one in still water with increased periodic motions (mainly heave, pitch and roll) and a component of added resistance due to waves, wind and, possibly, currents and other factors. These aspects, besides compromising the safety, comfort and work ability on board, due to slamming, green water, vertical and lateral accelerations and other dangerous effects, have a great influence on the propulsive system with an increased request of power in sub-optimal regimes, eventually being too demanding and out of the specific load diagram.

Since waves are commonly recognized as the major cause of degraded ship performances, only the sea-state has been considered so far, but the influence of other not negligible factors such as wind and currents will be taken into account in the future.

Ship modelling consists in the assessment of the sea-keeping responses and the propulsion performances in every possible weather condition (sea-state) and speed of advance. There are two ways to achieve this requirement: either to consider the specific sea-states described by means of the actual wave spectra (possibly directional) and apply them to the response amplitude operators (RAOs), or to compute off-line the ship responses for a set of more or less accurately pre-defined sea-states with a standard spectrum taken as a model for any encountered wave condition. The latter is widely adopted method cause it allows to describe the sea-states through few integral parameters simplifying the calculation and reducing the computational time. However it must be considered that different spectral shapes may have an effect on the short-term wave-induced ship responses, especially in case of multi-modal spectra (Guedes Soares, 1990; Orlandi and Bruzzone 2012). As a first approach, the first method has been chosen, but the good performances of the code in terms of computational time (see section 4) encourage to study more sophisticated ship models. In order to have a more reliable description of the ship performances they have been computed for many different sea-states (H_S from 0m to 15m with a step of 1m and T_P from 0s to 20s with a step of 2s), headings (from 0° to 180° with a step of 30°) and ship speed (from 0kn to 20kn with a step of 1kn) and saved in tables which are imported when an object of the "Ship" class is instantiated.

The hydrodynamic calculations are carried out off-line with a computer code based on the strip theory (Salvesen et al., 1979), which provides transfer functions of the motions and the added resistance.

2.1 Sea-keeping responses

Starting from the RAOs and the given wave spectrum, being the real or a standard one in a parametric form, several sea-keeping performances can be computed and their relevance depends on the type of ship and the mission of the trip many standards have been proposed in the literature (Nordforsk, 1987; Dubrovski, 2000). For a containership, important factors may be the structural load due to bottom or flare slamming or damage to the cargo due for instance to green water or lateral accelerations; the operability of a passenger ship, instead, will be more influenced by comfort criteria governed by the vertical acceleration such as the motion sickness incidence (MSI); in the case offshore supply vessels, navy vessels or any other ship that must guarantee the possibility to work on the deck, attention has to be paid on the motion induced interruptions (MII).

The code allows to adapt the optimization to the required strategy and to define appropriate limits for each effect. A risk coefficient is then computed, similarly to the concept adopted to define the seakeeping operability limit (see Nordforsk, 1987, fig. 3.3), through the following equation.

$$RISK_{coeff} = \left[\left| \max\left(\frac{Max\_Seakeeping\_eff_i}{Limit_i} \right) + \max\left(\frac{Mean\_Seakeeping\_eff_i}{Limit_i} \right) \right| \right] / 2 \quad (1)$$

The ratio aims to normalize all the seakeeping effects in a value in the range [0,1] depending on the distance from the respective limit, such as 1 correspond to the most extreme conditions that the ship can stand. All the effects are however stored and are available to the user in the set of optimal routes.

To run the test case (see section 4) three effects have been considered: the slamming probability, the green water and the vertical acceleration on the bridge.

A slamming event occurs when the keel emerges from the water, meaning that the relative motion is greater than the draft, and, later on, impacts on it with a speed higher than the critical value calculated as in equation 2, thus it implies the knowledge of the relative motions and velocity at bow:

$$V_{CR} = 0.093\sqrt{gL} \quad (2)$$

The probability of keel emergency is calculated assuming the Rayleigh distribution of the peaks as:

$$P_{ke} = \exp\left(-\frac{D_{ke}^2}{2C_S^2 D_{r\zeta}} \right) \quad (3)$$

where D_{ke} is the actual keel draft, which should take into account the trim, the sinkage and the ship's own wave and C_S is a coefficient to include the swell-up effect, here neglected. $D_{r\zeta}$ is the variance of the relative motion on the bow.

The probability to impact with a speed higher than the critical value is given by:

$$P_{\dot{r}_{CR}} = \exp\left(-\frac{V_{CR}^2}{2C_S^2 D_{\dot{r}\zeta}} \right) \quad (4)$$

where the variance of the relative velocity on the

bow is used.

From the previous considerations and assuming in first approximation the relative motion and velocity to be independent, the following equation yields:

$$P_{sl} = P_{ke} \cdot P_{\dot{r}_{CR}} = \exp\left(-\frac{D_{ke}^2}{2C_S^2 D_{r\zeta}} - \frac{\dot{r}_{CR}^2}{2C_S^2 D_{\dot{r}\zeta}} \right) \quad (5)$$

For the green water effect, the freeboard exceedance is a necessary, but not sufficient condition. If the speed is not too low, in fact, most of the time the exceedance of the freeboard does not imply a green water event. Nevertheless for this study the probability of deck submergence is considered to be a good indication and it is computed with the equation:

$$P_{ke} = \exp\left(-\frac{F_e^2}{2C_S^2 D_{r\zeta}} \right) \quad (6)$$

where F_e is the effective freeboard.

The criteria on the vertical acceleration, being a varying quantity, are given on the root mean square (rms) of the variance of the vertical acceleration on the bridge:

$$V\_acc_{rms} = \sqrt{D_{\ddot{v}\zeta}} \quad (7)$$

2.2 Propulsion system

The calculation of the total resistance in irregular waves for each sea-state derives from a time-domain method proposed by Prpić-Oršić and Faltinsen (2012) where the instantaneous ship speed is calculated according to the method of Journee (1976) and Journee & Meijers (1980) by taking into account propeller in-and-out-of-water effect on ship propulsion and the effect of mass inertia. The still water resistance is calculated according to Holtrop & Mannen method (Holtrop & Mannen 1982, Holtrop 1984), an approximate procedure, which is widely used at the initial design stage of a ship. The method is based on regression analysis of random model experiments and full-scale data, available at the Netherlands Model Basin. The added resistance in waves is computed according to direct pressure integration procedure developed by Faltinsen et al. (1980).

For the main engine a two-stroke electronically controlled low speed marine engine with main particular listed in table 1 is used.

Table 1. Characteristics of the main engine

Engine model	MAN B&W 9S50ME-C8-TII
MCR	14940kW at 127 rpm
SCR	12707kW at 120 rpm
Fuel	Heavy fuel oil

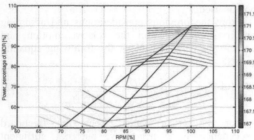

Figure 1. Main engine load diagram and SFOC contours.

The load diagram and the specific fuel oil consumption (SFOC) at an arbitrary load have been computer following the recommendation given by the constructor in the project guide (MAN B&W, 2010) and are shown in figure 1.

A B-series propeller is assumed and the open water propeller characteristics are obtained with the method of Oosterveld and van Oossanen (1975). The required brake power and the respective number of revolutions for all the considered conditions (in term of H_S, T_P, heading and ship speed) are then computed considering the thrust and the wake fractions computed with Holtrop method (1984).

From a simple comparison of the demand and the engine output, it is easy to obtain the fuel consumption per nautical mile for each condition, that is then used to compute the fuel oil consumption for each track and, finally for the whole route.

3 SOFTWARE DESCRIPTION

Due to the intrinsic nature of the problem, the best route does not exist. Instead a compromise among conflicting objectives has always to be found and it usually depends on the type of ship and the decisions of the captain. In this perspective, the multi-objective optimization is aimed to produce a set of favourable route variants making the user aware of their values and lacks enabling a conscious decision-making process for the final ranking and selection. This concept is the base of the Pareto theory. A solution is Pareto-optimal if an improvement of one objective result in the impairment of another one. The Pareto frontier is the border between the feasible and unfeasible solutions and contains the closest routes to the ideal optimal (and in the general case unfeasible) one. Among them the best possible compromise to satisfy the request can be found.

3.1 Code

A C++ code that integrates the multi-objective optimization performed by a robust genetic algorithm SPEA2 (Strength Pareto Evolutionary

Algorithm 2) proposed by Zitzler et al. (2002), has been developed. It takes care of the description of the sailing area, the weather condition as well as the modelling of the ship behaviour in the seaways, in terms of seakeeping responses and propulsive performances, and, ultimately, of the optimization of the route. The code consists in six classes: "Ship", "Journey", "Route", "Domain", "Land" and "Weather"; and five subclasses: "GeoPoint", "Time", "Seakeeping", "Engine" and "Environment". The architecture was designed to be flexible to future upgrades or modifications and to be completely controlled through simple input files.

3.2 Genetic operators

The initial population is made of random routes that are computed respecting the constrains. The ship performance is computed for many short tracks (30 miles in the case study in section 4) along the route, in order to ensure the weather condition to be stationary for each track.

The variation to produce the new generations is performed through three genetic operators: crossover, mutation and migration.

In the crossover two routes (parents) are randomly chosen from the selected set, as well as one of the waypoints (k^{th} waypoint). All the parameters (that is grid-point indices and ship speed) up to the kth waypoint are copied from the first to the new route (offspring), while from the $(k+1)^{th}$ to the nWP^{th} are copied from the second parent.

In the mutation one individual is randomly chosen from the selected set as well as one of the waypoints, then the parameters relative to the detected waypoint are randomly varied ensuring respect of the previously defined limitations. The mutation operator has usually a very low probability; nevertheless in this case it revealed a great capability of refining the routes in the proximity of the Pareto frontier, especially in an advanced stage of the optimization.

The migration simply consists in the generation of a new random route. This operator is often neglected in the genetic algorithms and can here be excluded.

3.3 Objectives

Although with a different extent in consideration of the ship, any navigation requires to reach the destination in a pre-determined or as short as possible time, in safe conditions and, especially when a weather routing system is adopted, to save fuel that likely also imply lower emissions. For this reason the classical objective in route optimization are the fuel oil consumption (FOC), the estimated time of arrival (ETA) and the safety (SAFETY).

Instead of using the actual values of the objectives, the following objective functions were introduced for the last two targets:

$$FOC_{obj} = \left(1 - FOC_{min}/FOC_{journey}\right)^2 \qquad (8a)$$

$$ETA_{obj} = \left(1 - ETA_{min}/ETA_{journey}\right)^2 \qquad (8b)$$

$$SAFETY_{obj} = k \cdot \left(RISK_{coeff}\right)^2 \qquad (8c)$$

$$k = \left(1 - ETA_{min}/ETA_{journey}\right) \qquad (8d)$$

where FOC_{min} is the minimum fuel oil consumption achieved in calm waters at the most efficient optimal engine load, ETA_{min} is the minimum possible duration of the journey if the shortest route is sailed at the design speed and k, given in equation 8d is a factor adopted to discard routes which are safe but practically useless for the optimization cause of the excessive duration.

These functions help to accelerate the propagation of the initial routes towards the optimal solutions, a skill that is especially important in case of rough sea or land avoidance when the respect of the constrain might slow down the creation and recombination of feasible individuals.

Furthermore the objective function relative to the safety reduces the impact of meaningless routes, which minimize the slamming probability, but imposing weird courses.

3.4 Selection

Once the stopping criterion is achieved, the post-processing starts. It consists in writing a series of files to allow the quick visualization of all the route parameters (time of departure, expected time of arrival, sea-states that the ship will encounter, sea-keeping responses, etc.) and the file for the plots. In this phase a ranking method is also applied. In this case the distance of each route from the utopia point in the objective space determines the rank.

4 EXAMPLE

A test case was run considering the S175 containership as a reference. The simulation is performed for a route between the port of Sines in Portugal and the entrance of the entrance of the Gulf of Saint Lawrence. The departure time is set at 9.00pm on January 31st, 2001 in order to include in the travel period the evolution of a storm occurred between February 4th and 5th, identified analysing the HIPOCAS (Hindcast of Dynamic Processes of the Ocean and Coastal Areas of Europe) hindcast database (Pilar et al., 2008). This database contains

44 years of wind and wave data in the North Atlantic in a grid of 2°x2° and every three hours.

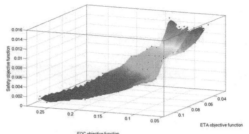

Figure 2. Resulting Pareto frontier after 1000 generations.

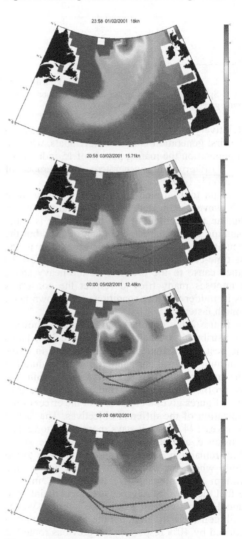

Figure 3. Paths of three of the Pareto frontier routes: smallest FOC (red stars), shortest time (red dots) and safest (red triangles).

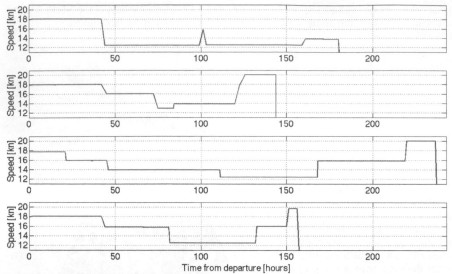

Figure 4. Speed profile of the routes depicted in figure 3: smallest FOC (first), shortest time (second) and safest (third), plus the first rank given in the selection process (fourth).

Relatively to the genetic algorithm, the dimension of the first generation is of 100 individuals, while in each generation the mating pool is of 15 individuals and 15 offspring are generated. The number of generation is 1000.

The run took around 15 minutes. Most of them are however dedicated to the initialization, in particular the reading of the bathymetry is quite demanding due to the large amount of data, but the computation of the first random generation is the most time-consuming routine, requiring almost 10 minutes cause to the presence of the heavy storm that imposes many tries to respect the constrains (runs in milder weather conditions, not shown here, are much faster).

In figure 2 the obtained Pareto frontier shows a good distribution of the results. In figure 3 the paths of the best routes among the selected ones in terms of minimum fuel oil consumption, shortest duration of the trip and smallest risk are shown, while the speed profiles are plotted in figure 4. A comparison of the 2 figures gives a clear view of the effect of the optimization of the different objectives. The fastest route takes 144 hours, it stays more north to reduce the distance and keeps an higher average speed 16.4kn, remaining out of the storm but close to the borders where waves can be quite heavy (the maximum encountered wave height is 5.5m). To save fuel the speed of advance is cut and maintained around 12.5kn for great part of the journey. Waves are still considerably high, but the fuel consumption is reduced by 40% respect on the previous route but taking one and a half day more to reach the destination. The safest modulates the speed in order

to reduce the ship motions but also circumnavigates the storm at a greater distance to avoid the tail.

The first rank route seems a good compromise. It takes 14 hours more than the fastest one, using 30% less fuel. It must be noted that even in this case rough weather condition are encountered, but always respecting the sea-keeping constrains. It is due to the presence of an heavy storm in the period of the passage. If more safety or comfort are required, once can set stricter constrains before the running or just chose another route in the rank. In the future a more flexible ranking method will be adopted.

5 CONCLUSIONS

The progresses in the development of an onboard decision support system have been described. Special attention was given to the modeling of the ship performance in terms of sea-keeping responses and propulsion system. Other important, but more critical sea-keeping effects must be included, in particular the roll motion in order to limit the occurrences of heavy beam sea. The optimization algorithm confirmed to be appropriate to deal with this problem, with a smooth distribution of the results in the Pareto frontier. The ranking method needs to be studied more accurately in order to be more flexible to the request of the decision-maker and to allow a re-selection if the first choice is not considered adequate.

The analyses of the computational time, deeper than the one discussed in section 4, indicates that the most critical part of the optimization is the construction of the first generation, while the calculation of the ship behavior is quite fast, in spite of the quite high resolution of the computation point along each route. This encourages to attempt an even more accurate modeling of the ship and the weather including comparison of the wave actual wave spectra and the RAOs in the code. This would allow to identify the possible dangerous situations due to the actual frequency distribution of the wave energy, especially in the cases of multi-modal spectra, and more in general to be more precise in the computation of the ship performances, thus in the optimization of the route.

AKNOWLEDGMENTS

This work was performed within the project SHOPERA-Energy Efficient Safe SHip OPERAtion, which was partially funded by the EU under contract 605221.

The first author was supported by the Portuguese Foundation for Science and Technology (FCT - Fundação para a Ciência e Tecnologia, Portugal) under the contract no. SFRH/BD/89476/2012.

REFERENCES

Amante, C. and B.W. Eakins, 2009. ETOPO1 1 Arc-Minute Global Relief Model: Procedures, Data Sources and Analysis. NOAA Technical Memorandum NESDIS NGDC-24. National Geophysical Data Center, NOAA.

Chen, H. (2011). Voyage optimization supersedes weather routing. Technical report, 2011.

Cormen, T.H., Leiserson, C.E., Rivest, R.L., Stein, C. (2001). "Section 24.3: Dijkstra's algorithm". Introduction to Algorithms (Second ed.). MIT Press and McGraw-Hill. pp. 595–601.

Dubrovaskiy, V.A. (2000). Complex Comparison of Seakeeping: Method and Example. Marine Technology, 37(4), 223-229.

Faltinsen, O.M. Minsaas, K.J. Liapis, N. Skjordal, S.O. (1980). Prediction of Resistance and Propulsion of a Ship in a Seaway, *Proc. 13th Symposium on Naval Hydrodynamics*, pp. 505-529.

Guedes Soares, C. (1990). Effect of Spectral Shape Uncertainty in the Short Term Wave-Induced Ship Responses. Applied Ocean Research, 12(2), 54-69.

Guedes Soares C. (1995) Effect of wave directionality on long-term wave-induced load effects in ships. *Journal of Ship Research*, Vol. 39, No. 2, pp. 150-159.

Guedes Soares, C., Fonseca, N., Ramos, J. (1998). Prediction of Voyage Duration with Weather Constraints. *Proceedings International Conference on Motions and Manoeuvrability*, Royal Institute of Naval Architects, London.

Hinnenthal, J., Saetra, O., (2005). Robust Pareto – optimal routing of ships utilizing ensemble weather forecasts. Maritime Transportation and Exploitation of Ocean and Coastal Resources, C. Guedes Soares, Y. Garbatov and N. Fonseca eds., Taylor & Francis Group, London, UK, pp 1045-1050.

Holtrop, J. A. & Mennen, G.G.J. (1982). An Approximate Power Prediction Method. *International Shipbuilding Progress*, Vol. 29, pp. 166-170.

Holtrop, J. A. (1984). Statistical Reanalysis of Resistance and Propulsion Data. *International Shipbuilding Progress*, Vol. 31, pp. 272-276.

IMO (1995). Guidance to the Master for Avoiding Dangerous Situations in Following and Quartering Seas. MSC/Circ.707.

IMO (2009). Guidelines for voluntary use of ship Energy Efficiency Operational Indicator (EEOI). MEPC.1/Circ.684.

IMO (2009). Report of the Marine Environment Protection Committee on its Fifty-ninth Session. MEPC 59/24.

James R.W. (1957). Application of wave forecast to marine navigation. Washington: US Navy Hydrographic Office.

Journee, J.M.J. (1976). Prediction of Speed and Behaviour of Ship in a Seaway. *International Shipbuilding Progress*, Vol. 23, No. 265, pp. 1-24.

Journee, J.M.J., Meijers, J.H.C. (1980). Ship Routeing for Optimal Performance. *Transactions IME, Rapport 0529_P*, Delft

Krata P. & Szłapczynska J. (2012). Weather Hazard Avoidance in Modeling Safety of Motor-Driven Ship for Multicriteria Weather Routing. TransNav - International Journal on Marine Navigation and Safety of Sea Transportation. Vol. 6. No. 1. pp. 71-78.

Kreysig, Erwin, Advanced Engineering Mathematics, Third Edition, John Wiley and Sons, Inc., 1972.

Mackie, G.V. & Houghton, J.F.T. (1992). Marine Meteorological Services to Shipping, Past, Present and Future. Journal of Navigation. Vol. 45, pp. 241-246.

Maki, A., Akimoto, Y., Nagata, Y., Kobayashi, S., Kobayashi, E., Shiotani, S., Ohsawa, T., Umeda, N. (2011). A new weather-routing system with real-coded genetic algorithm. Journal of Marine Science and Technology. 16:311–322.

MAN B&W (2010). S50ME-C8-TII – Project Guide. 198 75 86-7.1.

Marie S. & Courteille E. (2009). Multi-Objective Optimization of Motor Vessel Route. TransNav - International Journal on Marine Navigation and Safety of Sea Transportation. Vol. 3. No. 2. pp. 133-141.

Motte, R. (1985). Weather Routing of Ships for the Northern Oceans. Journal of Navigation. Vol. 38, pp. 274-282.

Nordforsk (1987). Assessment of Ship Performance in a Seaway. The Nordic Co-operative project: "Seakeeping Performance of Ships". Nordic Cooperative Organization for Applied Research.

Ocean Prediction Center, National Weather Service, National Oceanic and Atmospheric Administration. http://www.opc.ncep.noaa.gov/links.shtml.

Oosterveld, M.W.C. & van Oossanen, P. 1975. Further Computer-Analyzed Data of Wageningen B-Screw Series, I.S.P., Vol. 22, No. 251, Wageningen

Orlandi, A. & Bruzzone, D. (2012). Numerical weather and wave prediction models for weather routing, operation planning and ship designs: the relevance of multimodal wave spectra. Sustainable Maritime Transportation and Exploitation of Sea Resources. Rizzuto & Guedes Soares (eds). Taylor & Francis Group, London, pp. 817-826.

Pacheco M.B., Guedes Soares C. (2007). Ship Weather Routing Based on Seakeeping Performance. Advancements in Marine Structures. Guedes Soares, C. & Das, P.K. (Eds). London, U.K.: Taylor & Francis Group; pp. 71-78.

Padhy, C.P. , Sen, D., Bhaskaran, P.K. (2008). Application of wave model for weather routing of ships in the North Indian Ocean. Natural Hazards Vol. 44 (3), pp. 373-385.

Papatzanakis, G.I., Papanikolaou, A.D., Liu. S. (2012). Optimization of Routing with Uncertainties, Sustainable Maritime Transportation and Exploitation of Sea Resources, Rizzuto, E. & Guedes Soares, C (Eds), Taylor and Francis Group, pp 827-835.

Pilar P., Guedes Soares C., Carretero J.C. (2008). 44-year wave hindcast for the North East Atlantic European coast. *Coastal Engineering* 55, 861–871.

Prpić-Oršić, J., Faltinsen, O.M. (2012). Estimation of Ship Speed Loss and Associated CO_2 Emissions in a Seaway. *Ocean Engineering*, Vol. 44, No 1, 1-10.

Prpić-Oršić, J., Vettor, R., Guedes Soares, C., and Faltinsen, O.M. (2015) Influence of ship routes on fuel consumption and CO2 emission, Maritime Technology and Engineering. Guedes Soares, C. & Santos T.A. (Eds.). Taylor & Francis Group, London, UK, pp. 857-864.

Prpić-Oršić, J., Parunov, J., Šikić, I. (2014). Operation of ULCS – real life. International Journal of Naval Architecture and Ocean Engineering, Vol. 6, pp. 1014-1023.

Skoglund, L. (2012). A new method for robust route optimization in ensemble weather forecasts. KTH, School of Engineering Sciences, Master's thesis.

Spaans J.A. (1986). *Windship routeing.* Technical University of Delft.

Szłapczynska, J., Smierzchalski, R. (2007). Multiobjective Evolutionary Approach to Weather Routing For Vessels with Hybrid Propulsion, Proc. 14th Adv. Computer Systems.

Zitzler, E., Laumanns, M., Thiele, L. (2002). SPEA2: Improving the Strength Pareto Evolutionary Algorithm for Multiobjective Optimization. Evolutionary methods for design, optimizations and control with application to industrial problems (EUROGEN 2001), K.C. Giannakoglou and others, eds, International Center for Numerical Methods in Engineering (CIMNE), pp. 95-100.

Zoppoli, R. (1972) Minimum-time routing as an N-stage decision process. *Journal of Applied Meteorology,* 11, 429-43.

Simulation-Augmented Methods for Manoeuvring Support – On-Board Ships and from the Shore

K. Benedict, M. Kirchhoff, M. Gluch, S. Fischer & M. Schaub
Hochschule Wismar, University of Applied Sciences – Technology, Business and Design, Warnemünde, Germany

M. Baldauf
World Maritime University, Malmö, Sweden

ABSTRACT: The shipboard tasks and procedures on ships with high safety level and a high portion of manoeuvring activities have been changed to high back-up procedures as in air planes. For port manoeuvres e.g. the system of pilot/co-pilot was introduced on ferries in a sense that one officer is operating and the other is monitoring and checking the safe performance. In cruise-liner operation there are even new structures replacing the traditional rank-based system with a flexible system based on job functions to create a safety net around the person conning the vessel. Each operation is cross checked before execution by one or two persons depending on circumstances. The consequence is higher costs for double personnel on the one hand and the need for a technology to guarantee that the checking officer is able to monitor what the conning officer is doing on the other hand.

This opens up chances for the application of the new "Fast-Time Manoeuvring Simulation Technology" (FTS) developed at the Institute for Innovative Ship Simulation and Maritime Systems (ISSIMS). It calculates within one second of computing time up to 1000 seconds of manoeuvring time by a very complex ship-dynamic simulation model for rudder, engine and thruster manoeuvres. This enables the online prediction of all manoeuvres carried out by the conning officer for the observing officer, too. So it is easy for all officers to see whether the manoeuvring actions have at least the correct tendency and even more the effectiveness of the manoeuvres can be improved. This new type of support is called Simulation-Augmented Manoeuvring Design and Monitoring (SAMMON) – it allows not only overlooking the next manoeuvring segment ahead but also for the following or even for series of manoeuvring segments.

Currently, this technology is used within two new research projects: The Project COSINUS (Co-operative Ship Operation in Integrated Maritime Traffic Systems) sets out for implementing the FTS technology into integrated ship bridges and to also communicate the manoeuvre plans and display it to VTS centres. Within the European project MUNIN (Maritime Unmanned Navigation through Intelligence in Networks) this technology will be used to investigate if it is possible to steer autonomous ships where the information for manoeuvring the ship will be delayed due to the communication links.

1 INTRODUCTION

During the previous TRANSNAV conference in 2013 a fast-time simulation tool box was introduced to simulate the ships motion with complex dynamic models and to display the ships track immediately for the intended or actual rudder or engine manoeuvre in the ECDIS (Benedict et al., 2013). These "Simulation-Augmented Manoeuvring Design and Monitoring" - SAMMON tool box will allow for a new type of design of a manoeuvring plan as enhancement exceeding the common pure way-point planning – and it will play an important role in future education and training in simulators for ship handling.

During the INSLC 17 conference new concepts were presented for innovative organisational structures specifically for bridge management (Hederstrom, 2102).

This paper presents the potential of the new method specifically for the support of manoeuvring of ships both for the new manning concept and even for shore-based support or moreover for autonomous ships. Manoeuvring of ships is and will be a human-centred process despite of expected further technological developments. Most important elements of this process are the human itself and the technical equipment to support its task. However, most of the work is to be done manually because even today nearly no automation support is available for complex manoeuvres. Up to now there was nearly no electronic tool to demonstrate manoeuvring characteristics efficiently or moreover to design a manoeuvring plan effectively. However, due to the new demands there is a need to prepare harbour approaches with complete berth plans specifically in companies with high safety standards like cruise liners. These plans are necessary to agree on a concept within the bridge team and also for the discussion and briefing with the pilot.

For increasing the safety and efficiency for manoeuvring real ships, the method of Fast-Time Simulation will be used in future – even with standard computers it can be achieved to simulate in 1 second computing time a manoeuvre lasting about to 20 minutes using innovative simulation methods. These Fast-Time Simulation tools were initiated in research activities at the Maritime Simulation Centre Warnemuende (MSCW) which is part of the Department of Maritime Studies of Hochschule Wismar, University of Applied Sciences - Technology, Business and Design in Germany. They have been further developed by the start-up company Innovative Ship Simulation and Maritime Systems (ISSIMS GmbH)

A brief overview is given for the modules of the FTS tools and its potential application:

SAMMON is the brand name of the innovative system for "Simulation Augmented Manoeuvring Design & Monitoring". It is made for both:
– Application in maritime education and training to support lecturing for ship handling to demonstrate and explain more easily manoeuvring technology details and to prepare more specifically manoeuvring training in ship-handling simulators (SHS) environment and
– Application on-board to assist manoeuvring of real ships e.g. to prepare manoeuvring plans for challenging harbour approaches with complex manoeuvres up to the final berthing/unberthing of ships, to assist the steering by multiple prediction during the manoeuvring process and even to give support for analysing the result,

And SAMMON contains the following modules:
– Manoeuvring Design & Planning Module to design ships-manoeuvring concepts as "manoeuvring plan" for harbour approach and berthing manoeuvres (steered by virtual handles on screen by the mariner)
– Manoeuvring Monitoring & Multiple Dynamic-Prediction Module: monitoring of ships manoeuvres during simulator exercises or manoeuvres on a real ship using bridges handles, display of manoeuvring plan and predicted manoeuvres in parallel. It calculates various prediction tracks for full ships-dynamic simulation and simplified curved-headline presentation as look ahead for future ships motion.

– Manoeuvring Simulation Trial & Training Module: ship handling simulation on laptop display to check and train the manoeuvring concept (providing the same functions as monitoring tool; steered by virtual handles on screen)

SIMOPT is a simulation-optimiser software module based on FTS for optimising standard manoeuvres and modifying ship math model parameters both for simulator ships and for on board application of the SAMMON system.

SIMDAT is a software module for analysing simulation results both from simulations in SHS or SIMOPT and from real ship trials: the data for manoeuvring characteristics can be automatically retrieved and comfortable graphic tools are available for displaying, comparing and assessing the results.

The SIMOPT and SIMDAT modules were described in earlier papers (Benedict et. al: 2003, 2006) for tuning of simulator-ship model parameters. The modules for Multiple Dynamic Prediction & Control to be used on board as steering assistance tool and later the manoeuvring design and planning technology were described later (Benedict et. al: 2012, 2014).

In this paper, the focus will be laid on the potential of the SAMMON software supporting ship operations in a collaborative way on-board and ashore.

2 FUNCTION-BASED BRIDGE ORGANISATION

2.1 Functional Positions

The concept of Function-Based Bridge Organisation was introduced by Hans Hederstrom at the INSLC Conference in 2012. Acknowledging that all humans may make errors, the function-based bridge organization introduces organizational countermeasures to detect and manage human error before it leads to any negative consequence. It can help to remove hierarchical barriers and enhance teamwork and communication, if a traditional rank-based system has been replaced by a function-based bridge organization. The function-based bridge organization does not diminish the authority of the Master. The Master assigns officers to the particular functions based on watch-keeper competence and experience with the upcoming operation, making it a very adaptable system.

The system builds on the airline concept by introducing Navigator and Co-Navigator functions. The Navigator who is conning the ship is required to communicate intentions and orders to the Co-Navigator. This means that no course changes or engine orders will be carried out without a confirmation from the Co-Navigator. These new protocols also require a double watch-keeping system with a minimum of two bridge officers on watch at all times the ship is at sea.

For ships with a single watch-keeping officer and a lookout on watch, the system may be somewhat more difficult to introduce. However, with trained and engaged lookouts there are definitely advantages to gain. When the Captain joins the bridge team, there is no problem to use the function based system. The best way to apply the system in this situation would be if the Captain takes on the function as Co-Navigator, leaving the watch officer to continue conning the ship. The following definitions were given and the following assigned tasks are included in these procedures (only extracted items specifically for manoeuvring aspects) in Figure 1:

Operations Director: Overview of the entire bridge operation, ensuring that it is, at all times, carried out in accordance with these procedures; Direct monitoring of both the Navigator and Co-Navigator, ensuring that safe passage is maintained and that no internal or external influences are permitted to distract them from their primary tasks; Monitors workload and transfers tasks between functions as circumstances dictate; Unless directed otherwise by the officer with the charge, will conduct the Pilot exchange briefing; If the Operations Director takes the conn, then the position of Operations Director must be re-established as soon as possible.

Navigator: Responsible for conning, navigating the ship following the approved passage plan and collision avoidance. Ensure that the bridge team (including the Pilot) is aware of planned actions and intentions by "Thinking Aloud". If a pilot has the conn, the Navigator should ensure the Pilot's intentions and planned actions are understood in advance by all bridge team members and agreed upon by the Navigator. If s/he has the charge, the Navigator is responsible for taking back the conn from the Pilot whenever s/he determines that doing so is necessary or appropriate for the safe navigation of the vessel. .

Co-Navigator: Monitors and cross checks the actions of the Navigator. Supports, challenges, and recommends actions to the Navigator. Notifies the Master or Second in Command whenever s/he has reason to believe that the Navigator has taken or plans to take any action that violates the Master's orders or is inconsistent with the safe navigation of the vessel. Monitors and cross checks the ship's position against the passage plan using real time navigation methods. Monitors traffic and collision avoidance. Unless directed otherwise by the officer with the charge, is responsible for external VHF (may be delegated to the Pilot) and liaison with the ECR.

Administrator: Responsible for fixing the ship's position when paper charts are in use. Responsible for alarm management and actions. Alarms to be identified as either urgent or non-urgent alarm. Responsible for internal communications as directed. Responsible for logbook entries, checklist management and status board. Ancillary tasks as assigned.

Lookout: Maintains all around lookout by sight and by hearing, reporting all sightings and/or sound signals to the Navigator, unless otherwise directed. Maintains awareness of planned intentions and reports any necessary clearances before an alteration of course. Must be able to give full attention to the keeping of a proper lookout, and no other duties shall be undertaken or assigned which could interfere with the task. Be available to interchange duties with the Helmsman. The duties of the Lookout and the Helmsman are separate. The Helmsman shall not be considered the Lookout while steering.

Helmsman: Acknowledge and execute steering orders given by the person with the conn. Advise the person with the conn of any steering concerns.

2.2 *The Captain as a Leader instead of an Operator*

It is up to the Captain to decide who should fulfil any of the four functions. A Risk Factors Table and a Risk Analysis and Bridge Manning Level Table have been developed to assist the Captain in deciding what manning level to set. Those manning levels are to be seen in Figure 1.

The philosophy behind the system encourages the Captain to assume the role of Operations Director, acting as a leader while the team undertakes the operation.

By delegating the operational tasks, he demonstrates trust in his team. This has many positive effects, such as: enhanced learning; readiness to actively participate in problem solving; enthusiasm and motivation to work; and an engaged team directly leading to increased safety and efficiency.

As officers are allowed to conduct the vessel, they will be better prepared for their promotion when time comes. This will normally also increase job satisfaction, which facilitates officers'? retention rate.

Within this paper some elements are presented on how the communication with in the bridge team can be supported by the Fast Time Simulation Modules of the SAMMON System

3 SIMULATION-AUGMENTED SUPPORT FOR SHIP MANOEUVRING PROCEDURES

3.1 *Pre-planning with "Manoeuvre Planning & Design Module"*

As an example for creating a berth plan and briefing the navigational officer, a berthing scenario is chosen for a harbour area - the starting situation and the environmental conditions within this area on a sea chart is to be seen in Figure 2. The objective is to berth the ship with port side alongside Grasbrook Berth at Hamburg Port.

The respective harbour area is being divided into two manoeuvring sections following a specific aim:

1 Section 1: At the end of this section the speed over ground (SOG) should be around 3 kn and the heading slightly towards southeast as preparation for section 2.

2 Section 2: A state should be reached, where the ship can be held in the current at a position with constant heading and no speed. Then, the ship can then crab towards the berthing place mainly by means of thrusters. The current can be used as an additional supporting aid to go alongside.

In a conventional briefing only these rough indications of the manoeuvring status can be used to develop a potential strategy for berthing the ship. The manoeuvres and setting of engines, rudders and thrusters cannot be discussed in detail because no specific manoeuvring characteristics are available for the specific situations.

Required Functions at Each Bridge Manning Level							
		Bridge Functions					
		Navigator	Co-Navigator	Administrator	Operations Director	Lookout	Helmsman
Bridge Manning Level	Green	Yes	Yes		No	Yes	As required
	Yellow	Yes	Yes		Yes	Yes	Yes
	Red	Yes	Yes	Yes	Yes	Yes	Yes

Figure 1. Required Functions and Manning Concept for Functional Approach for Bridge Operation

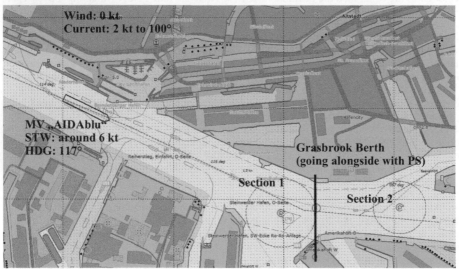

Figure 2. Exercise area and environmental conditions in Port of Hamburg for berthing scenario, divided into two sections for planning the manoeuvres

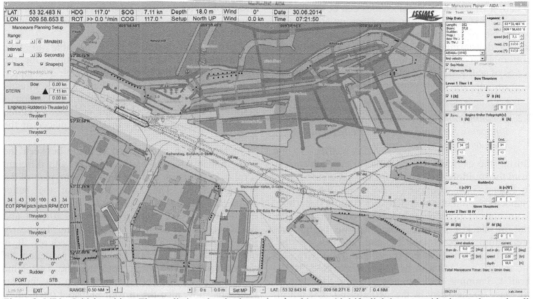

Figure 3. MP0 - Initial position: The prediction already shows that the ship would drift slightly to portside due to the set handle positions

With the new fast-time simulation there is the chance for designing a manoeuvre plan as a detailed strategy with the specific settings at distinguished positions called the Manoeuvring Points (MP). In the following, the course of actions is described in a series of figures to make a full manoeuvring plan by means of the control actions at the manoeuvring points, MP. In Figure 3 the initial position is to be seen where the instructor has set the ship in the centre of the fairway. The prediction already shows that the ship would drift slightly to port side due to the set handle positions. It can be learned that therefore the rudders have to be put slightly to starboard at the very beginning in order to follow the straight track until the next MP 1. At MP 2 the rudders are set amidships again and both propulsion units are used to slow down and to steer the ship: the starboard engine is kept at 34 %, resp. 43 rpm to allow for a certain rudder effectiveness for steering control, whilst at MP 3 the portside engine is set backwards in order to achieve about 3 kn SOG at the end of section 1.

In Figure 4 (top), the ship is stopped at MP4: The vessel's heading is chosen in that way, that all handles can be set in zero position, holding the ship with a minimum speed almost at the same position. At this moment, bow and stern thrusters can be applied to bring the ship safely to its berth. In the bottom figure the ship is already brought to the berth. The crabbing by means of bow and stern thrusters needs a further MP5 in order to reduce the transversal speed shortly before berthing at MP6.

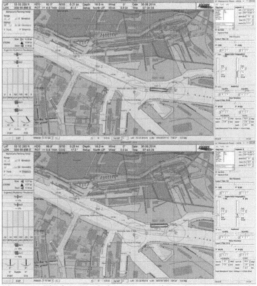

Figure 4. Final part of the manoeuvring plan: At MP4 the vessel is stopped and the heading is chosen in that way, that all handles can be set in zero position (Top); at MP5 / MP6 the ship is already brought to the berth (Dottom)

The complete manoeuvring plan can be saved to be used for the training or to be loaded again for editing the plan for an optimisation to achieve a better performance e.g. to do the whole manoeuvre in less time. For an in-depth discussion at the separate manoeuvring points and sections, there is the possibility to save the specific conditions as

111

situation files. These situation files can be useful for discussing strategies during the planning at different places where new challenges will come up as well as for the debriefing sessions. In Figure 4 at the right bottom corner the time is to be seen for the complete series of segments: the total manoeuvre time is about 17.5 minutes for this version of the plan.

3.2 Berthing making use of Simulation-augmented support in SHS and with SAMMON Monitoring Module and Training Tool

During the exercise it is possible to take advantage from the multiple predictions for the manoeuvres.

In Figure 5 an example is shown for the On-line manoeuvre prediction (dotted ship contours) starting from current position (black ship contours) at the end of black past track. On the ship bridge the prediction is controlled by the handles on the manoeuvring controls. For training and test purposes the manoeuvre can also be tested in the SAMMON Trail & Training Tool – in this software the controls are used to be seen in the Monitoring & Control Interface of the Training Tool presented on the right side in this figure .

For comparing the effectiveness of the simulation-augmented support tools a simulator test was made with trainees who have no support and trainees who have parts or even the full support of the SAMMON.

During the exercise it is possible to take advantage from the multiple predictions for the manoeuvres. The students can bring their own laptop onto the simulator bridge (where he has already developed the manoeuvring plan), the prediction is controlled via the bridge handles, and another laptop with the monitoring tool can also be placed at the instructor station.

In Figure 6 a comparison is made between the two simulator results of the trainees with different level of preparation and the manoeuvring plan of the second trainee. The achievements of the better prepared trainee are obvious – the planned manoeuvre is very close to the executed track and the actions of the controls has been done also nearly in accordance with the planned procedures. It is obvious that there is not just a reduction of manoeuvring time when applying the Fast-Time Simulation tool in briefing and training; the thruster diagrams show also that a well prepared manoeuvre can minimize the use of propulsion units and therefore be more efficient.

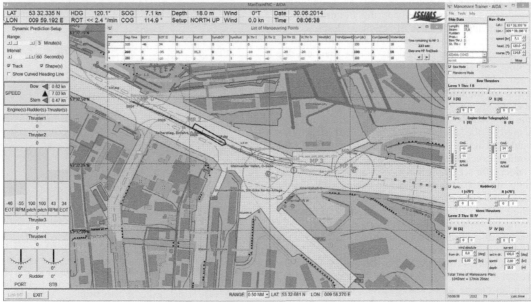

Figure 5. Example for overlay of a pre-planned manoeuvre plan (MP) as manoeuvring basis (blue) and manoeuvre prediction (dotted ship contours) starting from current position (black ship contours) at the end of black past track with her engines ordered in opposite direction presented in Monitoring & Control Interface of the Training Tool (right)

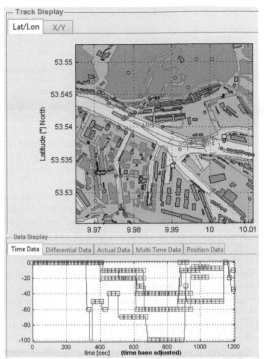

Figure 6. Results from two manoeuvring exercises in SIMDAT interface with ships track and time history of thruster activities. (Blue: run of the trainee without support by Fast-Time Simulation; Green: run of the trainee with full support by pre-planning with Design and Planning Module; Red: comparison to the prepared manoeuvring plan with manoeuvring points)

The benefit of using the FTS is to be seen for several purposes:
– The multiple dynamic predictions are always a great help for the Navigator steering the ship: They have a better overview on the current situation and the chances for the potential success of an action can immediately be seen; also for the Co-Navigator there is the chance to see both the manoeuvres and the success – this is a great situation because they can both share a better situation overview.
– Multiple dynamic predictions may be used to see both the current state of motion by the static path prediction and the future development of the ship motion caused by the current handle settings – it is expected that the static prediction changes into the dynamically predicted track, in this case the pre-diction is correct. If not then the handle settings can be slightly adjusted to correct for the tendency of the potential impact of environmental effect which might not have been considered by the dynamic prediction, e.g. a non-detected current.

4 RESEARCH PROJECT COSINUS - SIMULATION AUGMENTED MANOEUVRING FOR BRIDGE OPERATION AND FOR VTS

The goal of the project COSINUS ("cooperative operation of ships for nautical safety through integration of traffic safety systems) is to achieve the integration of maritime traffic safety systems on board and on shore. Therefore, novel concepts are investigated regarding the presentation of enhanced data to the operator and operation of new tools and services as well as decentralized data capturing, processing and storage. Processed data of land-based information systems will be visualized in such a way that a complete overview over the traffic and environmental situation is given in order to support the navigational operation of the vessel. This includes e.g. the representation of a shared route and manoeuvring plan, the operational interface to the VTS operator, and the depiction of weather-data along the voyage or at the destination port. The goal is to establish a cooperative picture which offers a dynamically enhanced view for the bridge crew going beyond traditional ship-based sensor information like own ship RADAR or AIS. This will improve the safety particularly in heavy traffic situations. A great deal of work will be carried out concerning the definition and establishment of new standards for the ship based navigation in cooperation with higher level traffic management systems. The main areas of work are the following:
– Visualization concept for representation of land-based information on ship bridges
– The proposal and the validation of modules and interfaces for autonomous communication between VTS and INS
– Combination of ECDIS representation of navigational data and VTS data to an integrated navigational and traffic picture
– Concept for cooperative route- and manoeuvre planning
– Investigation of communication channels and interfaces for exchange between VTS and INS
Specifically for the integration of the Simulation-Augmented Manoeuvring Support by SAMMON the new functions have to be interfaced:
– The results of the manoeuvre planning have to be made available into the Integrated Bridge System and
– Also the data transfer from ship data into the Monitoring and Control Module have to be adjusted.
– The data transfer from ship to shore into the VTS centre has to be established.
The concept for sharing the information between ship and shore is to be seen in Figure 7- both on the bridge and in the VTS the same display elements for planned routes and manoeuvres can be observed on

all screens in the same way. In Figure 8 a first result is to be seen for a ship station to display the manoeuvring plan together with a route plan of another vessel.

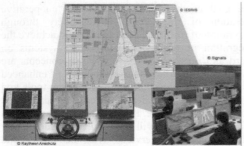

Figure 7. Project COSINUS – shared information on manoeuvring plans and multiple prediction in ECDIS between bridge and VTS

Figure 8. Project COSINUS results: Display of manoeuvring plan (green) of own ship together with route plan of another vessel (blue) presented in Integrated Navigation System of project partner Raytheon Anschütz transmitted from VTS station from SIGNALIS

5 RESEARCH PROJECT MUNIN - MANOEUVRING SUPPORT FOR AUTONOMOUS SHIPS

5.1 *Introduction & Objectives*

Maritime Unmanned Navigation through Intelligence in Networks (MUNIN) is a collaborative research project of eight partners from five European countries co-founded by the European Commission. MUNIN's aim is the development of an autonomous-ship concept and its simulation-augmented feasibility study. MUNIN Project coordinator is the Fraunhofer Center for Maritime Logistics and Services (CML) in Hamburg, Germany.

The Department of Maritime Studies at Hochschule Wismar (HSW), University of Technology, Business and Design in Rostock-Warnemünde, Germany, is involved in both parts of ship operation the navigational and technical systems.

– The ship-engineering department at HSW is responsible for the analysis and conceptual redesign of current engine-related tasks as well as for repair and maintenance optimisation for unmanned operation during the sea passage.
– The Institute for Innovative Ship Simulation and Maritime Systems (ISSIMS) at HSW develops a simulation augmented manoeuvring support systems for remote-controlled navigation in near coastal waters.
– The Maritime Training Centre Warnemünde at HSW serves with its simulation environment and partner's prototype integration for the feasibility study within the proof of concept.

The main idea behind the MUNIN concept is the autonomous sea passage of an unmanned vessel. Nevertheless, before the ship can be set to autonomous operation it has to put out at sea in the traditional way with a crew on board. For the unmanned voyage part the vessel is monitored by a Shore-Control Centre. When in autonomous mode, the vessel solves appearing problems with regard to weather and traffic situation by autonomous algorithms and follows its pre-defined voyage plan. If necessary, the operator takes over automatic control by commanding the vessels true heading and speed-over-ground. Furthermore, when exact manoeuvring is required, the operator enables a mock-up bridge to manually control the vessels manoeuvring systems like rudder and engine from a situation room within the Shore-Control Centre. Assuming that the connection fails, the vessel has to drift or, if possible, drop the anchor to maintain its position.

5.2 *Remote Manoeuvring Support System –On-line Prediction and presentation of operational limits*

The Remote Manoeuvring Support System envisages the improvement of the mental model of experienced ship officers on board sea-going vessels to a Shore-Control Centre. Since for the shore-based operators the feeling of the ship's motion is missing, a way must be found to transmit the impression and feeling of the ship's actual and future motion to the operators. The problem is: there is no scope for the conventional "trial and error corrections" or "touch and feel experiences" for vessels fully controlled by shore-side operators.

The remote manoeuvring support system's aim is to allow safe and efficient remote-controlled navigation in near-coastal waters. The innovative value of the Fast-Time Simulation technology is the look-ahead function of ship's motion by dynamic-prediction methods, so that a ship's officer or shore-side operator can foresee the vessels future path.

The Remote Manoeuvring Support System prototype contains three different modules - all based on Fast-Time Simulation und dynamic-prediction methods:

– Monitoring tool with visualisation of future ship track by means of dynamic-prediction methods
– Pre-planning tool to design safe and efficient manoeuvre plans for the upcoming manoeuvring
– Prediction of the operational limits visualising the required room to manoeuvre.

Not only for collision avoidance but also for navigation in narrow waters it is from high importance for a shore-side operator to know the operational limits of the vessels under his surveillance. The problem is that the manoeuvrability depends on many hard-to-estimate factors. High speed in shallow water e.g. causes squat effects, and the speed-through-the-water to speed-over-ground ratio increases/decreases rudder effectiveness as well as waves and gales affect the turning and stopping behaviour. The mariner aboard senses this and directly interprets the effect by the above named factors. He can feel and observe a squat effect way easier as an operator sitting in a control centre ashore in front his screens. He has trained his mental model of ship's motion by years of experience at sea.

To support the shore based operator by information on ship's motion dynamics, the Remote Manoeuvring Support System supplies the operator (and the collision avoidance system on board) with vessel data regarding its operational manoeuvring limits.

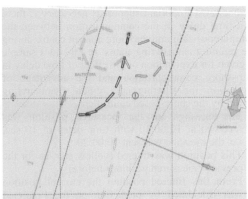

Figure 9. Sample for presentation of dynamic-manoeuvring prediction of actual manoeuvring track (black-dotted contours) and additional manoeuvring tracks for hard-to-STB (green) and PT (red) as well as for crash stop (black) from actual motion parameters - the ship has applied rudder amidships the contours of actual control are ahead of the ships position.

Figure 9 shows the monitoring concept with the prediction of the manoeuvring limits. All four manoeuvre predictions will be supplied in a 1 Hz update rate. This figure shows a situation for a collision threat: the own ship is the stand-on vessel and the ship on its port side is expected to do a course change to avoid a collision according to COLREG rule 15. In case the ship as not acting in proper time, the own ship is obliged to do an evasive manoeuvre according to COLREG rule 17. From the figure it is to be seen that a stopping manoeuvre would not help anymore but a turning circle to starboard would help.

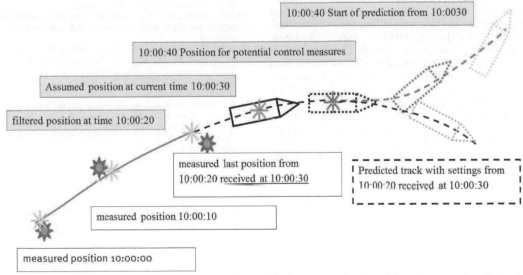

Figure 10. Sample for explanation of the effect of time delay in ship-shore-communication and the ad-vantage of prediction for filtering and remote manoeuvring at time point 10:00:30

The most important support is necessary if there is a time delay in the communication between the autonomous ship and the shore control centre during the remote manoeuvring status: in Fig. 13 a sample is given for explanation of the effect of time delay in ship-shore communication and the advantage of prediction for filtering and remote manoeuvring action by the shore-based controller.

- The message for the measured position was received at 10:00:30 with time delay of 10 sec, i.e. the message was sent 10:00:20.
- This position was filtered (yellow star, as for the previous measured positions before).
- From this filtered position the current position was calculated by prediction on the Predicted track (blue broken line) with control settings from 10:00:20. In the same way the position at 10:00:30 was found which the initial point for the new prediction is.
- From the assumed / predicted position at 10:00:30 the new prediction for new settings from 10:00:30 will take effect after another delay of 10 sec at the position at 10:00:40 – from there the red dotted contours and track are shown for the new predicted track.

It is obvious that it is very difficult to steer the ship if the time delay is increasing. Within the project it is planned to do some investigations into the maximum delay allowed to secure a safe control of the vessel from shore.

ACKNOWLEDGEMENTS

The research results presented in this paper were partly achieved in research projects "ADvanced Plan-ning for OPTimised Conduction of Coordinated MANoeuvres in Emergency Situations" (ADOPTMAN), COSINUS and Maritime Unmanned Navigation through Intelligence in Networks (MUNIN) funded by EU, by the German Federal Ministry of Economics and Technology (BMWi), Education and Research (BMBF), surveyed by Research Centre Juelich PTJ and DLR. Additionally it has to be mentioned that the professional version of the SAMMON software tools has been further developed by the start-up company Innovative Ship Simulation and Maritime Systems GmbH (ISSIMS GmbH; www.issims-gmbh.com).

REFERENCES

Benedict, K., Baldauf, M., Felsenstein, C., Kirchhoff, M.: " Computer-based support for the evaluation of ship handling simulator exercise results" MARSIM - International Conference on Marine Simulation and Ship Manoeuvrability, Kanazawa, Japan, August 25th – 28th 2003

Benedict, K., Baldauf, M., Kirchhoff, M., Koepnick, W., Eyrich, R. 2006: Combining Fast-Time Simulation and Automatic Assessment for Tuning Simulator Ship Models. MARSIM - International Conference on Marine Simulation and Ship Manoeuvrability, Terschelling, Netherlands, June 25– 30 2006. Proceedings, M-Paper 19 p. 1-9

Benedict, K.; Kirchhoff, M.; Gluch, M.; Fischer, S.; Baldauf, M. 2009: Manoeuvring Simulation on the Bridge for Predicting Motion of Real Ships and as Training Tool in Ship Handling Simulators. TransNav - the International Journal on Marine Navigation and Safety of Sea Transportation, Vol. 3 No. 1 - March 2009.

Benedict, K.; Kirchhoff, M.; Gluch, M.; Fischer, S.; Baldauf, M. 2013: Simulation Augmented Manoeuvring Design and Monitoring – a New Method for Advanced Ship Handling. TransNav - the International Journal on Marine Navigation and Safety of Sea Transportation, Vol. 3 No. 1 - May 2013.

Benedict, K.; Baldauf, M.; Fischer, S.; Gluch, M.; Kirchhoff, M.; Schaub, M.; M.; Klaes, S. 2012: Fast Time Manoeuvring Simulation as Decision Support for Planning and Monitoring of Ship Handling Processes for Ship Operation On-Board and Training in Simulators. MARSIM - International Conference on Marine Simulation and Ship Manoeuvrability, Singapore, 23 -27 April 2012.

Fischer, S., Benedict, K. 2009: "Analyses of manoeuvring procedures on ferry Mecklenburg-Vorpommern in Rostock Sea Port and potential improvements using alternative manoeuvring concepts with Dynamic Predictor" Internal research report (in German only), Hochschule Wismar, Dept. of Maritime Studies, Warnemuende

ISSIMS GmbH 2012: Web page for SIMOPT & SIMDAT: http://www.issims-gmbh.com/joomla/index.php/software-products

Benedict, K.; Baldauf, M.; Fischer, S.; Gluch, M.; Kirchhoff, M.; Schaub, M.; M.; Klaes, S.: APPLICATION OF FAST TIME MANOEUVRING SIMULATION FOR SHIP HANDLING IN SIMULATOR TRAINING AND ON-BOARD. Proceedings of INSLC 17. Sept 2014 at Hochschule Wismar, Rostock Warnemuende / Germany.

Hederstrom, Hans: MOVING FROM RANK TO FUNCTION BASED BRIDGE ORGANISATION. Proceedings of INSLC 17. Sept 2014 at Hochschule Wismar – Rostock Warnemuende / Germany

Decision Support System

Information, Communication and Environment – Marine Navigation and Safety of Sea Transportation – A. Weintrit & T. Neumann (eds.)

3D Navigator Decision Support System Using the Smartglasses Technology

A. Łebkowski

Gdynia Maritime University, Poland

ABSTRACT: The publication contains a proposal to use SMARTGLASSES technology as part of a 3D Decision Support System for Navigators (3DSSN). Information presented by the system is a fusion of normal navigator's vision from the bridge with the information presented in 3D technology using the Smartglasses. By using this technique it is possible for instance to steer the ship in conditions of limited visibility, or to select an optimal collision avoidance maneuver or route of passage.

1 INTRODUCTION

Seas and oceans surrounding the land were always inseparable part of mankind's history – they provided food, new land and allowed trade. On the other hand they were always dangerous and mysterious element, impossible to tame. The first basic problem was the difficulty in determination of ship's position. Navigators of the past were not always sure where they were headed and what was the destination port heading. One moment of inattention was enough to experience how sinister and hazardous the sea was. Treacherous tides, varying hydro meteorological conditions, maelstroms, shallows were very serious navigational obstacles that weighted on life and safety of transported goods. Over time, additional navigational hazards appeared. Following the growing number of traveling ships, their size and velocity also increased. No nation could grow without advancement in navigational and transportation technology.

Looking back in time, the biggest empires of their age built their might on their fleet – like the British Empire, which was controlling one quarter of land on whole Earth. We admire Scandinavian sailors' accomplishments – the Vikings, who as first Europeans set feet on America's shore. Scientific progress including astronavigation techniques using Sun's and stars' positions allowed them to explore high seas. Hungarian scientists have researched the theory of application of "sun stones" – natural polaroid filters in form of Calcite, Tourmaline and Mica minerals, which change their brightness and color depending on incident light angle. Usage of sun stones allowed the Vikings to ascertain their position regardless the atmospheric conditions. [1]. It is also known, that astronomical and geographical knowledge of Nordic people was many years ahead of their time. Awareness of Earth's sphericity and related terms such as longitude, latitude, arctic circle or the equator are among the examples. The exact system created and used by Viking people remains a mystery, however [9].

During the ages, the secrets of navigation were constantly evolved, and even today this process continues. Calculation of location using satellite based techniques is certainly not the limit of our capabilities, as the progress in this matter is constant.

Many new technologies take inspiration from the antiquity. Take for example the mythical Talos – a colossus forged out of bronze by the Hephaestus. It guarded Crete's shores and denied access to any foe. Having noticed an unidentified ship or other object on the sea, it heated up to cherry red glow and burned them in its fiery embrace. Legends of ignition and destruction of foreign objects were not uncommon in that era, however the historians' attention was attracted by one difference between Talos and other entities: what enabled him to identify friend from foe? This could be and a system of artificial reasoning and decision – a first artificial intelligence system. It could be said that Talos was the first robot. The issues of navigation and artificial intelligence are interconnected. Currently there are more and more systems developed using these techniques to make navigation safer, by helping foresee eventual occurrence of threats. Moreover, they possess a capability for self-learning, a feature recently reserved only to living organisms. Contemporary science implements many solutions based on artificial intelligence. Agent systems, evolutionary algorithms, neural networks, fuzzy logic or expert systems are among the best known methods of AI used in navigation systems. Listed methods are part of the e-Navigation strategy being currently developed by IMO. It can be described as a harmonized collection, integration, exchange, presentation and analysis of marine information on board and ashore by electronic means to enhance berth to berth navigation and related services for safety and security at sea and protection of the marine environment [10].

The obvious cause of this is a green light for development of new navigator decision support systems, such as one proposed by the author, which combines use of information available from ECDIS with new technical possibilities offered by Smartglasses technology (Figure 1) [19].

Figure 1. Assortment of Smartglasses [19].

Although current IMO regulations forbid the sailing of unmanned ships, doubtlessly the next step in ship construction will be development of such ships, capable of sailing without their crew present. Unmanned cargo ships – drone ships, controlled remotely via satellite link from a shore station are a near future. Preliminary designs of such ships are already proposed by Rolls-Royce (Figure 2) [13] and other designers also present models of their autonomous vessels [8,14,15,16].

Current IMO regulations do not allow presence, let alone proliferation of full-size autonomous, unmanned vessels in maritime traffic, however it is unknown what the final shape of e-Navigation strategy will be, as the proceedings will not be finished until 2019. Before we enter the phase of operation of unmanned ships, there will be a period of development of various technologies and systems supporting the navigator's decisions [3,4,5,6,7,8,11].

Figure 2. Views of proposed unmanned crafts [17].

These systems will undoubtedly increase the safety level on sea. They greatly reduce the time needed by the navigator to take an optimal decision given the navigational circumstances, they help by providing information on the situation around the ship as well as data on the ship itself. In 2010 researchers from Japan Institute of Navigation presented a system which displays data and information needed for safe navigation directly on the bridge windows [2]. Application of similar technology is proposed by scientists from Rolls-Royce and VTT in their visionary project presented in 2014, its final effects should be available around 2025. [12]. The project assumes displaying on the bridge's windows of all the data required for navigation. Information is to be additionally presented in 3D mode, which would make it more legible and comprehensible to the navigators.

The following publication is author's proposition of navigator decision support system using 3D Smartglasses technology.

2 DECISION SUPPORT SYSTEMS

Taking into account the number of collisions and accidents happening on sea each year, no matter what hydro meteorological conditions and water type, more and more advanced decision support systems are being used. Navigator commanding the cargo ship is required to constantly observe the dynamically forming navigational situation around his ship. Apart from that, he needs to pay attention to various instruments and devices which constantly flood him with information. Based on observations and readouts from auxiliary devices, in case of collision hazard he is required to designate an anti-collision maneuver in the shortest time possible. In such stressful situation, where single man has to make a quick and precise decision it is easy to make a mistake. An important matter from the point of cruise safety is the time between arising of collision situation and commanding an anti-collision maneuver. The shorter the time, the bigger the margin of error for navigator's eventual mistake. So given the long enough decision time, the navigator has opportunity to correct any eventual errors. Application of decision support system allows shortening of time between noticing a danger and issuing a command which neutralizes it. Apart from classic decision support systems, which mostly use the displays of existing systems or put additional ones, such as: Navigational Intension Exchange Support System (Passing Pattern); INT-NAV (Integrated Navigational System. Tracked Target Information and Video display); Collision Threat (Scheduled route based collision danger area display. Scheduled route and collision danger zone); Collision Avoidance (Display for automatic collision avoidance system - recommended collision avoidance route); NAVIEYE (Vessel surveillance equipment. Video image of target ship); Route Suggestion (Route suggestion from Shore to Ship); NAVDEC [11,18] and many others, there are systems which strive to dimensionally present parameters for given navigational situation.

One of such systems is Visual Lookout Support System proposed in 2009 by researchers from Japan Institute of Navigation. A team led by Kenjiro HIKIDA has implemented a solution which displays navigational parameters on transparent head-up display placed in the field of view of navigator standing on the bridge (Figure 3) [2].

Figure 3. View through the window with VLSS installed [2].

The system displays basic information on Head-Up Display such as object's name, bearing, range, heading, speed, CPA and TCPA. Results of experiments have shown that system's application has shortened the time for interpretation of navigational situation when using radar system display. Using VLCC, navigator can save 2 to 10 seconds only on interpretation time. Therefore the

correct manner of presenting the data can influence the delay before issuing anti-collision maneuver.

Another solution could the system presented by Rolls-Royce and VTT [12,13]. It assumes displaying of necessary information on specially crafted bridge window, including parameters of passing objects, data on hydro meteorological conditions, digital maps projection or showing the distance to nearby objects. Some of the information provided can be displayed three-dimensionally. Project of the future bridge is shown in Figure 4.

Figure 4. The concept of the navigation bridge by Rolls-Royce and VTT [17].

3 TECHNOLOGY

First Smartglasses models [19] had integrated Optical Head Mounted Displays (OHMD) or transparent Heads-Up Displays (HUD). Next generations had transparent pieces allowing the user to see projections of virtual video (Augmented Reality – AR) with possibility of 3D elements display.

Using Smartglasses technology, which combines a function of glasses with display and a computer, all in one package, it will be possible to present navigation parameters and display of navigational situation around on ship. State of the art Smartglasses, as proposed in this article, have extended functions of data processing, like in smartphones and tablet computers. (Figure 5).

Figure 5. View of META Smartglasses [20].

Navigation between applications can be realized using navigator's voice or with his hands. Pointing at any element, the navigator will have possibility to select various display modes and parameters. Using "pinching" and "spreading" in the air it will be possible to mark objects or browse through their parameters essential from the point of navigation. Additionally, modern Smartglasses can launch mobile applications, recognize and interpret objects in environment. They also have built-in components like multiaxial accelerometers, gyros, Bluetooth, Wi-Fi and GPS modules. They will allow the navigator to pan and tilt his head and observe through transparent displays virtual holograms of other objects, his ship route, other ships routes, potential crossing or approach points or other information such as boundaries of allowed navigation areas, actual meteo condition messages and important messages from VTS station or AIS system with satellite modules.

Figure 6. Example view through Smartglasses.

An ideal set for the navigator on a cargo ship's bridge would be a combination of Smartglasses technology with land CCTV and VTS systems and onboard CCTV system (equipped with infrared and thermal cameras).

4 SYSTEM STRUCTURE

The constant drive to make sailing better, raise the efficiency of transportation services and provide for crew and transported cargo safety is the reason for installation of modern anti-collision devices and systems on ships of all types. Presence of these systems not only gives the awareness of increased safety, but also increases the efficiency of vessel motion from source to destination ports. These systems realize informational and computational functions as well as they provide warnings and alarming in case of dangerous situations with probable grim consequences to both crew and environment.

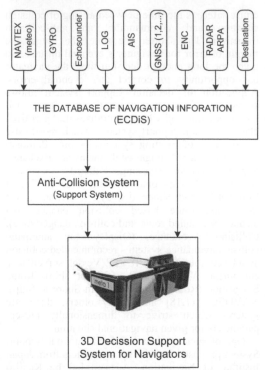

Figure 7. The structure of 3D Navigator Decision Support System using the Smartglasses Technology.

In Figure 7 a structure of 3D Decision Support System for Navigators is shown. The definition of "System" includes a set of some elements, isolated from their neighborhood as a relatively autonomous entity, whose elements possess internal connections and feedbacks. All the elements work together, towards common goal. In this case, the goal is to safely navigate a ship. Therefore system is an organized object, acting in a defined environment and consisting of smaller elements called sub-systems. Sub-systems are systems within greater system, which are a part of cooperative relationships so that each can realize the main goal and can contribute to common success. System is also called a set of elements connected into common dependencies, created to fulfill certain goals. Summing up all these statements it can be said, that a system is a set of interconnected elements, which operate as a unity and have a single goal, which they constantly seek.

Modern anti-collision systems allows precise tracking of even up to 300 various dynamic objects. Similarly, the goals for 3D Decision Support System for Navigators will include:
- indication of safe space for ship navigation or determination of ship route (fairways, ice fields, passage over shallows, harbor approach zones, navigation exclusion zones),

- detection and tracking of objects potentially dangerous to own ship,
- indication and display of parameters (bearing, range, heading, speed, CPA, TCPA) of objects being tracked and indicated by the navigator,
- alarming in the condition of possible collision,
- indication of collision parameters for dangerous and indicated by navigator objects (CPA, TCPA),
- graphical imaging of relative and absolute motion of tracked objects,
- signaling in the event of rapid change in course or speed of tracked objects,
- decision support on safe maneuvers including suggested course and speed in order to maintain safe distance from passed objects,
- presentation of data on parameters of passed objects (mostly AIS data: name, call sign, ship type, size, etc.),
- presentation of data on hydro meteorological conditions (weather forecast, present weather on given waters, etc.),
- graphical simulation of trial anti-collision maneuver along with projected consequences.

3DSSN in its structure relies mainly on data collected prior by ECDIS, which include data from: log, gyrocompass, AIS, radar system with ARPA, electronic map, satellite navigation system and additional systems such as: NavTex, local Meteorological system, NOAA, NWS, NOS and others. Moreover, 3DSSN can interoperate with other decision support systems which could calculate, based on information available to them, the optimal ship route [3,4,5,6,7,8,18]. Thanks to its capabilities, 3DSSN can be used almost on any ship. System properties allow it to be used during passage through regions of limited visibility, regions where sea bottom raises dangerously and constitutes a danger to the ship, or in regions with possibility of icebergs presence.

5 CONCLUSIONS

Application of 3D Decision Support System for Navigators could greatly increase the safety level on sea.
- Due to possibilities of holographically presentation of calculation results and decision support could be a powerful tool for navigational data implementation by the navigator. Thanks to presentation of navigational situation by 3D holograms and digital markers it is possible to plot passage routed as well as other traffic parameters for both the own vessel, and other vessels present in the vicinity. It allows the navigator to know well in advance the other ships navigators' desired maneuvers and avoid collision.

- The system appears to be much cheaper both to purchase and to operate than the presently proposed and developed 3D display systems [2,12,13]. Its simple construction and low cost could make it find lots of users in the near future. The only limit in application of this system seems to be the imagination of designers of displayed data.
- 3DSSN could be used on many ship types, its practical applications pretty unlimited. It could be used for example by: navigators during adverse weather conditions, harbor pilots, and (after software upgrade) by other maritime workers such as platform crew, mechanics, electricians or deck crew. In the near future, a prototype will be constructed and its usefulness will be evaluated on training ships. An interesting solution would be an application of Smartglasses technology on submarine vessels.

REFERENCES

[1] Bernáth B., Farkas A., Száz D., Blahó M., Egri A., Barta A., Akesson S., Horváth G., How could the Viking Sun compass be used with sunstones before and after sunset? Twilight board as a new interpretation of the Uunartoq artefact fragment 10.1098/rspa.2013.0787 Published 26 March 2014.

[2] HIKIDA K., Development of a Shipboard Visual Lookout Support System with Head-up Display, Navigation System Research Group, Navigation and Logistics Engineering Department, National Maritime Research Institute, CiNii 2010.

[3] Lisowski J.: Optimal and game control algorithms of ship in collision situations at sea. Control and Cybernetics, No. 4, Vol. 42, 2013, p. 773-792.

[4] Lisowski J.: Optimization-supported decision-making in the marine game environment. Solid State Phenomena, Trans Tech Publications, Switzerland, Vol. 210, 2014, p. 215-222.

[5] Lisowski J.: Game strategies of ship in the collision situations. TransNav – The International Journal on Marine Navigation and Safety of Sea Transportation, Vol. 8, No. 1, 2014, pp. 69-77.

[6] Lisowski J.: Computational intelligence methods of a safe ship control. 18th International Conference in Knowledge Based and Intelligent Information and Engineering Systems KES2014, Gdynia, Elsevier Procedia Computer Science, No 35, 2014, pp. 634-643.

[7] Łebkowski A., Control of ship movement by the agent system, Polish Journal of Environmental Studies Vol.17, No. 3C, 2008

[8] Łebkowski A., Negotiations between the agent platforms. Scientific Papers of Gdynia Maritime University, Gdynia 2013.

[9] Majek M., Comparative analysis of selected systems and control systems - artificial intelligence methods currently used for process control of the ship. Master's Thesis, supervisor dr inż. Andrzej Łebkowski, AMG 2013.

[10] IMO, NAV 59/6. Development of an e-Navigation Strategy Implementation Plan, Report of the Correspondence Group on e-Navigation, submitted by Norway. IMO (2013).

[11] http://www.e-navigation.net/index.php?page=portrayal-examples

[12] http://www.vtt.fi/files/events/ITSEurope/UXUS_pressreleaseFINALprintversion.pdf

[13] http://www.rolls-royce.com/marine/customer_focus/

[14] http://www.unmanned-ship.org/munin/

[15] http://gotransat.com/

[16] http://www.wired.com/2014/10/navy-self-driving-swarmboats/

[17] https://www.flickr.com/photos/rolls-royceplc/sets/72157647334399764

[18] http://www.navdec.com

[19] http://www.Smartglassesnews.org/smart-glasses/

[20] https://www.spaceglasses.com/products.

Information, Communication and Environment – Marine Navigation and Safety of Sea Transportation – A. Weintrit & T. Neumann (eds.)

Decision Support System

Neuroevolutionary Ship Maneuvering Prediction System

M. Łącki
Gdynia Maritime University, Gdynia, Poland

ABSTRACT: The paper presents a concept of the advanced ship maneuvering prediction system with neuroevolutionary method. This ship handling system simulates a learning process of an autonomous control unit, created with artificial neural network. This unit observes input signals and calculates the values of predicted parameters of the vessel maneuvering in confined waters. In neuroevolution such units are treated as individuals in population of artificial neural networks, which through environmental sensing and evolutionary algorithms learn to perform given task efficiently. The task is: continuous online learning and prediction of the values of a navigational parameters of the vessel after certain amount of time, regarding an influence of its environment.

1 INTRODUCTION

Prediction of vessel movement during maneuvers in restricted waters is of primary meaning regarding safety of people, equipment and the environment. Increase of computational power of standard computers allow to implement complex algorithms into advanced decision support systems also in the field of marine navigation.

Such a system should include the following main functions:
- analysis of the navigational situation online, in continuous mode,
- warning before the dangerous situation may take place, e.g. possible collision or exit from a particular limited area in an undesirable direction,
- providing transparent information that can be used in co-operation with local authorities and other auxiliary units of the area,
- it shall give the answers to "what if" and "when" questions, in the field of ship maneuvering actions and navigational situations.

All these requirements may be obtained with neuroevolutionary methods. These methods are intensively studied and implemented in different fields of science, including robotics (Haasdijk et al., 2010; Lee et al., 2013), automation processes (Bagnell and Schneider, 2001; Kenneth et al., 2005; Zhao et al., 1995), multi-agent systems (Nowak et al., 2008), designing and diagnostic (Larkin et al., 2006) and many others. Neuroevolutionary algorithms are successful methods for optimizing neural networks topologies, especially in dynamic continuous reinforcement learning tasks. Their significant advantage over gradient-based algorithms is the capability to modify network topologies along with connection weights.

The basic concept of the system is presented in figure 1.

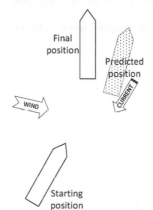

Figure 1. Main task of advanced prediction system is to calculate a position of the vessel as accurate as possible to her final position

The system shall, through continuous learning process, predict the ship position and state of the environment after specified time interval as accurate as possible in comparison to final real position of the vessel. It is possible to calculate a probable position when there is a simulation model of the vessel available. It is required that simulation model includes the equations and coefficients for wind, current and waves. But in most cases such advanced non-linear simulation model is not available. And again, a good solution for this problem is neuroevolution.

2 NEUROEVOLUTIONARY METHODS

Neuroevolutionary methods are part of intelligent computing methods that are capable of finding solution of complex tasks with artificial neural networks (ANN) created with evolutionary algorithms (EA). Such combination gives an advantage of flexibility and adaptation, which allows to adjust computational structures to dynamically changing tasks.

In neuroevolution ANN is treated as an individual in a population of multiple networks. Basic topologies of the initial population are randomly determined at the beginning of learning process. Each individual begins the process of finding a solution with the same starting parameters. The action of each individual is usually assessed with the reinforcement learning algorithms (Stanley et al., 2005) and evolutionary stage of the system shall select individuals best suited to the task during selection stage, which determines the whole population to improve its genetic material over time.

Evolutionary stage of the system consist three main processes:
– selection of the best individuals,
– reproduction (with cross-over and mutation sub-processes),
– replacement (offspring replaces worst individuals).

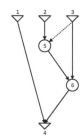

From	1	2	3	3	4	6
To	4	4	4	6	6	4
Weight	-0.4	0.02	-0.11	0.9	-1.0	0.75
Innov. num.	1	2	4	5	9	10
Disabled?	No	No	Yes	No	No	No

Figure 2. An example of encoded artificial neural network topology (phenotype) into a connection genome (genotype), due to NEAT method

In case of evolutionary method the genetic encoding of neural network topology is very important. Author of this paper implemented the modified NEAT method to this system (Figure 2). NEAT (NeuroEvolution of Augmenting Topologies) adjust the topology of ANN's with EA (Stanley and Risto, 2002a) gradually to given task, allows to obtain a set of ANN's that are best fitted to this task (Łącki, 2009).

Each node represents a neuron that produces a real value between 0 and 1 as a result of normalized weighted sum of its inputs. Normalization of weighted sum is performed with sigmoid function, as in Equation 1.

$$o_j = \frac{1}{1 + e^{-(S_j \beta + \theta_j)}} \tag{1}$$

where:
o_j – output value of an neuron,
S_j – weighted sum of input values x_{nj} with weights w_{nj},
β – slope coefficient,
θ_j – bias.

Adding the bias signal of constant value 1, allows to shift the output value of the activation function (Figure 3). Influence of bias may be adjusted through changing weight of this signal, when the mutation stage is performed in evolutionary process during creation of an offspring in the reproduction stage.

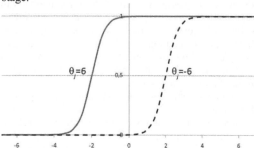

Figure 3. Influence of bias coefficient θ_j to value of sigmoid function

In this stage two best neural networks are chosen and its genetic material is crossed-over to create two new individuals. Cross-over of disparate topologies is processed in a meaningful way by pairing up genes with the same historical markings, called innovation numbers. With this approach the offspring may be formed in one of three ways:
– In uniform crossover, matching genes are randomly chosen for the offspring genome, with higher probability for better fitted parent.
– In blended crossover, the connection weights of matching genes are averaged.
– In elite crossover disjoints and excesses are taken from more fit parent only, all redundant genes

from less fit parent are discarded. All matching genes are averaged.

Genes that do not match with the range of the other parent's innovation numbers are called disjoints (when they occur within the genome) or excesses (when they occur outside of the genome).

These three types of crossover were found to be most effective in neuroevolutionary algorithms in comparison to other crossover methods (Stanley and Risto, 2002b).

Genes that have been disabled in previous generations have a small chance of being re-enabled during new offspring creation, allowing ANNs to make use of older solutions once again (Łącki, 2012).

Evolutionary neural network can keep historic trails of the origin of every gene in the population, allowing matching genes to be found and identified even in different genome structures. Old behaviors encoded in the pre-existing network structure have a chance to not to be destroyed and pass their properties through evolution to the new structures, thus provide an opportunity to elaborate on these original behaviors.

The number of inputs and outputs is fixed. During evolution, in mutation stage, the number of internal neurons and connections may change. In classic NEAT method the number of nodes and connections may only increase over time, with possibility to temporary disable the connection. This guaranties to transfer learning experience from ancestors to new offspring and fast learning of new tasks for new population but it may be disadvantageous in such dynamic environments as ship maneuvering in restricted waters. In this case an experience of old population may be insufficient and its learning ability to slow, due to size of experienced ANN's. Through mutation, the genomes in modified NEAT will gradually get larger for complex tasks and lower their size in simpler ones. Genomes of varying sizes will result, sometimes with different connections at the same positions.

Historical markings represented by innovation numbers allow neuroevolutionary algorithm to perform crossover operation without analyzing topologies. Genomes of different organizations and sizes stay compatible throughout evolution, and the variable-length genome problem is essentially solved. This procedure allows for used method to increase complexity of the structure while different networks still remain compatible.

During elite selection process the system eliminates the lowest performing members of every specialized group of individuals from the population. In the next step the offspring replaces eliminated worst individuals. Thus the quantity of the population remains the same while its quality shall improve according to assumed goals and restrictions of the task.

3 INPUTS AND OUTPUTS OF ANN'S

Input and output signals of ANN's must be determined at the beginning of designing phase of the system. Proper set of signals considered in the model is crucial for efficient performance of the system and for its fidelity and accuracy in comparison to the real navigational situation.

Input signals in the system, with three degrees of freedom vessel movement, are as follows:
- Ships' course over ground,
- Ships' angular velocity,
- Ships' speed over ground,
- Ships' position,
- Angle and velocity of a current,
- Angle and velocity of a wind.
- Main propeller revolutions (current and preset),
- Rudders' angle (current and preset).

In future research other signals from environment may be taken into account, i.e. waves, cargo, trim and roll.

Output signals of ANNs shall generate the values for important parameters that may change after certain amount of time. Most important signals in prediction process are:
- Ships' position,
- Ships' course over ground,
- Ships' speed over ground,
- Ships' angular velocity.

All of the input and output signals are encoded as real values between 0 and 1.

Computational flexibility and ability to adapt a network topology to a given task allows to design complex sets of inputs and outputs of ANN's. Since the neural network with multiple outputs learns slower than one with only one output, the proposal of the author is to divide a population of ANNs into different specialized groups of networks, designed to calculate a predicted value of a single particular output signal (Figure 4). This is a very sophisticated neuroevolution method that can deal with premature convergence that preserves diversity and gradual complexity of explored solutions.

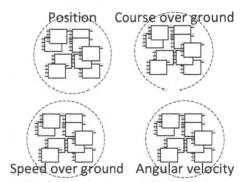

Figure 4. Division of population into specialized groups of ANNs

Each group has separate ranking list and individuals compete only within their own group. This approach requires much more memory allocation for higher amount of genetic material.

Performance of each individual is measured in predicted time interval and its fitness value is calculated as a difference between real and predicted value (Formula 2).

$$f_i = (r_n - p_i)^2 \leftarrow min \qquad (2)$$

where: f_i – fitness value of an individual i (a cost criteria), r_n – real value of a signal of n-group, p_i – predicted value of a signal of an individual i.

In this case the individuals with least fitness value are more likely to reproduce their genetic material in next generation.

General algorithm for prediction system is presented on figure 5.

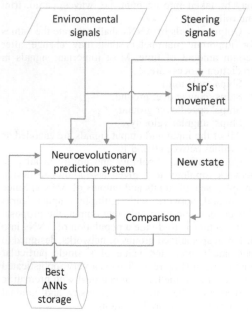

Figure 5. General algorithm of prediction system with neuroevolutionary method

Input signals have been divided into two groups – environmental and steering signals. Group of environmental signals consist all data incoming from vessels surroundings (i.e. winds and currents speed and direction) which creates an input state vector for the system.

Implementation of mathematical model of wind to the motion control in neuroevolutionary prediction system increases its performance and robustness in simulated environment.

Under pressure of wind force, depending of the ships' design (location of the superstructure, the deployment of on-board equipment and cargo, etc.)

she tends to deviate from the course, with the wind or into the wind. The smaller the speed and draft of the ship, the greater the influence of wind. Of course, the size of the side surface exposed to wind is essential to the ships movement.

When the ship moves forward the center of effort of the wind (wind point, WP) is generally close to amidships, away from pivot point (PP). This difference creates a substantial turning lever between PP and WP, thus making the ship, with the superstructure deployment at stern, to swing the bow into the wind.

For ship moving forward there are defined terms of relative wind speed V_{rw} and angle of attack γ_{rw} (Isherwood, 1973). Wind forces acting on symmetrical ship are in general calculated from data as ship's overall length, surfaces affected by the wind, air density and coefficients calculated from available characteristics of ships model, i.e. from wind loads data of Oil Companies International Marine Forum (OCIMF, 1977). This organization identifies safety and environmental issues facing oil tankers, barges, terminals and offshore marine operations, and develops and publishes recommended standards that will serve as technical benchmarks for regional and worldwide exploitation.

Additional forces that affect ships movement are water flows from current and tides. In this case the water moves in relation to the bottom of a river, sea or an ocean.

Position of the Moon and the Sun in relation to the Earth affects mostly waters in the oceans. Many naval ports of the world have their locations at a river estuaries. Currents of that rivers may be often affected by ocean tides. Tidal force is determined by the difference in forces at the Earth's center and surface. Tidal forces are mostly available from local authorities if a form of timetable sets and charts.

Tidal charts are grouped in sets, each set covers the time interval between two consecutive high tides in the area. These charts they give the average tidal flow information for desired time span for the ship maneuvering area. Information on a tidal chart consists of direction and average flow speed in knots for specified tides.

An alternative to the tidal charts is the tidal diamonds method, which may be found on most nautical charts. With this method one can relate certain points on the chart to a table that will calculate the direction and velocity of tidal flow and according to this flow the estimated position of the vessel can be calculated regarding the direction of the tide.

The gravitational tides are not the only one affecting flow of the water, there are also terrestrial tides, atmospheric tides and thermal tides. Atmospheric tides are caused by the radical effects of weather and solar thermal tides and in most cases they reinforce oceanic tides.

Under the pressure of a current a ship is drifting together with the water, relative to the ground and any fixed objects. When the ship is moving in current the speed over ground is resultant velocity of ship speed and velocity of the current.

Stronger water flows may cause the tendency to swing the ship with the stern or bow towards the side, which may lead to dangerous situation. This might be particularly difficult to overcome when working in following tide because of the small effectiveness of the rudder at low speed.

Steering signals consist data that may be changed by a user of the system (i.e. a navigator or a commander on the bridge). Steering signals include propellers revolutions or thrust and rudders angles.

All these input signals affect ship's movement which creates a new state of the environment with the moving vessel in it.

At the same time the similar new state parameters are being calculated in the neuroevolutionary system, regarding the same input signals. The real and predicted values are compared and the result of comparison provides substantial information for the system that allows to elaborate the quality of created ANN's and overall performance of whole population. During evaluation the ranking of ANNs it created and the best networks are stored for future exploitation.

4 SIMULATION MODELS AND RESULTS

For the purpose of a ship movement simulation an application has been created by the author (Figure 6).

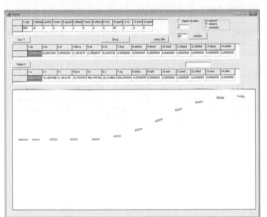

Figure 6. An application for testing behavior of simulation models of different vessels

The designed application allows to choose specific model of the vessel, to set a starting parameters of navigational situation in restricted waters, including speed and direction of a wind and a water current, and run a simulation with partially observable prediction values that can be saved to a file and analyzed after the simulation.

Two simulation models of ships with three-degrees-of-freedom had been used in the system for the purpose of systems performance test. Main parameters of ships has been compared in table 1.

Table 1. Main parameters of simulation ship models.

Name	Blue Lady	Cape Norman
Type	VLCC	Container ship
Scale	1:24	1:1
Length	13,78 [m]	175 [m]
Beam	2,38 [m]	26,5 [m]
Draft	0,86 [m]	14,2 [m]
Capacity/Tonnage	22,83 [T]	1504 [TEU]
Max. speed	3,1 [kn]	20,4 [kn]

The sets of output data of ANN's has been calculated and recorded during task evaluation in every generation as the results of simulation. The population consist four separate groups of ANNs, with 50 individuals each, resulting in 200 ANNs in total.

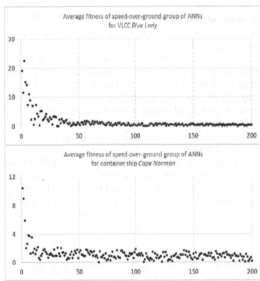

Figure 7. The examples of average fitness values of specialized groups of ANNs

The chart presents evaluation of the simulation for 200 generations (Figure 7). Since there is always only one optimal solution for every parameter in the prediction system, the algorithm tends to converge quickly to this solution. These simulation results prove a good performance of learning process of a single output neural network.

127

For two different models of the vessels the average fitness values become at stable levels after about 20-40 generations, which took about few minutes of simulation on a standard PC.

Spread of average fitness for speed over ground is greater for container ship, regarding its greater range of possible predicted speed values.

5 REMARKS

Intelligent maneuvering prediction system for maritime transport has some valuable benefits:

- increases of the safety of navigation in a restricted water area by improving the data analysis for decision-maker during maneuvers,
- improvement of the operation of ships in port, due to the increased bandwidth,
- reduction of operating costs of vessels,
- minimization of the occurrence of human errors,
- reduction of the harmful impact of transport on the environment.

It is important to notice that all these benefits strictly depend on proper adjustment of evolutionary parameters, the size of ANNs population and the encoding methods of signals considered in serviced environment.

Neuroevolutionary approach to ship handling in confined waters improves a quality of maneuvers and safety of navigation effectively (Łącki, 2008). For the simulation study, mathematical model of three-degrees-of-freedom maneuvering VLCC tank ship with the single-propeller and single-rudder was applied to test the prediction performance of the system. Artificial neural networks based on modified NEAT method increase complexity and performance of considered model of ship maneuvering in confined waters.

Implementation of additional input signals related to influence of wind and current allows to simulate complex behavior of the vessel (Łącki, 2012) in the environment with much larger state space than it was possible in a classic state machine learning algorithms (Łącki, 2007). Promising simulation results of maneuvers in variable wind and current conditions of different ship models encourage to further research of the neuroevolutionary methods which may finally be implemented into advanced navigational prediction systems to increase the safety of navigation.

REFERENCES

Bagnell, D. & Schneider, J. 2001. *Autonomous helicopter control using reinforcement learning policy search methods.*

Haasdijk, E., Rusu, A.A. & Eiben, A.E. 2010. *HyperNEAT for Locomotion Control in Modular Robots.*

Isherwood, J.W. 1973. Transactions of Royal Institution on Naval Architects, *Wind resistance of merchant ships*, vol. 115, 327–332.

Kenneth, S., Nate, K., Rini, S. & Risto, M. 2005. *Neuroevolution of an automobile crash warning system*, Washington DC, USA.

Larkin, D., Kinane, A. & O'Connor, N. 2006. *Towards hardware acceleration of neuroevolution for multimedia processing applications on mobile devices*, Hong Kong, China.

Lee, S., Yosinski, J., Glette, K., Lipson, H. & Clune J 2013. Applications of Evolutionary Computing, *Evolving gaits for physical robots with the HyperNEAT generative encoding: the benefits of simulation.*

Łącki, M. 2007. *Machine Learning Algorithms in Decision Making Support in Ship Handling.* , Katowice-Ustroń, WKŁ.

Łącki, M. 2008. *Neuroevolutionary approach towards ship handling.* , Katowice-Ustroń, WKŁ.

Łącki, M. 2009. *Ewolucyjne sieci NEAT w sterowaniu statkiem.* Inżynieria Wiedzy i Systemy Ekspertowe, Warszawa: Akademicka Oficyna Wydawnicza EXIT, pp. 535–544.

Łącki, M. 2012. TransNav - International Journal on Marine Navigation and Safety of Sea Transportation, *Neuroevolutionary Ship Handling System in a Windy Environment*, vol. 6.

Nowak, A., Praczyk, T. & Szymak, P. 2008. Zeszyty Naukowe Akademii Marynarki Wojennej, *Multi-agent system of autonomous underwater vehicles - preliminary report*, vol. 4, 99–108.

OCIMF 1977. *Prediction of Wind and Current Loads on VLCCs* , Oil Companies International Marine Forum.

Stanley, K.O. & Risto, M. 2002a. *Efficient evolution of neural network topologies.*

Stanley, K.O. & Risto, M. 2002b. *Efficient Reinforcement Learning Through Evolving Neural Network Topologies.*

Stanley, K.O., Bryant, B.D. & Risto, M. 2005. IEEE Transaction on Evolutionary Computing, *Real-time neuroevolution in the NERO video game*, vol. 9, 653–668.

Zhao, J., Price, W.G. & Wilson, P.A. 1995. *Automatic collision avoidance: towards the 21st century.*

Geoinformation Systems and Maritime Spatian Planning

Information and Communication Technologies in the Area with a Complex Spatial Structure

A. Kuśmińska-Fijałkowska & Z. Łukasik
University of Technology and Humanities in Radom, Poland

ABSTRACT: Typically, communication systems in land terminals are selected without a detailed analysis of the specificity of transmission within that object. As a result, due to noise and various constraints, this system usually does not work properly. In order to avoid this, the authors of this article proposed a DECT wireless telephone system. As a result, as the standard architecture of the system was analyzed, attention was paid to the parameters that should be considered when choosing this type of systems in order to provide continuous connectivity in a complex spatial structure.

1 INTRODUCTION

Reloading terminals in transport chains provide space, equipment and environment ready to translocate the units of intermodal transport between various transport connections. (Brill J., Łukasik Z. 2014; Kozyra J. 2009; Kisilowski J., Krzyszkowski A. 2014). Therefore, there is a need to implement a system of *Information and Communication Technologies* (ICT) which will enable to increase effectiveness of functioning of a reloading terminal in the chains of intermodal transport.

Applied systems of a communication in the reloading terminals are chosen without detailed analysis, specificity of transmission within such structure.(Craven P., Wong R., Fedora N. 2013; Eugen R., Serban R., Augustin R. M. 2014; Zhu Bo O., Chen K., Jia N. 2014) They are usually short-wave transmitters. As a result, due to interferences and various limitations, this system usually doesn't work properly. (Van Vorst D. G.; Yedlin Matthew J.; Virieux J. 2014) As a result of analysis of available systems, a DECT system can best fulfil given requirements in the areas with a complex spatial structure, in which parameters relevant for a specific reloading terminal were taken into consideration. (Kuśmińska-Fijałkowska A. Łukasik Z. 2009; Li T., Wang X., Zheng Ch. 2014)We should consider many aspects concerning a DECT standard, which is proposed by the authors. Firstly, standard structure of a DECT system must be precisely analysed.

2 DIGITAL ENHANCED CORDLESS TELECOMMUNICATION DECT

System of digital enhanced cordless telecommunication DECT was designed to integrate different technologies of data transfer. With an application of appropriate devices, DECT systems enable integration with networks GSM, WLAN, ISDN and X.25 (Tuttlebee W.H.W. 1996). Apart from standard voice transmission, data transmission between various subscribers with a binary flowability to 384kbi/s is also possible. (Jackowski S. 2005). Therefore, this system can be applied in the areas with a complex spatial structure – land reloading terminal.

System DECT consists of FP *(Fixed Part)* and some portable parts *(PP)* performing a role of transmitters and receivers of information. FP consists of one central element CCFP *(Central Control Fixed Part)* performing control functions, for example, commutations and a certain number of the so-called RFPs *(Radio Fixed Part)* grouped in the so-called clusters. Each RFP is equipped with groups of radio transceivers, using one antennal system creates an individual cell. (Jackowski S. 2005). PP connect with RFP. (Fig. 1) Two or more RFPs is connected with CCFP. Access to external networks is realized through Interworking Unit (IWU). DB *(Data Base)* cooperates with CCFP in order to ensure management of connections of system users (Kuśmińska-Fijałkowska A., Łukasik Z. 2013)

DECT system is mainly used to ensure the so-called basic connections serving to sound transmission *(basic connections)* and connections serving to data transmissions, faxes etc. *(advanced connections)*. It is possible thanks to cooperation of the groups of protocols defined as a DECT protocol. (Wesołowski K. 2003).

This protocol is divided into 4 layers:
– PHL (Physical Layer),
– MAC (Medium Access Control). This layer is internally divided into 2 sublayers: CCF (Cluster Control Functions) and CSF (Cell Site Functions),
– DLC (Data Link Control),
– NWK (Network Layer).

Figure 2 presents a comparison of layers of DECT protocols with a model ISO OSI. System DECT is also described as the so-called access system, that is system enabling cordless access to services offered by other systems, for example, ISDN (Fig. 1)

Therefore, a set of protocols is usually extended by an element called IWU *(Inter Working Unit)*, of which task is to adjust services offered by NWK to the needs of an appropriate layer of a system, to which access is enabled. An example of cooperation of protocols with ISDN network is presented on Figure 3.

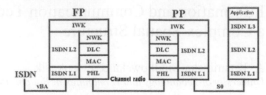

Figure 3. Cooperation protocols ISDN DECT

Coverage of RFP according to DECT specification is maximum 500m in an open area, although most of producers say that maximum distance is 300m. When we use telephone inside a building, coverage usually doesn't exceed 50m. Apart from good quality of sound, DECT systems also offer relatively high possibility of extension. Standard RFP can serve to 12 mobile devices, while one mobile can be registered in maximum 10 RFPs. It means that when we use a telephone in an open area, maximum 6 users switched to one telephone line can speak in the area of maximum 2.5 square kilometres. Using specialist subscriber exchanges, DECT systems can serve more users. Some sources even say that maximum number of users of DECT systems can even reach 10 000 per square kilometre. Apart from connections to a fixed network, RFPs also enable connections with mobile parts or between these parts. Starting from 1997, a possibility of extension of DECT systems increased even more. GAP *(Generic Access Profile)* - standard created by ETSI enables extension of DECT systems also with the use of devices from various producers. Buying another receiver to RFP, we can be sure that it will work properly.

In case of DECT technology, band 1880-1900 MHz divided into 120 channels is used to transmission. It allows to improve considerably the quality of sound and reduce interferences of signals coming from RFPs of various subscribers or various receivers. When there are interferences of signals, channel of transmission can be changed. For some time, information is transmitted on two channels, then electronic systems of RFP check, which channel offers better quality and choose it to further data transmission. The situation is similar in case of translocation of users within a coverage of various RFPs. When user comes closer to RFP with a stronger signal, then (provided that receiver has appropriate access authorization) transmission taken over by a station with a stronger signal. User does not see any change. However, we should add that not all DECT systems offer a possibility of change of RFP during conversation. Most of house telephones do not have such a function.

DECT system divides radio channel into three dimensions: spatial, frequency and time. (Wesołowski K. 2003).
– Spatial division is a limited coverage of propagation of radio-waves around RFPs, thus

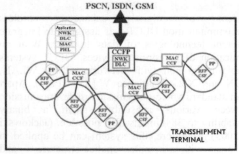

Figure 1. Realization of DECT system in the area with a complex spatial structure – land transshipment terminal. Layers of protocol realized in blocks are marked. CCFP –central control fixed part, RFP – Radio Fixed Part

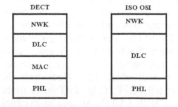

Figure 2. Comparison of DECT protocol and ISO OSI model

132

creating cells, in which a given RFP is heard by PP. Standard allows two levels of power of transmitters - 250 mW and 2,5 mW.

– Frequency division is realized by applying a dozen or so carrier frequencies. Basic standard uses 10 carrier frequencies distant from each other by 1,728 MHz within a coverage of 1880 – 1900 MHz with frequencies:

$$f_c = 1897,344MHz - c \cdot 1,728MHz \qquad (1)$$

where c is a number of carrier with a coverage $0,1,2....9$. Standard allows to increase a number of carriers by the so-called *extended carriers* with a frequency:

$$f_c = f_9 + c \cdot 1,728MHz \qquad (2)$$

while $c>9$. Inaccuracy of carrier frequency used by RFP shouldn't exceed 50kHz. Carriers are modulated with the use of modulation GFSK *(Gaussian Frequency Shift Keying)* with a nominal deviation of frequencies 288kHz. Binary flowability of integrated channel in one band is 1152kbit/s.

In order to get time division, TDMA technique is applied. One frame lasts 10ms and consists of 11520 bits. Frame is divided into 24 time slots *(full slot)* In first 12 time slots, FP transmission usually occurs (Fig. 4). The last 12 slots of frame are designated to PP transmission. These slots are paired, that is, for example, when FP conducts a transmission in slot 3, PP conducts transmission in a slot 15, with whom FP realizes this connection. Corresponding to each other pairs of time slots in a given frequency are called physical channels. (Jackowski S. 2005).

Time slots, in the extended connections, can be (Fig. 5, Fig. 6) paired (double slot) or divided further into two parts *(half slot)*. (Wesołowski K. 2003).

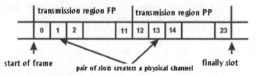

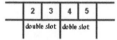

Figure 4. DECT frame structure

Figure 5. Time slots can be combined two by two, creating gaps double

Figure 6. Slots can be divided into two half-slots (half slot)

In the first case, number of physical channels will be two times smaller, whereas, in the second case

two times higher. It enables to increase binary flowability two times to a given subscriber, or decrease basic flowability two times.

Inside time slots, radio terminals can transmit data packets. Standard distinguishes 4 sizes of data packets: short, normal, small capacity and large capacity. (Wesołowski K. 2003). Their features are the following:

– Short packet, 96 bits in length, is used only by RFP

to spread with the use of the so-called dummy bearers *(beacokn)*.

– Full packet, 424 bits in length, is used in the basic connections.

– Large capacity packet *(double packet)*, 904 bits in length, is used in the extended connections. This packet occupies double time slot.

– Small capacity packet, 184 bits in length, occupies half slot, can be applied in the extended connections.

Apart from data, each packet must contain synchronization bits (different for FP and PP), field Z (except for short packet). Packets with other slots are separated by protective field, in which there is no transmission (Fig. 6). Field Z, which is a repetition of last bits of packet, enables to detect and counteract of the so-called *slipping collisions*, that is, a situation when as a result of lack of synchronization of transmission in different slots, it starts overlapping. Protective field provides space between packets transmitted by different radio terminals. (EN 300 175-1. 1999).

In case of DECT system, it is also possible to obtain channels with an increased flowability. In a borderline case, using all slots to data transfer from a one subscriber, we can obtain effective flowability $12 \times 32 = 384$ kbit/s.

Power of transmitter is approximately 10mW. Total number of channels in a DECT system is 120, therefore, with such a low power of transmitter, system can serve high telecommunications traffic of 10000 erlangs/km^2/level. (Jackowski S. 2005). Block arrangement of PP of a system is presented on Figure 7.

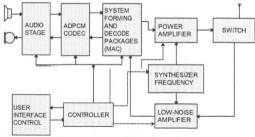

Figure 7. Block arrangement of PP of DECT system

3 CONCLUSIONS

ICT System implemented in the areas with a complex spatial structure, which is a reloading terminal should meet the following conditions:

1 To provide constant communications in the area of terminal dislocation;
2 To enable realization of cumulative digital communications of the short packets of information from any object that can be found in the area of terminal;
3 Selective bidirectional contact with external users should be possible, both communicative and in a form of short cables, from any external system;
4 Information exchanged in a system of communications should be protected from access to it by the unauthorized people.

The analysis, in which parameters relevant for a reloading terminal were taken into consideration, proved that proposed system of cordless telecommunication DECT can fulfil given requirements the best. Many aspects concerning proposed standard should be considered parameters, telecommunications traffic, specific area for propagation of electromagnetic wave, quality of radio transmission between RFP and PP.

REFERENCES

Brill J., Łukasik Z.: „Economic, technical and strategic intermodal transport" Rail Transport Technology magazine 3/2014, str. 12-24, ISSN 1232-3829

Craven P., Wong R., Fedora N.; et al. Book Group Author(s): ION Studying the Effects of Interference on GNSS Signals Craven, Paul; Wong, Ronald; Fedora, Neal; et al. Conference: International Technical Meeting of the Institute-of-Navigation Location: San Diego, CA Date: JAN 27-29, 2013 Sponsor(s): Inst Nav Proceedings of the 2013 International Technical Meeting of the Institute of Navigation Pages: 893-900 Published: 2013

EN 300 175-1, Digital Enhanced Cordless Telecommunications (DECT); Common Interface (CI); part 1 Overview, V.1 4.2., June 1999.

EN 300 175-1, Digital Enhanced Cordless Telecommunications (DECT); Common Interface (CI); part 3 Medium Access Control (MAC) Layer, V.1 4.2., June 1999.

Eugen, Rosca; Serban, Raicu; Augustin, Rosca Mircea; et al. Transshipment modelingand simulation of of container port terminals Edited by: Carausu C; Cohal V; Doroftei I; et al. Conference: ModTech International Conference - Modern Technologies in Industrial Engineering Location: Profess Assoc Modern Mfg Technol, Sinaia, Romania Date: JUN 27-29, 2013 Sponsor(s): Silesian Univ Technol; Maritime Univ Constanta Modern Technologies in Industrial Engineering Book Series: Advanced Materials Research Volume: 837 Pages: 786-791 Published: 2014

Jackowski S.: „Telecommunication", cz. 2. Technical University of Radom 2005

Li T., Wang X., Zheng Ch. et al.Investigation on the Placement Effect of UHF Sensor and Propagation Characteristics of PD-induced Electromagnetic Wave in GIS Based on FDTD Method IEEE Transactions on Dielectrics and Electrical Insulation Volume: 21 Issue: 3 Pages: 1015-1025 Published: JUN 2014

Kisilowski J., Krzyszkowski A.: „A fast train air transport" Logistics 3/2014

Kuśmińska-Fijałkowska A. Łukasik Z.: „The transmission of the information of the system of telecommunicational DECT in the trans-shipping terminal". 8 International Navigational Symposium TransNav'09, Gdynia 2009, Poland. Monograph „Marine Navigation and Safety of Sea Transportation", p. 317-323, ISBN 978-0-415-80479-0

Kuśmińska-Fijałkowska A., Łukasik Z.: "Management of a transshipment terminal supported by ICT systems", 13th International Conference „Transport Systems Telematics TST'13", Conference Proceedings, October 23-26, Katowice-Ustroń, Poland 2013, ISBN 978-83-927504-5-1, pp. 60 PUBLIKACJA International Scientific Journal: Archives of Transport System Telematics, Volume 6, Issue 4, November 2013, ISSN 1899-8208, pp 17-20

Kozyra J.: „The use biofuels in transport". Logistyka Nr 3/2009.

Tuttlebee W.H.W.: Cordless telephones and cellular radio: Synergies of DECT and GSM (Reprinted from Cordless Telecommunications Worldwide) Electronics & Communication Engineering Journal, Volume: 8 Issue: 5, October 1996, ISSN: 0954-0695, pp 213-223

Van Vorst, D. G., Yedlin Ma. J., Virieux J; et al. Three-dimensional to two-dimensional data conversion for electromagnetic wave propagation using an acoustic transfer function: application to cross-hole GPR data. Geophysical Journal International, Volume: 198, Issue: 1, July 2014, ISSN 0956-540X, pp 474-483

Wesołowski K.: Mobile radio systems, 2003.

Zhu Bo O., Chen K., Jia, N. et al. Dynamic control of electromagnetic wave propagation with the equivalent principle inspired tunable metasurface Scientific Reports Volume: 4 Article Number: 4971 Published: MAY 15 2014

Establishing a Framework for Maritime Spatial Planning in Europe

A. Kuśmińska-Fijałkowska & Z. Łukasik
University of Technology and Humanities in Radom, Poland

ABSTRACT: In this paper authors present a European framework for planning for EU countries that have access to the sea. Maritime spatial planning will contribute to the effective management of marine activities and the sustainable use of marine and coastal resources, by creating a framework. Through maritime spatial plans, Member States can reduce the administrative burden and costs in support of their action to implement other relevant Union legislation. The Member States of the European Union are required to develop plans for zoning marine areas, until 31 March 2021.

1 INTRODUCTION

The high and rapidly increasing demand for maritime space for different purposes, such as installations for the production of energy from renewable sources, oil and gas exploration and exploitation, maritime shipping and fishing activities, ecosystem conservation, the extraction of raw materials, tourism, require an integrated planning and management approach.

In order to promote the sustainable coexistence of uses and, where applicable, the appropriate apportionment of relevant uses in the maritime space, UE (Directive 2014/89/UE 2014) introduced a framework should be put in place that consists of the establishment and implementation by Member States of maritime spatial planning, resulting in plans.

Maritime spatial planning will contribute to the effective management of marine activities and the sustainable use of marine and coastal resources, by creating a framework for consistent, transparent, sustainable and evidence- based decision-making. In order to achieve its objectives, this Directive (2014/89/UE 2014) should lay down obligations to establish a maritime planning process, resulting in a maritime spatial plans; such a planning process should take into account land-sea interactions and promote cooperation among Member States. Without prejudice to the existing Union acquis in the areas of energy, transport, fisheries and the environment, this Directive should not impose any other new obligations, notably in relation to the concrete choices of the Member States about how to pursue the sectored policies in those areas, but should rather aim to contribute to those policies through the planning process.

Maritime spatial planning will contribute, inter alia, to achieving the aims of Directive 2009/28/EC of the European Parliament and of the Council, Council Regulation (EC) No 2371/2002, Directive 2009/147/EC of the European Parliament and of the Council, Council Directive 92/43/EEC, Decision No 884/2004/EC of the European Parliament and of the Council, Directive 2000/60/EC of the European Parliament and of the Council, Directive 2008/56/EC, recalling the Commission communication of 3 May 2011 entitled 'Our life insurance, our natural capital: an EU biodiversity strategy to 2020', the Commission communication of 20 September 2011 entitled 'Roadmap to a Resource Efficient Europe', the Commission communication of 16 April 2013 entitled 'An EU Strategy on Adaptation to Climate Change' and the Commission communication of 21 January 2009 entitled 'Strategic goals and recommendations for the EU's maritime transport policy until 2018', as well as, where appropriate, those of the Union's Regional Policy, including the sea-basin and macro-regional strategies. (Directive 2014/89/UE 2014).

2 MARITIME SPATIAL PLANNING

Within the Integrated Maritime Policy of the Union, that framework provides for the establishment and implementation by Member States of maritime spatial planning, with the aim of contributing to the objectives specified in point 5, taking into account land-sea interactions and enhanced cross-border cooperation, in accordance with relevant Unclos provisions.

In the framework of the European Spatial Planning definitions used (Directive 2014/89/UE 2014).

'**Integrated Maritime Policy**' (IMP) means a Union policy whose aim is to foster coordinated and coherent decision- making to maximize the sustainable development, economic growth and social cohesion of Member States, and notably the coastal, insular and outermost regions in the Union, as well as maritime sectors, through coherent maritime-related policies and relevant international cooperation;

'**maritime spatial planning**' means a process by which the relevant Member State's authorities analyze and organize human activities in marine areas to achieve ecological, economic and social objectives;

'**marine region**' means the marine region referred to in Article 4 of Directive 2008/56/EC;

'**marine waters**' means the waters, the seabed and subsoil as defined in point (1)(a) of Article 3 of Directive 2008/56/EC and coastal waters as defined in point 7 of Article 2 of Directive 2000/60/EC and their seabed and their subsoil.

3 OBJECTIVES OF MARINE SPATIAL PLANNING

Europe's maritime spatial planning shall aim to contribute to the objectives (Directive 2014/89/UE 2014):
1 When establishing and implementing maritime spatial planning, Member States shall consider economic, social and environmental aspects to support sustainable development and growth in the maritime sector, applying an ecosystem-based approach, and to promote the coexistence of relevant activities and uses.
2 Through their maritime spatial plans, Member States shall aim to contribute to the sustainable development of energy sectors at sea, of maritime transport, and of the fisheries and aquaculture sectors, and to the preservation, protection and improvement of the environment, including resilience to climate change impacts. In addition, Member States may pursue other objectives such as the promotion of sustainable tourism and the sustainable extraction of raw materials.
3 Directive (2014/89/UE) is without prejudice to the competence of Member States to determine how the different objectives are reflected and weighted in their maritime spatial plan or plans.

4 MINIMUM REQUIREMENTS FOR MARITIME SPATIAL PLANNING

Member States shall establish procedural steps to contribute to the objectives listed in point 2, taking into account relevant activities and uses in marine waters.

Member States:
– take into account land-sea interactions;
– take into account environmental, economic and social aspects, as well as safety aspects; (Carcamo Francisco, Garay-Fluehmann P, Squeo Rosa, A Francisco., et al. 2014; Gonzalez-Mirelis, Genoveva; Lindegarth, Mats; Skold, Mattias 2014; Michelle V., William G., Heather G., 2012).
– aim to promote coherence between maritime spatial planning and the resulting plan or plans and other processes, such as integrated coastal management or equivalent formal or informal practices;
– ensure the involvement of stakeholders in accordance with Article 9 (Directive 2014/89/UE 2014);
– organize the use of the best available data in accordance with Article 10 (Directive 2014/89/UE 2014);
– ensure trans-boundary cooperation between Member States in accordance with Article 11 (Directive 2014/89/UE 2014);
– promote cooperation with third countries in accordance with Article 12 (Directive 2014/89/UE 2014).

Maritime spatial plans shall be reviewed by Member States as decided by them but at least every ten years.

5 MARITIME SPATIAL PLANNING RP

When establishing and implementing maritime spatial planning, Member States shall set up maritime spatial plans which identify the spatial and temporal distribution of relevant existing and future activities and uses in their marine waters, in order to contribute to the objectives set out in point 2.

Member States shall take into consideration relevant interactions of activities and uses. Without prejudice to Member States' competences, possible activities and uses and interests may include:
– aquaculture areas,
– fishing areas,
– installations and infrastructures for the exploration, exploitation and extraction of oil, of gas and other energy resources, of minerals and aggregates, and for the production of energy from renewable sources,
– maritime transport routes and traffic flows,
– military training areas,

- nature and species conservation sites and protected areas,
- raw material extraction areas,
- scientific research,
- submarine cable and pipeline routes,
- tourism,
- underwater cultural heritage.

The main purpose of maritime spatial planning is to promote sustainable development and to identify the utilization of maritime space for different sea uses as well as to manage spatial uses and conflicts in marine areas. Maritime spatial planning also aims at identifying and encouraging multi-purpose uses, in accordance with the relevant national policies and legislation.

In order to achieve that purpose, Member States need at least to ensure that the planning processes result in a comprehensive planning identifying the different uses of maritime space and taking into consideration long-term changes due to climate change. Figure 1.

Phases of the project plan, zoning Polish marine areas.

PHASE I - study of the conditions to the plan.

Development of spatial study of the conditions of Polish sea areas with spatial analysis (03/07/2014 agreement Maritime Institute in Gdansk).
- As part of Phase I is scheduled:
- to collect data on marine waters;
- to collect information about how the development of the coastal strip in order to ensure consistency between sea and land plan;
- collection and analysis of international and national legislation and policy documents affecting the spatial aspect of the use of the sea;
- for the collection and analysis of project results and available research work;
- to gather information about the planned method of management of marine areas;
- to perform spatial analysis.

Deadline for completion of the first stage is 07. 12. 2014

PHASE II- develop a strategic plan

As part of Phase II is planned to develop a plan and forecasts as well as public consultation.

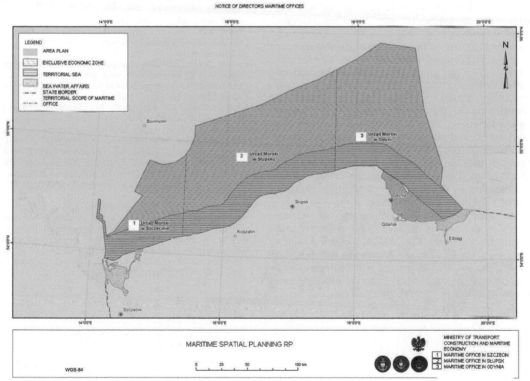

Figure 1. The Maritime spatial planning RP [11]

PHASE III - Transboundary consultations (planned completion date 2016-2017)
NEXT PHASE

Complementary research resulting from the indications contained in the strategic plan;
Develop plans for the Lagoons, port areas and waters requiring detailed plans identified in the strategic plan (scheduled for completion in 2021)

Problems encountered in drafting zoning plan Polish marine areas:
- lack of expected changes in the law on maritime areas of the Republic of Polish and Maritime Administration (lack of procedures to develop a project plan);
- lack of funds (ongoing work on the multi-annual program, which provides funding for the years 2015 to 2021) (Stelmaszyk-Świerczyńska A, Mostowiec A. 2014).

6 CONCLUSIONS

Through maritime spatial plans, Member States can reduce the administrative burden and costs in support of their action to implement other relevant Union legislation. The timelines for maritime spatial plans should therefore, where possible, be coherent with the timetables set out in other relevant legislation, especially: Directive 2009/28/EC, which requires the share of energy from renewable sources in gross final consumption of energy in 2020 to be at least 20 % and which identifies coordination of authorization, certification and planning procedures, including spatial planning, as an important contribution to the achievement of the Union's targets for energy from renewable sources; Directive 2008/56/EC and point 6 of Part A of the Annex to Decision 2010/477/EU, which require Member States to take the necessary measures to achieve or maintain good environmental status in the marine environment by 2020 and which identify maritime spatial planning as a tool to support the ecosystem-based approach to the management of human activities in order to achieve good environmental status; Decision No 884/2004/EC, which requires that the trans-European transport network be established by 2020 by means of the integration of Europe's land, sea and air transport infrastructure networks. (Directive 2014/89/UE 2014).

Member States are required to develop plans for zoning marine areas until 31 March 2021.

REFERENCES

Carcamo F., Garay-Fluehmann P, Squeo R.,.; "Using stakeholders' perspective of ecosystem services and biodiversity features to plan a marine protected area" Environmental Science & Policy, Volume: 40, Pages: 116-131 Published: Jun 2014

Directive 2014/89/EU of the European Parliament and of the Council of 23 July 2014 establishing a framework for maritime spatial planning

Directive 2008/56/EC of the European Parliament and of the Council of 17 June 2008 establishing a framework for Community action in the field of marine environmental policy (Marine Strategy Framework Directive) (OJ L 164, 25.6.2008, p. 19).

Directive 2009/28/EC of the European Parliament and of the Council of 23 April 2009 on the promotion of the use of energy from renewable sources and amending and subsequently repealing Directives 2001/77/EC and 2003/30/EC (OJ L 140, 5.6.2009, p. 16).

Directive 2009/147/EC of the European Parliament and of the Council of 30 November 2009 on the conservation of wild birds (OJ L 20, 26.1.2010, p. 7).

Council Directive 92/43/EEC of 21 May 1992 on the conservation of natural habitats and of wild fauna and flora (OJ L 206, 22.7.1992, p. 7).

Decision No 884/2004/EC of the European Parliament and of the Council of 29 April 2004 amending Decision No 1692/96/EC on Community guidelines for the development of the trans-European transport network (OJ L 167, 30.4.2004, p. 1).

Directive 2000/60/EC of the European Parliament and of the Council of 23 October 2000 establishing a framework for Community action in the field of water policy (OJ L 327, 22.12.2000, p. 1).

Gonzalez-Mirelis, G.; Lindegarth, M.; Skold, M. "Using Vessel Monitoring System Data to Improve Systematic Conservation Planning of a Multiple-Use Marine Protected Area, the Kosterhavet National Park (Sweden)" AMBIO Volume: 43 Issue: 2 Pages: 162-174 Published: MAR 2014

Council Regulation (EC) No 2371/2002 of 20 December 2002 on the conservation and sustainable exploitation of fisheries resources under the Common Fisheries Policy (OJ L 358, 31.12.2002, p. 59).

Stelmaszyk-Świerczyńska A, Mostowiec A. "Zoning plans Polish maritime areas-problems and challenges Maritime Offices". Maritime Office in Gdynia. Jastarnia 04/09/2014

Voyer M., Gladstone W.., Goodall H..: "Methods of social assessment in Marine Protected Area planning: Is public participation enough?" Marine Policy, Volume: 36 Issue: 2, Pages: 432-439 Published: Mar 2012.

Application of Intelligent Geoinformation Systems for Integrated Safety Assessment of Marine Activities

V.V. Popovich, O.V. Smirnova, M.V. Tsvetkov & R.P. Sorokin
SPIIRAS Hi Tech Research and Development Office Ltd, St. Petersburg, Russia

ABSTRACT: The intelligent geoinformation systems are essential for successful analysis of the complicated situations connected with the safety of marine activities. Currently, different dangerous unforeseen situations can develop in offshore zone, such as ship collisions, ecological catastrophes, terrorism, piracy and others. Existing offshore monitoring systems do not allow us to solve problem related to integrated safety and situation assessment and also their modernization will require significant costs. Thus, development of new technologies and tools in intelligent geoinformation systems will allow to operationally estimate current dangerous situations and to predict their progress. This paper considers tools and technologies, intellectual data processing capabilities and complicated situation assessment.

1 INTRODUCTION

Developing of maritime situation monitoring systems is without doubt relevant issue. Its actuality first of all is dictated by growing number of personal vessels, increasing complexity of vessel's navigation due to large variety of installed navigational aids, poor quality of special training. Moreover, increases the rate of various dangerous situations at sea such as ship collision, ecological disasters, natural hazards (storm, snow fall, fog etc.), piracy, terrorism etc.

Many scientific researches, technological developments for building effective maritime situation monitoring systems are oriented on the increase of navigation safety. However, most of these systems possess a limited functionality and low level of data analysis automatization. Global and local maritime situation monitoring systems provide data collection (from radar, AIS and other sources), data preprocessing, complex maritime route planning, monitoring and analysis of current maritime situation in navigation area.

To increase the quality of visualization and processing of maritime situation data, modern monitoring systems are being developed based on GIS-technologies.

GIS-technologies support:
– precise binding, systematization, selection and integration of all incoming and stored data (interconnect address space);
– composite nature and illustration of data representation for decision making;
– capability to dynamically model processes and effects;
– capability to automatically solve tasks, associated with location areas' peculiarities analysis;
– capability for operational analysis in case of dangerous situation arising.

However, application of traditional GIS (ArcGIS, MapInfo) for solving such tasks is highly laborious and costly. The user of such GIS has to analyse and process data about unfolding maritime situation on his own and make decisions about further actions relying on his intuition and experience.

In this paper we present new approach for development and building maritime situation monitoring systems and estimating safety of maritime activities. Main factor of this approach is application of intelligent GIS, that allows to solve tasks, related to automatized control over development of complex situations, interpretation of analysis results and intelligent identification of location and nature of activities of maritime objects.

2 RELATED WORKS

In paper (Claramunt et al. 2007) the issue of application of GIS-technologies in maritime situation monitoring systems is closely examined, capabilities of modern GIS for decision making support of operator is analysed. Also, in the paper peculiarities of collecting data from sensors (AIS, ECDIS) are discussed and distributed architecture of building is maritime situation monitoring systems, operating in real-time mode, is suggested.

In paper (Tetreault 2010) the issue of utilization of AIS in perspective navigation control systems is considered, perspective directions of application of such sensors in coast surveillance systems are disclosed.

In paper (Goralski, Gold, 2007) is given example of safe navigation system development based on GIS. Some attention is given to the question of navigation monitoring in vessel's location area, aspects of system functioning in real-time mode are regarded. It is necessary to note that given work is dedicated to the problem of shipborne system development. Integration of solution suggested in (Goralski&Gold 2007) with external intelligent navigation system is presented as one of the new lines of research.

Paper (Baylon&Santos 2013) describes capabilities of modern GIS for 'intelligent decision making' in information systems associated with safety of maritime activities. Authors pay close attention to perspectives of cooperation of different organizations that possess important information about various fields of maritime activities. Also, aspects related to training personnel to operate monitoring and safe navigation systems are considered in this work.

Papers (Popovich, Potapichev et al. 2006, Popovich, Claramunt et al. 2009) deal with concept of intelligent GIS and questions concerning application of IGIS for solving safe navigation problems. In work (Popovich, Claramunt et al. 2009) formalized concept of tactical situation, associated with vessels' location in sea area in given along with general architecture of IGIS that allows to solve vast spectrum of issues, related to safe navigation.

In full paper (Popovich 2013) questions about development of intelligent GIS for maritime situation monitoring are studied in detail. Also, this work focuses on issues of integration, harmonization and fusion of data, obtained from various sources (sensors). As a basis for building IGIS ontology approach is offered.

3 MARITIME SITUATION ASSESSMENT

Estimation and detection of dangerous situations is one of the key problems of maritime activities monitoring and decision support. The term maritime situation denotes a combination of some parameters, directly or indirectly defining observable system's state in given point of time. Maritime situation assessment represents analysis of certain characteristics for obtaining conclusion about current state of system and its possible state in near future (Blasch 2002). The term situation management, in turn, denotes purposeful influence on system aiming at changing the situation to our benefit. Stated influence can be accomplished by executing certain actions, oriented on changing system's properties, characterized by detected features.

Maritime situation assessment suggests:
- detection of factors (features) that influence development of dangerous situations in maritime activities location;
- visualized mathematical (computer) modelling of current situation in real-time mode;
- prediction of further development of dangerous situations and their analysis;
- generation and execution of decisions regarding further actions concerning arising dangerous situations;

Solving these tasks must be constructed on mathematical analysis methods, efficient numerical methods and algorithms. We believe that efficient tool for solving such tasks can be intelligent GIS (IGIS).

4 IGIS FOR ESTIMATION OF MARITIME ACTIVITIES' SAFETY

The term intelligent GIS is defined here as GIS that includes integrated tools and/or systems of artificial intelligence (AI) (Popovich 2003).

Figure 1 illustrates IGIS architecture. The central part of the IGIS is knowledge base including ontology. Ontology presents the 'framework' for representation concepts and relations between them in subject domain. Another part of knowledge base is storage of subject domain real object instances.

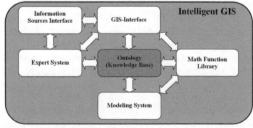

Figure 1. Intelligent GIS architecture.

Various active and passive sensors (AIS, radar) as well as information systems of global and local level can act as sources of information for maritime navigation system. While obtaining data from different sources the following problems may arise. First problem is duplicated information i.e. obtained information from different sources about same objects. However, similar data about same object is not necessarily a disadvantage since it increases credibility of data and consequently increases quality of specific decisions. One of the means for solving duplication problem and increasing quality of data obtained from various sources is application of concept of integration, harmonization and fusion of data which is described in (Popovich 2013).

Second problem is the issue of discrepancy between concepts in different systems. To format initial data to one standard we have developed a unified information interoperability model on basis of ontology database. Information interoperability model includes three ontology levels: domain ontology, geographical ontology and upper ontology.

GIS-interface is a program component for visual representation of geo-spatial data in various digital formats and of objects stored in knowledge base (Fig. 2). It combines different sources of geo-spatial data and program components that execute data processing using traditional methods.

GIS-interface allows:

- to update and display data in real-time mode along with processed results, prognosed and modelled data;
- to display all infrastructure of observed area of maritime activities;
- to set combinations of algorithms for execution of all stages of dangerous situation modelling (verification, interpolation, prediction);

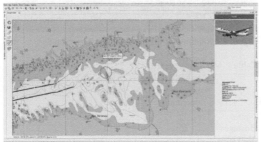

Figure 2. GIS Interface.

Library of mathematical functions is one of the important parts of intelligent GIS. Set of functions has to be open for access by any subsystem of IGIS, support changeability and expansion.

For example, for modelling spatial processes associated with dangerous situation development in maritime activities location following functions from library can be applied:

- mathematical model of different dangerous situations (e.g. oil spill, seizure of vessel by pirates, terrorist and etc.);
- vessel navigation on given route;
- search in location of rescue operation and etc.

To increase the quality of specific decisions made by user it is necessary to include prognostic models in mathematical functions library, i.e. such models, that allow to obtain estimation of dangerous situation development in future instances of time, based on data obtained to the current point of time.

It should be noted that any functions from mathematical functions library can be used in creating production rules for expert system.

The term expert system (ES) denotes a system that utilizes expert knowledge (knowledge form specialists) to provide qualified solution to tasks in given field of study (Popovich 2013).

Such systems are able to present knowledge, to clarify (examine) their processes-reasoning and are intended for fields of study where a person can reach professional level only after years of special education and training.

Expert knowledge is represented in form of condition-action rules and is saved in IGIS knowledge database. Let us note that any of mathematical functions given in library can be used in creating rules for IGIS expert system.

Expert system allows to execute integrated assessment of maritime activities' safety and suggest further actions to user in case of dangerous situation arising. For example, if system's user views successive triggering of rules: "Dangerous closing-in of vessels 1 and 2" and "Vessel 2 has left channel", evident is situation: "Vessel 2 has violated regime of navigation".

IGIS modelling system is intended for computer modelling of various spatial processes and also for visual generation of corresponding scenarios of processes development based on expert systems technology and represented as ontology database. Visual representation of the modelling allows the user to effectively estimate occurring process.

Modelling system allows us to solve the following tasks:

1 Building models of complex spatial processes based on their description in form of visual scenarios that are represented as two-dimensional digraph (block-scheme) where nodes are separate scenario tasks and decision making points in which scenario branches on various execution routes depending on satisfaction of specified conditions (Fig. 3). Scenario tasks can be executed both sequentially and in parallel, depending on block-scheme. For merging parallel branches special nodes "connectors" are used. Scenario tasks consist of individual atomic actions and are as well represented as block-schemes connecting atomic actions. Task's actions can also be executed both sequentially and in parallel, depending on task's block-scheme.

2 Construction, debugging and testing of scenarios by field of study specialists mostly with use of visual drag-and-drop of icons, corresponding to tasks, solutions and actions, to the scenario and

task scheme form and connecting them in accordance with scenario logic.

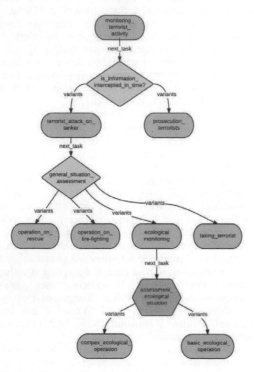

Figure 3. Example of terrorist situation assessment.

3 Scenario execution of complex spatial processes in optional time scale against digital map background, represented as moving simulated point objects with changing form, size, location, colour, transparency and etc. extended objects along with messages on natural professional language.
4 Interaction of number of complex processes, modelled on one as well as on several machines in network.
5 Manual object and process control option, modelling process in general (start/stop, time scale change, maps and etc.), scenario replay from control point, time jumps and etc.

5 CASE STUDY

As an explanation, consider the following situations that demonstrate maritime dangerous situation assessment using the proposed IGIS-technologies.

Situation 1. The cargo ship was captured by pirates in the Gulf of Aden. The search and rescue operation scenario on release of the seized cargo ship was developed. Figure 4 illustrates the example of this scenario. Release of the cargo ship (on the Figure 5 cargo ship is marked by blue color) is carried out by group including helicopter and two boats. During implementation of the developed scenario cargo ship seized by pirates was detected in 0 hours 50 minutes in point with coordinates {latitude 12 37'19"; longitude 47 0'6"} (data are displayed in information window of expert system). Results of modelling of search and rescue operation on release of the cargo ship seized by pirates are given in Table 1. Also these data are displayed in information window of expert system (Fig. 5).

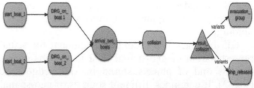

Figure 4. Scenario of the seized cargo ship release.

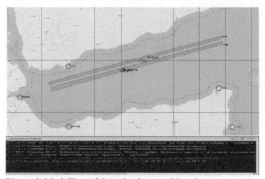

Figure 5. Modelling of the seized cargo ship release.

Table 1. Modelling results

	Number at the beginning operation	Number of killed	Number of wounded
Pirates	32	4	3
Hostages	25	2	5
Members of release group	29	1	1

Situation 2. In the Barents Sea at crash of the tanker there was an oil spill up to 5000 tons within 3 days. Figure 6 shows the scenario of oil spill taking into account navigation conditions around a dangerous situation, hydrographic and hydrometeorological situation. Also the assessment and the analysis of an ice situation are done.

The mathematical model on the basis of which the scenario was developed considers coastline, ice situation in given area, water flows and weather conditions. Figure 7 shows modelling of the ecological search and rescue operation using IGIS.

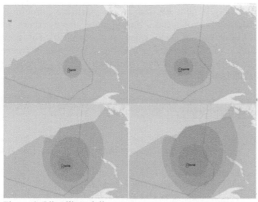

Figure 6. Oil spill modeling.

Figure 7.1. Modelling of search and rescue operation.

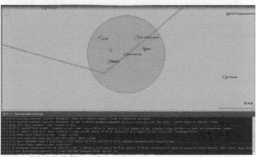

Figure 7.2. Modelling of search and rescue operation.

Modelling of search and rescue operations allows to carry out the analysis of the estimated action plan, to estimate efficiency of rescue forces' actions, to consider various weather conditions and features of the area.

6 MOBILE IGIS

On the base of stationary maritime situation monitoring system mobile version was created. The developed mobile IGIS is indented for mobile devices such as smartphones, laptops, and tablets with Android operating system. To update the information model and download the function modules Internet access is required.

The generalized structure of the mobile IGIS is shown on Figure 8. System includes the following elements:
- Cloud Services Platform – network of distributed servers that provides the set of necessary services;
- Client Integration Platform – application, that provides necessary interfaces for connecting modules that implement required features and that are installed on the mobile device.
- Function Modules Repository – storage of functional modules, capable to extend functionality of client's integration platform.

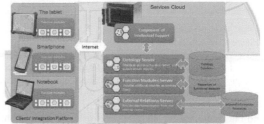

Figure 8. General structure of the mobile IGIS for vessels.

By means of these structure components mobile IGIS supports a wide range of functions and services such as:
- displaying the nautical chart showing the location of user's own ship and other maneuvering vessels;
- option of manual and automatic (AIS, radar) targets' drawing, target movement calculation;
- ship's routing between predetermined ports/points, accounting for the navigational area features and desired duration of movement/recreation;
- display and registration of weather data at a point/on the route, broadcast of storm warnings;
- calculation of maneuvers aimed at storm escape; speed and course calculations;
- solving tasks of maneuvering; positions' gaining and targets' passing;
- warning about dangerous maneuvers; entry into the closed/restricted areas, areas with special conditions for navigation, shallow water areas;
- warning of maneuvers on the route in advance (time and turning point, speed changes, etc.);

– retrieving data about locations of other vessels equipped with mobile IGIS.

Figure 9 shows a map where vessels' icons, routes (blue line) and additional information are plotted. The data on the map are updated in real time mode. By clicking on the vessel's icon additional information about the vessels, such as vessel's location (latitude and longitude), their lengths, widths, speeds, courses and names can be obtained. Various colors are used for displaying various vessels' types: green color for cargo vessels, blue – for passenger vessels, orange – for high speed crafts, grey – for unspecified vessels.

Figure 9. Vessels' routes.

It is necessary to notice that the system is rather simple and easy in use; the user needs no additional knowledge and skills to start working with the system. Furthermore, the system can work in autonomous mode that allows using the system without a permanent connection with the server during a certain period of time.

7 CONCLUSIONS

The safety of marine activities is the world priority, demanding solution on a global scale. Currently, not a single country is able to solve this problem independently, without interaction with the world communities. Every year the number of various dangerous maritime situations of both natural, and technogenic character increases.

The proposed new tools and technologies of assessment and modelling of dangerous maritime situations on the basis of IGIS aim to increase safety of sea activities. We have presented new mathematical methods for modelling of complex geo-spatial processes. Main elements of suggested IGIS architecture for maritime situations assessment are library of mathematical functions and modelling system.

As further research direction we suggest considerable expansion of services for the solution of a bigger class of tasks and problems connected with assessment of dangerous maritime situations, extensions of the knowledge base due to introduction of new rules and scenarios for new types of dangerous situations. We are actively developing user services and instruments for mobile IGIS as well.

REFERENCES

Baylon, A. M., Santos, E. M. R. 2013. Introducing GIS to TransNav and its Extensive Maritime Application: An Innovative Tool for intelligent Decision Making? *TransNav the International Journal on Marine Navigation and Safety of Sea Transportation*, Vol. 7, Num. 4: 557–566.

Blasch, E. 2002. Fundamentals of Information Fusion and Applications. *Tutorial, TD2. Proceedings of the International Conference on Information Fusion.*

Claramunt, C., Devogele, T., Fournier, S., Noyon, V., Petit, M., Ray C. 2007. Maritime GIS: From Monitoring to Simulation Systems. In V. Popovich, M. Schrenk and K. Korolenko (Eds.) *Proceedings of Information Fusion and Geographic Information Systems (IF&GIS'07), Springer-Verlag, LN series in Geoinformation and Cartography*, May 27–29, 2007, St. Petersburg, Russia: 34–44.

Goralski, R.I., Gold, C.M. 2007. The development of a dynamic GIS for maritime navigation safety. *ISPRS Workshop on Updating Geo-spatial Databases with Imagery & The 5th ISPRS Workshop on DMGISs, 28–29 August 2007, Urumchi, China:* 47–50.

Popovich, V., Claramunt, C., Osipov, V., Ray, C., Wang, T., Berbenev, D. 2009. Integration of Vessel Traffic Control Systems and Geographical Information Systems. *Proceedings REAL CORP 2009 Tagungsband 22–25 April 2009, Sitges:* 271–284.

Popovich, V., Potapichev, S., Pankin A., Shaida, S., Voronin, M., 2006. *Intelligent GIS for Monitoring' Systems, SPIIRAS, Serial 3, Vol. 1, St. Petersburg, Russia:* 172–184.

Popovich, V., Pankin, A.V., Voronin, M.N., Sokolova, L.A. 2006. Intelligent Situation Awareness on a GIS Basis. *In Proceedings of MILCOM 2006, October 24–26, Washington, USA.*

Popovich V, 2003. Concept of geoinformatic systems for information fusion. *In: Proceedings of the 1st international workshop on information fusion and geographic information system, September 17–20, St. Petersburg, Russia.*

Popovich, V. 2013. Intelligent GIS Conceptualization, Information Fusion and Geographic Information Systems. In V. Popovich, M. Schrenk and K. Korolenko (Eds.) *Proceedings of Information Fusion and Geographic Information Systems (IF&GIS'13), Springer-Verlag, LN series in Geoinformation and Cartography, St. Petersburg, Russia:* 17–44.

Smirnova, O., Tsvetkov, M., Sorokin, R. 2014. Intelligent GIS for monitoring and prediction of potentially dangerous situations. *In 14th International Multidisciplinary scientific geoconference SGEM 2014, Conference proceedings, Vol. 1, June 17–26, 2014:* 659–666.

Tetreault, B. J. 2010. Expanded use of Automatic Identification System (AIS) navigation technology in Vessel Traffic Services (VTS). *In Ports 2010: Building on the Past, Respecting the Future, Proceedings of the 12th Triennial International Conference, Jacksonville, Florida, April 25– 28, 2010:* 789–796.

Hydrometeorological Aspects

Design Tide and Wave for Santos Offshore Port (Brazil) Considering Extreme Events in a Climate Changing Scenario

P. Alfredini & E. Arasaki
Polytechnic School of São Paulo University, São Paulo, São Paulo State, Brazil

A.S. Moreira
Companhia Docas do Estado de São Paulo, Santos, São Paulo State, Brazil

ABSTRACT: Santos Port, in São Paulo State Coastline (Brazil), is the most important maritime cargo transfer terminal in the Southern Hemisphere. The increasing number of larger vessels is the setting for the future. The port is located in an estuarine area with the depth maintained by high dredging rates, limiting possible choices for new estuarine expansions. Like many other important ports in estuarine areas that have already expanded seaward, Santos is planning its offshore port. The awareness on climate changing in maritime projects led Agencies from different countries, e.g. The Netherlands, UK and USA, to issue in the latest decades recommendations or requirements about sea level rise trend, which, associated with the tidal trend recorded, guided the tidal forecasting until 2100. The tidal dataset of Santos Port is the longest in Brazil (1940 - 2014). The hindcasting meteorological mathematical models, calibrated and validated with the modelling of two different dataset of wave buoy, provided the wave climate assessment.

1 INTRODUCTION

Santos Port (Figs. 1 to 3) throughputs approximately 15% of Brazilian maritime cargo, more than 110 million tons per year (Fig. 4), and is the most important maritime cargo transfer terminal in the Southern Hemisphere. The increasing number of larger container vessels and the important oil and gas reserves in the Offshore Basin of Santos is the economic scenario for the future. The port is located in a confined estuarine area, comprising small inland fluvial deltas, with the hydrodynamical pattern induced by tides (Figs. 2 and 3) and the depth maintained by high dredging rates (7 million m³/year to maintain depth – 15.0 m Chart Datum (CD). To receive larger vessels there is the increasing necessity to expand the nautical areas (access channel, turning basins and berths) out off the estuarine area.

Like many other important estuarine ports that have already expanded seaward (e.g. Rotterdam, Le Havre and Shanghai), Santos is planning its first offshore port, with jetties (Moreira 2009). Due to the high values involved of the facilities and infrastructures, it will be necessary minimize the risks of natural disasters. Hence, the conceptual planning for the new port must consider the climate changing in extreme events of tides and waves with a recommended recurrence period greater than 100 years.

The tidal dataset of Santos Port is the longest in Brazil (1940 - 2014). The awareness on climate changing in maritime projects led Agencies from different countries, e.g. The Netherlands, UK and USA, to issue in the latest decades recommendations or requirements about sea level rise (PIANC 2014a). These recommendations associated with the tidal trend recorded of Santos Port in the last 75 years, guided the tidal forecasting for 2050 and 2100.

The hindcasting meteorological mathematical models, calibrated and validated with the modelling of two different dataset of wave buoy, provided the wave climate assessment.

2 MATERIAL AND METHODS

Data sets for this assessment were obtained from a physical model of Santos Bay, Estuary and nearby beaches showing the impact of maritime climate changes (Alfredini et al. 2008), wave climate and tidal forcing (Alfredini et al. 2012, 2013 and 2014, Fournier 2012).

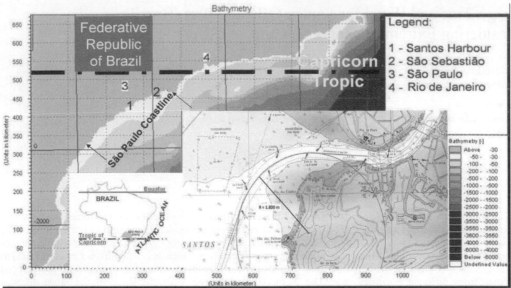

Figure 1. Santos Port site location

The long term tidal level variability assessment of CODESP, the Santos Dock Company, from 1940 until 2007 and of the Santos Pilots-Zenith, 2008 until now, showed the possibility to have exactly four lunar declination periods (1940 – 2014) of 18.61 years each one (Fig. 5). The forecasting trends of Higher High Water (HHW) and Lower Low Water (LLW) depend largely upon the meteorological tide, beyond sea level rise (Alfredini et al. 2013). Comparing with PIANC 2014a, the tidal trend of Santos Port is similar to the moderate trend of the UK recommendations, meaning a mean sea level rise around 50 cm from 1950 until 2050, and 110 cm from 1950 until 2100. For HHW and LLW the trends are similar, as showed in Figure 5. Hence, in the future dramatic changes in tidal range and in the currents would not occur.

Based upon water speed surveys, it was possible to calibrate and validate the mathematical hydrodynamical model of the tidal current of Santos Bay using the Mike 21 software (Baptistelli et al. 2008). Figures 2 and 3 show charts for a spring tide in July. The flood and ebb conditions are maxima, respectively. It is possible to observe that the maximum current velocities reaches values around 0.8 knots (deep averaged) in the estuarine port area. In equinoctial spring tides, the values may arrive up to 2 knots. However, in Santos Bay the tidal current velocities are roughly half those figures.

An assessment of the characteristics and historical frequency of extreme storm events employed the long-term (Jan 1st 1980 – August 6th 2012) a wave climate database (www.OndasdoBrasil.com). The database include definitions of significant wave height (H_{m0}), spectral peak wave period (T_p) and spectral directions as required. Hindcasting deep-water was developed with the aid of the WaveWatch III, the third generation wave model, and was calibrated with Topex Satellite data along Brazilian coastline, and subsequently validated with a directional sea buoy. The deep-water wave climate transferred to shallow water, in the Santos Bay showed the rose in Figure 6 (Fournier 2012). The statistical assessment of extreme wave considering a lognormal fitting gives the following significant heights estimated for a 100 years recurrence period:

– Upper Estimate: 8.92 m
– Average Estimate: 7.94 m
– Lower Estimate: 6.96 m

Statistically, for a lifespan of 100 years and one recurrence period of 100 years, the probability of occurrence of these numbers is 63%.

In the immediate future, the Port of Santos aims to receive Post Panamax Plus (5th generation container vessels) 9,000 TEUs with L_{pp}:335 m; B: 46.00 m and T: 15.00 m. These vessels will require a channel depth of - 17.0 m (CD). Huge amounts of capital dredging, maintenance dredging and some operations should be cancelled, especially during April to September, period of the highest waves (Alfredini 2013).

The other issue is about the width of two-way channel of 220 m, suitable for Panamax vessels, but would be better at least 315 m (equal to 6.8 times the beam) to accommodate vessels Post Panamax Plus and an appropriate enlargement of the curve (Alfredini 2013).

The last issue to consider relates to the channel's curvature radius presently set at 1800 m (Fig. 1). It is small for the Panamax vessels and too small for the Post Panamax Plus. Without tugboats aid, it would have to be redesigned to at least 3300 m, unless if the use of tugboat would begin in the Outer Access Channel. That curve is the most accreted area of the Outer Access Channel due to the longshore sand transport along Santos beaches (Alfredini 2013).

3 THE SEAWARD CONCEPTUAL PLANNING PHASE 1

Considering the necessity to enlarge the geometric dimensions (depth, width and radius) of the Outer Access Channel of Santos Port with the purpose of navigating in a two way with Post Panamax Plus vessels of 9000 TEUs, training walls are the best engineering solution (Alfredini 2013). This solution is present in the Master Plan of Santos Port proposed in the 1960 decade and it is the Phase 1 of the Conceptual Planning for the Offshore Port.

Figure 2. Tidal currents chart in the maximum spring tide flood current in July 21st 2005.

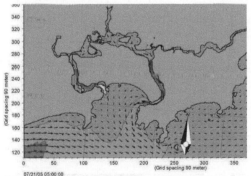

Figure 3. Tidal currents chart in the maximum spring tide ebb current in July 21st 2005.

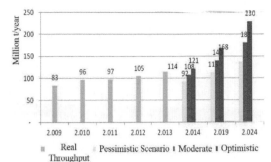

Figure 4. Throughput trends forecasting for Santos Port.

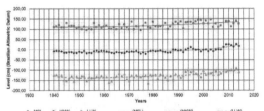

Figure 5. Santos Port annual tidal level variability from 1940 until 2014.

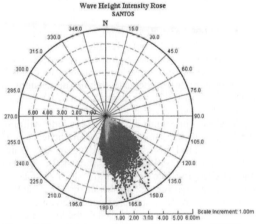

Figure 6. Wave height rose in Santos Bay (1980 until 2012).

Considering an increasing of 10% in the offshore extreme waves, as suggested in PIANC (2009) due to the climate changing, the design value of significant wave suggested for the training walls heads is 9.0 m, according to the upper estimate previously mentioned. Considering the sea level rise, the HWN (High Water Neap) and the wave uprush, the altitude of the crest of the training walls heads would be around + 13.0 m (Brazilian Altimetry Datum) with an armor of precast concrete blocks, or + 8.0 m , using deflector concrete cap to avoid overtopping the crest level.

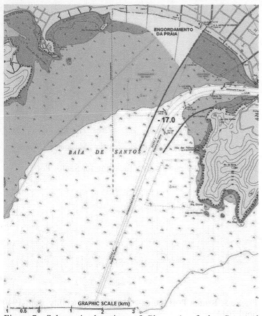

Figure 7. Schematic location of Phase 1 of the Seaward Planning for the Offshore Port with probable beach nourishment.

A rubble mound structure, such as the two training walls with a total length of 9 km (Fig. 7) will provide a solution. Significantly reducing the OPEX requirements (dredging maintenance cost), and redesigning the adequate channel dimensions to depth -17.0 m (CD), the channel would be suitable for the larger vessels.

As shown in Figure 7, the Western Jetty will trap the longshore sand transport along the beaches' surf zone, a natural nourishment, reducing the Outer Access Channel sedimentation. Indeed, the present annual dredging rate, with depth – 15.0 m (CD) in the channel is around 2.5 million m³. The training walls will provide a sheltered access channel. That process will enhance the coastal defenses against flooding due to sea level rise, storm surges, increasing the coastal resilience against these phenomena. Indeed, in the last decade increased the frequency of inundation episodes in the areas of Santos municipality.

4 THE SEAWARD CONCEPTUAL PLANNING PHASE 2

Figure 8 shows the layout for Seaward Conceptual Planning Phase 2 for Santos Offshore Port. The best area for this expansion is the southeastern extreme of Santos Bay, where the average depth is around – 14.5 m (CD). However, it will be necessary to provide an external jetty of 7.5 km length to protect the berths from the waves ESE to SSW (see Fig. 6).

The maintenance of a deeper access channel will enlarge the Phase 1 of Western Jetty more 4.5 km, which will protect additionally the beaches westward, growing their resilience against the increasing frequency and impact of storm surges foreseen. At last, the Phase 1 of Eastern Jetty will be enlarged with more than 3.5 km, used for a coastal closure and providing a reclamation land.

For the training walls crest level it will be used the same figures explained in item 3.

Considering the sea level rise for HWN (High Water Spring) and the residual wave, the quay apron platform level will be around + 4.0 m (Brazilian Altimetry Datum). The storage yards retro-areas will be reclaimed at the same level.

The Seaward Conceptual Planning Phase 2 for Santos Offshore Port will be the construction of the deep berths for the following vessels type:

– Triple E – Ultra Large Container – up to 22,000 TEUs with L_{pp}: 470 m; B: 59.0 m andT: 17.0 m.
– NPX – New Panamax – up to 15,000 TEUs with L_{pp}: 400 m; B: 56.0 m and T: 16.0 m.
– Post Panamax Plus – up to 9,000 TEUs with L_{pp}: 335 m; B: 46.0 m and T: 15.0 m.
– Ore Capesize – up to 150,000 DWT with L_{pp}: 276 m; B: 44.0 m and T: 17.0 m.
– Suezmax for ethanol – up to 150.000 DWT with L_{pp}: 270 m; B: 49.5 m and T: 17.0 m.

Considering the larger draught of 17.0 m, it will be necessary to provide the depth of – 19.0 m (CD) for the access channel extension of Phase 2.

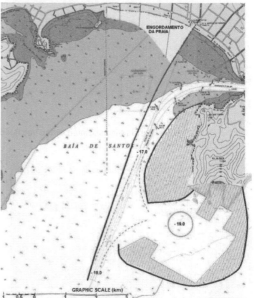

Figure 8. Schematic location of Seaward Planning Phase 2 for Santos Offshore Port.

150

The exact geometry of these jetties and the quays' alignment will be dependent of an accurate study of fill and dredging volumes, associated with the retroarea necessary, according to the cargo to be handled.

The Seaward Conceptual Planning Phase 2 for Santos Offshore Port will follow the sustainable port concept (PIANC 2014b), including the climate adaptation.

5 CONCLUSIONS

It is possible, based upon the international guidelines suggested by PIANC (2009, 2014a and 2014b), propose the Seaward Conceptual Planning for Santos Offshore Port considering the future sea level rise and wave climate that will be predictable until the end of this century. Indeed, the tidal database of Santos Port is one of the longest and confident in the Southern Hemisphere and its sea level rise trend is similar to the UK recommendations for a moderate scenario of climate changes. Hence, this validation gives the best estimative for altimetry design parameters for the jetties crest and retroarea. The estimative of the wave climate has considered a well-known hindcasting technique, based upon the last three decades to forecast the 100 years' design wave, which is statistically justified. Considering the increasing wave heights due to the climate changing scenario, the recommendation is to use the upper estimate.

ACKNOWLEDGEMENTS

This paper has the financial support of CAPES, Human Resources Improvement Agency of Brazilian Government. The authors also thank the support of São Paulo University, CODESP, Santos Pilots, Zenith, Baird & Associates Coastal Engineers Ltd.

REFERENCES

Alfredini, P., Arasaki, E. & Amaral, R. F. 2008. Mean sea level rise impacts on Santos Bay, Southeastern Brazil – Physical modelling study. *Environmental Monitoring and Assessment* vol.144: 377-387. Amsterdam: Springer.

Alfredini, P., Arasaki, E., Pezzoli, A., Arcorace, M., Cristofori, E. I. & Sousa Jr., W. C. 2014. Exposure of Santos Harbor Metropolitan Area (Brazil) to wave and storm surge climate changes. *Water Quality, Exposure and Health* vol. 6, Issue 1-2, pp 73-88. Springer Netherlands.

Alfredini, P., Arasaki, E., Pezzoli, A., Fournier C. P. 2013. Impact of climate changes on Santos Harbor, São Paulo State (Brazil). . In A. Weintrit & T. Neumann (Org.), *The International Journal on Marine Navigation and Safety of Sea Transportation TransNav* vol. 7(4). London: CRC Press/Balkema.

Alfredini, P., Pezzoli, A., Cristofori, E. I., Dovetta, A. & Arasaki, E. 2012. Wave and tidal level analysis, maritime climate changes, navigation's strategy and impact on the coastal defences – Study case of São Paulo State coastline harbour areas. *Geophysical Research Abstracts*: vol. 14, EGU2012-10735. Vienna.

Baptistelli, S. C., Harari, J. & Alfredini, P. (2008). Using "Mike 21" and "POM" for numerical simulations of the hydrodynamics in Baixada Santista region (São Paulo, Brazil). *Afro-America Global Sea Level Observing System News*: 12 (1).

Fournier, C. P. 2012. Ondas Santos. *Workshop Molhes Guias-Correntes na Barra de Santos – O futuro da acessibilidade marítima ao Porto.* Santos: Associação de Engenheiros e Arquitetos de Santos.

Moreira, A. S. 2009. *Sistema de informações de indicadores do modelo porto concentrador de carga.* PhD Thesis in Hydraulic Engineering, presented to the Hydraulic Department of the Polytechnic School, University of São Paulo, São Paulo (Brazil).

PIANC – The World Association for Waterborne Transport Infrastructure 2009. *Waterborne transport, ports and waterways: A review of climate change drivers, impacts, responses and mitigation.* Bruxelles: EnviCom – Task Group 3 Climate Change and Navigation.

PIANC – The World Association for Waterborne Transport Infrastructure 2014a *Countries in Transition (CIT) Coastal Erosion Mitigation Guidelines.* Bruxelles: Cooperation Commission (CoCom), PIANC Report 123.

PIANC – The World Association for Waterborne Transport Infrastructure 2014b *Sustainable Ports A Guide for Port Authorities.* Bruxelles: IAPH EnviCom Report Working Group 150, PIANC Report 150.

Mathematical Modeling of Wave Situation for Creation of Protective Hydrotechnical Constructions in Port Kulevi

A. Gegenava, I. Sharabidze & A. Kakhidze
Batumi Maritime Academy, Batumi, Georgia

ABSTRACT: The presented paper continues the previous two papers - "New Black Sea Terminal of Port Kulevi and it Navigating Features" and "Analysis of Hydrometeorological Characteristics in Port of Kulevi Zone", which was published in "8th and 9th INTERNATIONAL NAVIGATIONAL SYMPOSIUM ON MARINE NAVIGATION AND SAFETY OF SEA TRANSPORTATION. GDYNIA, POLAND, 2009 and 2011" and deals with the aspects of safety navigation provision in Port Kulevi by means of conduction of mathematical modelling of wave situation for creation of protective hydrotechnical constructions in port Kulevi.
As it was said in the previous papers, the port of Kulevi from the point of view of hydrometeorological condtions is a difficult one. The influence of prevailing wind directions – East and West, constant sea and curents of river Khobi, as well as the absence of the special protective hydrotechnical constructions allows the waves and sediments from the rivers Rioni and Khobi to enter to the channel. This paper presents the matmatical modeling of wave situations in different variants of protective measures for creation of protective hydrotechnical constructions which can be conducted in Port Kulevi having the aim the decreasing of hydro meteorological conditions to provide the safety navigation.

1 INTRODUCTION

The analysis of hydrometeorological characteristics shows that the main factor, which influences upon the level of safety of navigation in water area of Port of Kulevi is heavy sea. The analysis of wave conditions was conducted for the provision of the necessary level of safety of navigation as well as the provision of effective and safe maintenance of hydrotechnical constructions. The automatical metcorological station which is a part of Port of Kulevi VTS equipment gave possibility of every day (per each 30 minutes) keeping of the wind condition (direction, speed, maximal and minimal indicators). The complex of measures of Port of Kulevi water area protection was worked out on the basis of that analysis.

2 MATHEMATICAL MODELING OF WAVE SITUATION

2.1 *Theoretical justification*

Mathematical modeling was carried out using the theory of frequency-angular spectrum using the coefficients of surface and of bottom friction.

To describe the fields of excitement uses the concept of frequency-angular spectrum

$E(\sigma, \theta, x, y, t)$,

where:

σ, θ - frequency and angular coordinates;

x, y, t - horizontal coordinates and time.

Appropriate equation for the determination of E is given by the law of conservation of energy:

$$\frac{\partial}{\partial t}N + \frac{\partial}{\partial x}(c_x N) + \frac{\partial}{\partial y}(c_y N) + \frac{\partial}{\partial \omega}(c_\omega N) + \frac{\partial}{\partial \theta}(c_\theta N) = \frac{S}{\omega} \quad (1)$$

where,

$N = E/\omega$ - density of wave action;

Magnitudes, c_X, c_Y, c_ω, c_θ - transport velocity along the appropriate spatial and frequency-angular coordinates.

For numerical solution of equation (1) is applied a modified version of the model, in which the source function is defined as

$$S=S_{in}+S_{nl}+S_{wc}+S_{bf}+S_{dib}, \qquad (2)$$

where:

S_{in} - the source of generation of waves by the wind;

S_{nl} - non- linear interactions spectral harmonics;

S_{wc} - dissipation of the energy due to the caving of the wave crests;

S_{bf} - dissipation of the energy due to of bottom friction;

S_{dib} - caving waves on the critical depths.

In the original version the source of wave generation S_{in} - is determined by assuming that the friction coefficient C_d only depends on the wind speed.

Solution of problem (1), (2) makes it possible to obtain estimates a number of spectral wave characteristics, in particular the significant wave height and root mean square wave height.

A characteristic feature of this model is the use of the coefficients of surface and of the bottom friction, characteristic only of the area of Kulevi port.

2.2 Coefficient of surface friction

In most cases, the modeling of wind flows coefficient of surface friction is assumed constant or linearly dependent on wind speed.

A more realistic approach is based on the theory of the boundary layer. Under the assumption that the profile of the wind speed near the surface of the sea has a logarithmic form, the value of C_d determined by the expression

$$C_d = \kappa^2 / \ln^2(z / z_0) \qquad (3)$$

where:

k = 0.4 - Karman constant,

z = 10 m, and

z_0 - sea surface roughness parameter characterizing the hydrodynamic properties of the underlying surface and determines the mode of turbulent mixing at the interface of atmosphere-sea.

2.3 Coefficient of bottom friction

Under the assumption of logarithmic depending profile current speed near the -of bottom for the coefficient of bottom friction C_b we have expression similar to (3):

$$C_b = \kappa^2 / \ln^2(z / z_0) \qquad (4)$$

where:

z_b - parameter roughness of bottom surface;

z - distance from the bottom up to the point at which determined the coefficient of friction.

3 TERMS OF MODELLING

The main physiographic characteristics that influence the wave processes are wind strength, direction, which determines the overclocking wave (the distance at which the wave is generated the energy), the duration of the wind and the depth terrain.

Calculations were carried for dangerous wind conditions the western part. For the initial condition takes steady excitement with certain characteristics, which were set at the boundary investigated area. The boundaries were taken as liquid boundaries flow (boundary conditions) with the calculated oceanographic tables wave characteristics established excitement (the period and wave height). At the NW wind overclocking length was set 550 km, at SW - 180 km, at W - 1,100 km. The depth of area was set field of depths obtained by conversion data hydrographic observations. Depth values are tied to the local reference system (local coordinates) and interpolated on a computational grid 80x40 nodes in increments of 50 and 75 meters respectively.

The results of modeling are shown in Fig. 1-3.

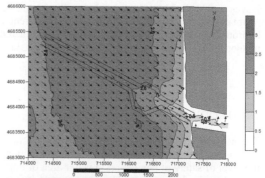

Figure 1. Direction of wind - NW, speed of wind – 30 m/s.

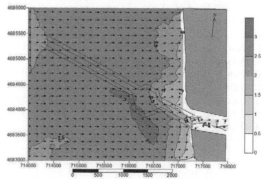

Figure 2. Direction of wind - W, speed of wind – 30 m/s.

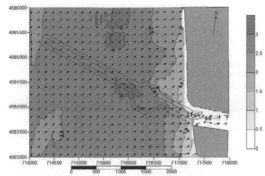

Figure 3. Direction of wind - SW, speed of wind – 30 m/s.

4 ANALYSIS OF MODELLING

The analysis shows that the largest area of maximum wave fields is observed from West and South-West directions.

As a result of this work on mathematical modelling have been analysed several options of protective hydrotechnical constructions. A characteristic feature of the chosen protective hydrotechnical construction (see fig. 4) is the output of the extremities of protective hydrotechnical constructions into natural depths 7-8 m, resulting entry of the sediment from the river Rioni in the approaching canal. Approximate volume - about 40-50 thousand cubic meters per year. The remaining sediments from the river Rioni will form the beach. Extremities of protective hydrotechnical constructions are located on a natural depth of the order of 7-8m. Dimensions of protective hydrotechnical construction given in table 1.

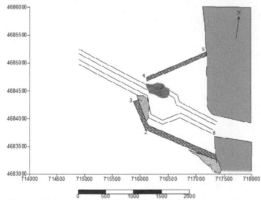

Figure 4. Scheme of protective hydrotechnical construction.

Table 1. Dimensions of protective hydrotechnical construction

The plot	1-2	2-3	3-4	4-5	6-1	5-1
Distance (m)	1400	500	450	1150	500	1900
Azimuth (deg)	118	167	--	90	--	--

5 MODELLING OF WAVE SITUATION WITH PROTECTIVE ACTIONS

Have been selected initial wind conditions: wind North-West, West and South-West directions at the speed of 30 m/s.

The results of mathematical modeling represented of the fields in the form of wave loads in Fig. 5-7.

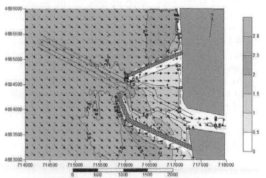

Figure 5. Wawe loads (NW), speed of wind – 30 m/s.

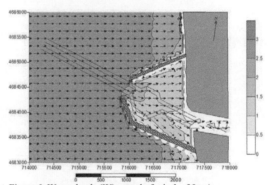

Figure 6. Wawe loads (W), speed of wind – 30 m/s.

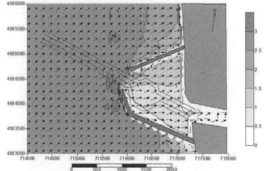

Figure 7. Wawe loads (SW), speed of wind – 30 m/s.

155

6 CONCLUSIONS

From the positions of safety of navigation, the prospects development of port, the economic feasibility and efficient operation of protective hydrotechnical constructions preferred is protective hydrotechnical construction shown in the figure 4. Results of modelling show:

− At the north-west wind speeds up to 30 m/s in port aquatory will go wave height not exceeding 2,0-0,8m, at the berths - no more 0,4-0,2m;
− At the westward wind speeds up to 30 m/s in port aquatory will go wave height not exceeding 2,0-0,8m, at the berths - no more than 0.5 m;
− At the south-west wind speeds up to 30 m/s in port aquatory will go wave height not exceeding 1.0 m, at the berths - no more 0,2-0,1m;
− Wave loads on the hydrotechnical constructions at rated wind speed of western rhumb 30 m/s will not exceed 0.5 t/m2;
− At calculations of the wave loads recommended to take speed of winds from the western rhumb - 33 m/s.

REFERENCES

Georgian Government Order N57 on "Establishment of ships movement separating schemes, sea corridors and special sea regions in Georgian territorial Water area". Tbilisi, 2014;

Maritime Transport Agency of Georgia, Director Order on approval of Port Rules, №019. Batumi, 2013;

Dzhaoshvili S. "Hydrologic-morphological processes in mouth zone river Rioni and their anthropogenesis changes". Water resources, w.25, №2, 1998;

Varazashvili N. "Geological processes and the phenomena in a zone of construction of sea hydraulic engineering constructions and actions on improvement of a coastal situation". Engineering geology (excerpt). Academy of Sciences the USSR. Moscow, 1983;

Dzhaoshvili S. "New data about beach formed sediments of a coastal zone of Georgia". Water resources (excerpt). Academy of Sciences the USSR. Moscow, 1984;

Materials of supervision over elements of a hydrometeorologcal mode for the period 1971-1999. (Information of Poti mouth stations);

Bettes P. "Diffraction and refraction of surface waves using finite and infinite elements". Numerical Meth. London, 1997;

Bach H., Christiansen P. "Numerical investigations of creeping waves in water theory". ZAMM. New York, 2004;

Mathematical modelling of hydrological characteristics in the port of Kulevi. LLC "Anchor". Kulevi, 2009;

Black Sea Terminal of Port Kulevi. http://www.kulevioilterminal.com;

Gegenava A, Varshanidze N, Khaidarov G. " New Black Sea Terminal of Port Kulevi and it Navigating Features". TransNav 2009, Gdynia, 2009;

Gegenava A, Khaidarov G. "Analysis of Hydrometeorological Characteristics in Port of Kulevi Zone". TransNav 2011, Gdynia, 2011.

Hydrometeorological Aspects
Information, Communication and Environment – Marine Navigation and Safety of Sea Transportation – A. Weintrit & T. Neumann (eds.)

The Northerly Summer Wind off the West Coast of the Iberian Peninsula

N. Rijo & A. Semedo
Escola Naval-CINAV, Lisbon, Portugal

D.C.A. Lima, P. Miranda, R.M. Cardoso & P.M.M. Soares
Instituto Dom Luiz, University of Lisbon, Lisbon, Portugal

ABSTRACT: Winds are a natural hazard for mariners and for offshore infrastructures. In spite of having been used as a propulsion system throughout times, nowadays it can be a source of danger in extreme situations. High wind speeds can also represent an economic driver, since, on the one hand, wind energy can be harvested from off-shore wind farms, but on the other hand fishing activity or, for example, off-shore fish farms are affected. The predominant winds in the North Atlantic (the Westerlies) have a basin-wide scale, and are highest during winter. During summer, when the large scale forcing is weaker, mesoscale and local wind features prevail. That is the case of coastal jets.

During summer the Iberian Peninsula is under the effect of the semi-permanent Azores High and of a thermal low pressure system in-land. This synoptic pattern drives a seasonal northerly wind along the west coast of the Iberian Peninsula. This regional wind is an example of a coastal jet.

An insight on the climatology of the Iberian Peninsula coastal jet (IPCJ) is presented, based on a regional reanalysis produced with the WRF mesoscale model for the period 1989-2007. It is shown that although the IPCJ are more frequent in the south area of the west coast of Iberia, in the north area coastal jets are stronger (with higher wind speeds). The dynamic structure of the IPCJ is studied through the analysis of a strong summer IPCJ event along west coasts of Galicia and north of Portugal (at 3 km horizontal resolution). This case study was identified in vicinity of Aveiro on 14th July 2002 at 1900 local time.

1 INTRODUCTION

The regional climates along the mid-latitude western continental Coastal regions are affected by the ocean and atmospheric circulations and their interactions. At these regions, mostly in summer, the contrast between the cold oceanic waters and the warm land are strong, which is fundamentally responsible for the existence of a coast parallel wind designated as the coastal low-level jets (CLLJ) or as "coastal Jets". In spite of being a mesoscale feature, CLLJ occurrences have a typical large scale synoptic forcing behind them, characterized by a high pressure systems over the ocean and a thermal low in-land. This pressure gradient is responsible for driving a coastal parallel wind pattern along the coast. For that reason CLLJ occurrence regions coincide with cold equator-ward eastern boundary currents, located along the east flanks of the atmospheric semi-permanent high pressure cells (Winant et al. 1988) in the mid-latitudes. During summer these coast parallel winds features are responsible for upwelling processes, bringing deep cold nutrient-rich water to the shelf and to the surface (Vallis 2012). As a result sharp temperature and pressure gradients develop at the coast (Chao 1985), and consequently an intensification of coastal winds occur, which means that CLLJ are synoptically forced but mesoscale intensified.

CLLJ have a small vertical extend with wind speed maxima (jet core) located at low altitudes, of the order of hundreds of meters above sea level (ASL), confined at the marine atmospheric boundary layer (MABL) height (Beardsley et al. 1987, Burk and Thompson 1996, Söderberg and Tjernström 2002). The MABL is capped by a sharp temperature inversion, which is strengthened by the warm subsiding air above and the cold air close to the surface, that cools in contact with the low sea surface temperature (SST) due to upwelling. The horizontal extent of CLLJ can exceed hundreds of kilometres, limited in by the Rossby radius of deformation (Winant et al. 1988, Garreaud and Muñoz 2004, Ranjha et al. 2013, Soares et al. 2014).

The presence of coastal topography, capes, and headlands, can result in additional local enhancement of the wind speed, and changes in the flow direction if the mountains are higher than the sloping capping inversion of the MABL. The cross-coast flow is blocked and the wind becomes semi-geosptrophic. In these situations the flow is channelled, and the wind speed increases even more at the coast (Winant et al. 1988, Burk and Thompson 1996, Tjernström and Grisogono, 2000).

Coastal jets are essentially a summer phenomena, which occur when the synoptic pattern prevail. Their occurrence, as mentioned before, is strongly linked to upwelling and to low SST along the coast. The major upwelling areas, coinciding with the CLLJ areas (Ranjha et al. 2013) are where the most productive regions of the World Ocean in terms of fishing: about 17% of the worldwide fish captures take place along the coastal upwelling regions (Pauly and Christensen 1995), which in turn cover less than 1% of the world ocean area. The regional climates of the mid-latitude western continental areas are related CLLJ and to the associated cold SST. The low SST at the coast lowers the evaporation over the ocean, and the strong coast parallel winds substantially reduce the advection of marine air in-land. For this reason the atmospheric water vapour content in-land is low and some of the mid-latitude western coastal regions are arid or desert regions (Warner 2004). Due to shear (mechanically driven) turbulence the MABL at the coast in CLLJ areas is usually stable. After some consecutive days of coastal jet occurrences, if the synoptic forcing changes and favours the advection of warm moist air towards the along coast low SST areas, coastal fog or stratus clouds can occur (Tjernström and Koracin 1995, Burk and Haak 1999). Since CLLJ have an impact on cloud cover, visibility, SST, and wind structure in coastal areas, their importance for mariners, fishing, and shipping activity can be considerable. Hence a more in deep knowledge of this wind feature is of paramount importance.

The Iberian Peninsula Coastal Jet (IPCJ) is an example of a CLLJ, developed mostly during the summer season, due to the effect of the semi-permanent Azores High in North Atlantic sub-basin, and to the presence of a thermal low pressure system inland over the Iberian Peninsula. This synoptic pattern drives a seasonal northerly wind along the west coast of the Iberian Peninsula often called the Nortada (Lopes et al. 2009).

The climatology of the Iberian Peninsula CLLJ was recently presented by Soares et al. (2014). They have found that, in of spite not being as prevalent as other coastal jets (Ranjha et al. 2013, 2014), during summer (particularly in July) it has frequencies of occurrence of the order of 34-35%. In this study we present a detailed climatology of the IPCJ as well as an insight the mesoscale structure, spatial variability and temporal characteristics of the IPCJ,

complementing the study of Soares et al. (2014). The results were obtained through the Weather Research and Forecasting (WRF) model downscaling data produced in Soares et al., 2012, at 9km resolution, for a period 1989 to 2008. A detail analysis of a IPCJ case study (at 3km resolution) along the west coasts of Galicia and north of Portugal, occurred in vicinity of Aveiro on 14th July 2002 at 1900 Lt is also presented.

2 DATA AND METHODOLOGY

2.1 Filtering algorithm

The calculation of the occurrence of the CLLJ, and in particularly the IPCJ, follows the CLLJ detection algorithm proposed by Ranjha et al. (2013). This algorithm is based on the inspection of the vertical wind speed and temperature profiles. According with this criteria the IPCJ is considered to occur when:

- The height of the jet maximum (maximum wind speed) is within the lowest 1 km in the vertical;
- The wind speed maxima is at least 20% higher than the wind speed at the surface;
- The wind speed above the jet maximum decreases to below 80% of the wind speed at the surface (i.e. a 20% falloff) before reaching 5 km above its maximum;
- The temperature at the jet maximum height is lower than the temperature at two model levels above it (inversion detection criterion); and

These criteria were then used to scan the regional reanalysis at each grid point and to identify the locations and occurrences of coastal jets.

2.2 Regional reanalysis and case study simulations

A climate simulation (regional reanalysis), produced with the WRF model (version 3.1.1), covering the period 1989-2007, is used in the present study (Soares et al. 2012), namely the results from the innermost domain, with a 9 km horizontal resolution. The WRF model (Skamarock et al. 2008) is a non-hydrostatic mesoscale model, suitable for simulations at a wide range of scales, with a large number of available options concerning the model core and its physical parameterizations, making it appropriate for weather forecasting, and mesoscale meteorological studies. More recently the WRF model has also been extensively for dynamical downscaling in regional climate studies. The regional reanalysis used here was produced by downscaling the ERA-Interim global reanalysis from the European Centre for Medium-Range Weather Forecasts (ECMWF). A complete and more detailed description of the model set-up can be found in Soares et al. (2012), Cardoso et al. (2013) and

Soares et al. (2014), where the simulation results were validated for inland maximum and minimum temperatures and precipitation, and to marine surface wind speeds, showing a good agreement with observations.

The case study presented here used, as boundary and initial conditions, data from the regional reanalysis at 9 km horizontal resolution mentioned above. These boundary and initial conditions were used to force an inner domain at 3 km resolution (see Figure 1). The case study WRF simulation was parameterized with 49 vertical levels, placing roughly 20 vertical levels in the boundary layer, with the lowest model sigma level at approximately 10 m of height and model top at 50 hPa. The lateral boundary conditions and SST were both updated every 6 h, from the 9 km regional reanalysis. The WRF model run was set to start at 0000 UTC July 10, 2000, and end at 0000 UTC July 19, 2000).

3 DETAILED CLIMATOLOGY OF THE IBERIAN PENINSULA COASTAL JET

A statistical analysis of the global occurrences of coastal jets was performed by Ranjha et al. (2013), and more recently the climatology of an IPCJ was presented by Soares et al. (2014). Their results showed two main areas of occurrence of the IPCJ along the west coast of Iberia. These areas are located south of Cape Finisterre, and south of Cape Roca, the west-most point of continental Europe. In complement of their study we have recomputed the frequencies of occurrence of IPCJ for a strip along the west coast of the Iberian Peninsula, and separately for each of the North (A1) and South (A2) areas, represented in Figure 1.

The intra-annual variability of IPCJ occurrences and the mean wind speed, (mean monthly occurrences) for the west strip (Figure 2a,b) and for areas A1 (Figure 2c,d) and A2 (Figure 2e,f) are shown. The mean wind speed (ms-1), was computed at each grid point when jet occurs, at the height of the maximum wind speed (jet core).

From the frequencies of occurrence it is clearly the seasonality of the IPCJ occurrence displaying the maximum values in the summer months, particularly in July. The geographical distribution of the IPCJ occurrence is also clear where area A2 (south) show the mean frequency of occurrence of the IPCJ highest (close to 30% in July, i.e., 1 out of 3 days in that month comprises a coastal jet occurrence). In area A1 (north) the mean frequency of occurrence of the IPCJ is lower (slightly higher than 20% also in July, when the occurrences peak), but the wind speed, in July, at the jet height, is now higher than in the south area (close to 15 ms-1 and about 12 ms-1 in areas A1 and A2, respectively).

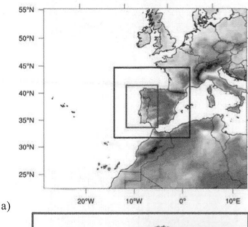

a)

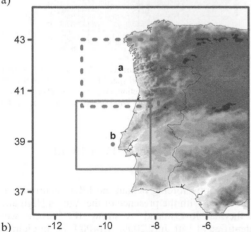

b)

Figure 1. (a) WRF domains used in the downscaling, from outermost to innermost domain with horizontal resolution of 9 km (black line) and 3 km (blue line), and (b) areas of statistical analysis, west coast strip (green line), and sub areas A1 (dust red line) and A2 (solid red line) showing the key points a (41.132N 9.42W) and b (38.436N 9.71W), respectively.

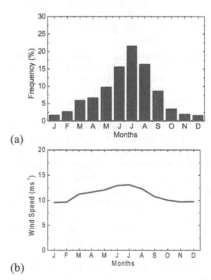

(a)

(b)

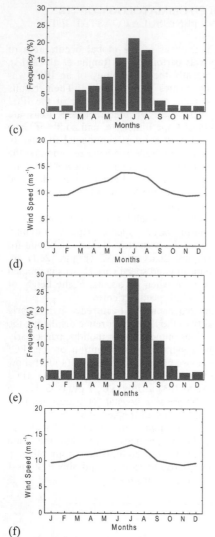

(c)

(d)

(e)

(f)

Figure 2. Monthly mean (Jan to Dec) frequency of occurrence (%) histograms of IPCJ at (a) west strip, (c) area A1, and (e) area A2, and monthly mean (Jan to Dec) wind speed (ms-1), when jet occurs, at jet height (height of the maximum wind speed) at (b) west strip, (d) area A1, and (f) area A2.

The analysis of the IPCJ characteristics of the Jet-height, wind speed maximum and direction for summer season are illustrated in figure 3, for each area. From the wind rose plot (Figure 3a and 3b) during summer months, in both areas, the predominant IPCJ wind direction at jet core is northerly (30%) to north-north-easterly (35%) during JJA, almost parallel to the coast of Iberia. In transition months May, June and September the pattern is slight different, with wind directions slightly more easterly (not shown).

The distributional pattern for the jet height (Figure 3b and 3d) is similar in both areas, revealing

that IPCJ occur at 200-400 m, and have a wind speed maxima around 15 ms-1. In area A1, nevertheless, the wind speed at the jet height in JJA is higher (wind speeds of the order of 20-24 ms-1 occurring ~7% of the time; Figure 3c) than in areas A2 (Figure 3a).

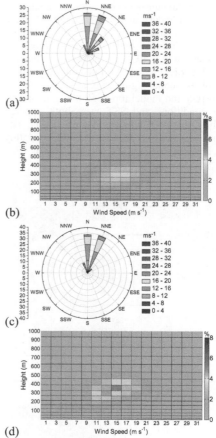

(a)

(b)

(c)

(d)

Figure 3. IPCJ statistics for summer (JJA), from WRF (9 km), (a,c) wind roses of mean wind speed (ms-1), when jet occurs, at jet height with prevail direction, and (b, d) the 2D histogram (%) of the jet height with the wind speed maxima (ms-1), for for area A1 and A2, respectively.

4 CASE STUDY ANALYSIS OF IBERIAN PENINSULA CLLJ

The synoptic forcing behind the IPCJ occurrence is associated with the presence of the Azores High and a thermal low inland. A strong IPCJ event was identified at 14th July 2000, at 1400 UTC in vicinity of Aveiro. The synoptic pattern of this case study is characterized by a deeper high pressure system north of the Azores Islands, and consequently the zonal pressure gradient strait to central Iberia.

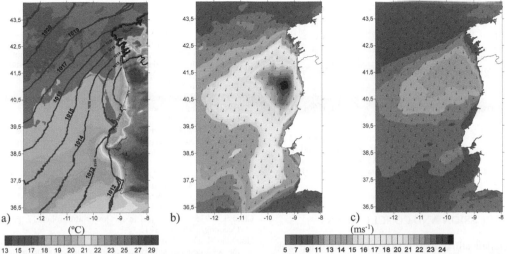

a) ... b) ... c) ...

(°C)   (ms⁻¹)

13 15 17 18 19 20 21 22 23 25 27 29   5   7   9 11 13 14 15 16 17 18 20 21 22 23 24

Figure 4. Horizontal fields from WRF at 3 km of (a) Surface temperature ~2m (°C) overlaid with the MSLP contours, wind speed (b) at lowest model level ~10 m (ms-1), and (c) at jet height (ms-1) overlaid with wind direction, for July 14th 2000 at 1900 (Lt).

The contrast between land-sea temperatures associated with this zonal pressure gradient provides the baroclinic structures behind the IPCJ case study event as it is seen in Figure 4a from the WRF simulations at 3 km resolution. The horizontal wind speed maps for this case study simulation at the lowest model level (~10 m ASL) and at the jet height are represented in Figure 4b and 4c, respectively. The close to surface wind speed pattern clearly shows the interaction of the flow with Cape Finisterre ~43°N, and the flow acceleration southward, with an area of increased wind speed, in excess of 14 ms-1 west of Porto to Aveiro, in Portugal. At the jet height the wind speed increase is even better defined, with a jet core with wind speeds higher than 25 ms-1.

The ~10 m height surface wind speed pattern (Figure 4b) also shows a clear interaction of the flow with Cape Finisterre. Actually the flow is accelerated at Cape Finisterre, and this acceleration dominates the wind speed pattern along the whole west coast of Iberia. The changes in the wind direction, due to the topography effect are marginal. Nevertheless slight coast-ward wind direction changes can be seen, at the surface, south of Cape Finisterre, and south of Cape Roca and Cape St. Vincent. Wind speed maxima at jet height occurs at a considerable lower altitude (~200 m).

To investigate the vertical structure of the IPCJ the vertical across- and along-flow sections are show in Figures 5 through lines passing on key point a (Figure 1b). The potential temperature lines (isentropes) are also shown in these figures. From the cross-flow cross-section (Figure 5a) of case a high wind-speed core with wind speeds in excess of 25 ms-1, is clearly visible near the coast. The offshore extension of this coastal jet occurrence is

considerable, with the jet core spanning to distances higher than 250 km from the coast. The tilting of the isentropes towards the coast and the location of the jet core within the MABL capping inversion close to the coast are well defined, and typical of the characteristics of a CLLJ (Beardsley et al. 1987). This slope provides the thermal wind that forces the jet. The potential-temperature contours also show the capping inversion separating the warm air above and the relatively cool MABL at the surface. In the along-flow section shown in Figure 5b a very pronounced lowering of the MABL height (capping inversion) is seen. In less than 150 km the MABL height goes from more than 500 m to less than 200 m. As a consequence the wind speed, at the lower MABL height area south of Cape Finisterre, increases significantly through a Bernoulli like effect, where the flow is constrained to a considerably lower section.

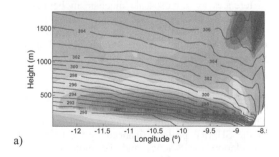

a)

161

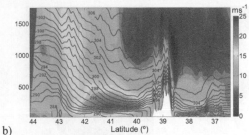
b)

Figure 5. Case-study vertical cross-sections from WRF at 3 km for case study 1 on July 14th 2000 at 1900 (Lt) of (a) cross-coast and (b) along-coast (see lines in Figure 1) of wind speed (ms-1; shading contours) and potential temperature (K; blue full lines).

5 DISCUSSION AND CONCLUSIONS

A detailed climatology of the IPCJ along the west coast of Iberia is presented, based on a regional reanalysis (Soares et al. 2012), produced with the WRF model for the period 1989-2007, complementing the more general climatology from Soares et al. (2014).

The IPCJ is a summer event whereas July is the month where the IPCJ is more frequent, and also the month where the wind speeds are higher. It is shown that although the IPCJ are more frequent in the south area of the west coast of Iberia (area A2), the coastal jets in the north area (area A1) are stronger, since the mean wind speed at the jet height is higher. The IPCJ occur in low troposphere in firs 200-400 m ASL, which could be considered as a relative shallow coastal jet.

The dynamic structure of the IPCJ is also studied trough the analysis of a particular IPCJ event. It is shown that the Cape of Finisterre is the main flow "accelerator" in the north of west coast of the Iberian Peninsula. It is analyzed the costal baroclinicity structures through the cross-coast pressure gradient and the land-sea temperature differences as a mesoscale mechanism behind the IPCJ event.

The extension offshore of the IPCJ is shown to be of the order of 250 km although limited in the vertical extension to less than 400m.

ACKNOWLEDGEMENTS

The work on this study was pursuit in the framework of the SHARE project, financed by the Portuguese Foundation for Science and Technology (FCT – Fundação para a Ciência e Tecnologia, Portugal). All authors, with the exception of Nadia Rijo, are part of the SHARE project (RECI/GEO-MET/0380/2012). Nadia Rijo is financed by the Portuguese Navy.

REFERENCE LIST

Beardsley R. C., Dorman C. E., Friehe C. A., Rosenfield L. K., and Winant C. D., 1987: Local atmospheric forcing during the Coastal Ocean Dynamics Experiment 1: A description of the marine boundary layer and atmospheric conditions over a northern California upwelling region. J. Geophys. Res., 92, 1467-1488.

Burk S. D., and Thompson W. T., 1996: The summertime low-level jet and marine boundary layer structure along the California coast. Mon. Wea. Rev., 124, 668–686.

Burk S. D., Haack T., and. Samelson R. M, 1999: Mesoscale simulation of supercritical, subcritical, and transcritical flow along coastal topography. J. Atmos. Sci.,56, 2780-2795.

Cardoso R.M., Soares P.M.M., Miranda P.M.A., Belo-Pereira M., 2013. WRF high resolution 627 simulation of Iberian mean and extreme precipitation climate. Int J. Climatol. 33, 2591-2608. DOI: 10.1002/joc.3616.

Chao S., 1985: Coastal jets in the lower atmosphere. J. Phys. Oceanogr., 15, 361-371.

Garreaud R. D. and Munõz, R. C. 2005. The low-level jet off the west coast of subtropical South America: structure and variability. Mon. Wea. Rev. 133, 2246-2261.

Lopes A., Oliveira S., Fragoso M., Andrade J. A., and Pedro P., 2009. Wind risk assessment in urban environments: the case of falling trees during windstorm windstorm events in Lisbon, Bio. And Nat. Hazards. International Scientific Conference, Poľana nad Detvou, Slovakia, ISBN 978-80-228-17-60-8

Pauly D., and Christensen V., 1995. Primary production required to sustain global fisheries. Nature. 374, 255-257.

Ranjha R., Svensson G., Tjernstro" M., and Semedo A., 2013. Global distribution and seasonal variability of coastal low-level jets derived from ERA-Interim reanalysis. Tellus A. 65, 20412.

Ranjha, R., Tjernström M., Semedo A., and Svensson G., 2014: Structure and Variability of the Oman Coastal Low-Level Jet. Under review - Tellus A.

Skamarock W. C., Klemp J. B., Dudhia J., Gill D. O., Barker D. M., and co-authors. 2008. A Description of the Advanced Research WRF Version 3. NCAR Tech. Note TN-468 + STR, 113 pp.

Soares P. M. M., Cardoso R. M., Miranda P. M. A., Medeiros J., De, Belo-Pereira M., and co-authors. 2012. WRF high resolution dynamical downscaling of ERA-Interim for Portugal. Clim. Dyn. 39, 2497_2522.

Soares P.M.M., R.M. Cardoso, Semedo A., Chinita M.J., Ranjha R., 2014: Climatology of the Iberia coastal low-level wind jet: weather research forecasting model high-resolution results. Tellus A, 66, 22377.

Söderberg S., and Tjernström M., 2002: Diurnal Cycle of Supercritical Along-Coast Flows', J. Atmos. Sci. 59, 2615–2624.

Tjernström M., and Koracin D., 1995: Modeling the im act of marine stratocumulus on boundary layer structure. J. Atmos. Sci., 52, 863–878.

Tjernström M., and Grisogono, B. 2000. Simulations of supercritical flow around points and capes in a coastal atmosphere. J. Atmos. Sci. 57, 108-135.

Vallis, G.K. 2012. Climate and the Oceans. Princeton University Press, p244.

Warner T. T., 2004. Desert Meteorology. Cambridge University Press, Boston, MA, 595 pp.

Winant C. D., Dorman C. E., Friehe C. A. and Beardsley R. C., 1988. The marine layer off northern California: an example of supercritical channel flow. J. Atmos. Sci. 45, 3588-3605.

Inland Shipping

Emergency Group Decision-Making with Multidivisional Cooperation for Inland Maritime Accident

B. Wu, X.P. Yan & Y. Wang
Intelligent Transport Systems Research Center (ITSC). National Engineering Research Center for Water Transport Safety. Engineering Research Center for Transportation Safety (ERCTS, MoE). Wuhan University of Technology, Wuhan, China

J.F. Zhang
Centre for Marine Technology and Engineering (CENTEC), Instituto Superior Técnico, Universidade de Lisboa, Portugal

ABSTRACT: This paper aims to address the decision-making problem of emergency response for inland maritime accident. Since a maritime accident usually involves the behaviors of the ships in trouble and the ships passing-by, as well as the actions of the maritime authorities, the decision-making model should take into account the reactivity from multiple parties. Unlike the traditional group decision-making model which concerns more about the aggregation of opinions given by different experts, this paper focuses on the cooperative decision-making for experts from different departments. A cooperative group decision-making model is introduced after some basic assumptions are made. The presented model is then applied to a case study in the predefined scenario. Compared with the traditional group decision-making model, the combination option made by utilizing the cooperative model is more feasible and sensible for maritime accident. Finally, conclusions and remarks on the employed model are made and future research work is also proposed.

1 INTRODUCTION

Group decision-making plays an essential role in choosing a best or compromise option, and has been a keen research topic in the field of multiple criteria decision-making (MCDM). Unlike MCDM, group decision-making focuses on the coordination of the experts when they make different or even conflict choices. This is very common because each expert may have their preferred orderings of the attributes. Moreover, the choice of a single expert may be not persuasive or feasible due to the inherent complexity or incomplete information in the process of decision-making. Thus group decision-making is widely used in order to integrate the experts' viewpoints and overcome limitation of individuals.

One of the critical problems is how to deal with conflict information among experts in group decision making. This conflict may result from the following factors: (1) the experts assess the situation diversely due to preference orderings which are influenced by their educational backgrounds or professional experience such as the personalities, attitude and others. (2) The experts pay different extent of attentions to the attributes according to their own interests. For example, Maritime Safety Administration (MSA) closely concerns about the organization of search and rescue (SAR) and the

safety of the whole waterway traffic, while ship operators focus on ship damage control and crew safety. (3) The high level of stress may influence the ability of decision makers. The accident may suddenly lead to entirely different results according to different intervention strategy, thus the decision makers should make the best choice in very short time. (4) The option made under uncertainty may be not be reliable and trustable. The uncertainty involves incomplete or imprecise information in emergency response.

In order to overcome the vagueness and fuzziness, many effective and feasible methods have been introduced. To list some, Xu (2007) summarized that all group decision-making methods should involve the following information: (1) Collection of information about attributes weights and values; (2) weighted aggregation of the values across all attributes for each alternative to obtain an overall value; (3) Selecting the best alternative according to the orders of overall values.

As for collecting information, linguistic variables (Xu, 2007) have been widely used to demonstrate the importance of linguistic variables. Specifically, Fuzzy rules (Ölçer & Ödabaşi 2005, Yeh & Chang 2009, Yu et al., 2009), Evidential Reasoning (Xu et al., 2006), Utility Function (Huang et al., 2013) are the mostly utilized methods for illustrating the preference orderings, such as fuzzy preference relations (Fan et al., 2006), interval preference orderings (Liu et al., 2012,), Xu et al., 2006, Xu, 2013) and linguistic preference orderings (Dong et al., 2006). While in the aggregation of attribute values for incomplete information (Mateos et al., 2006, Xu et al., 2006, Xu, 2007), the group ranking problem can be roughly classified into two major approaches, the total ranking approach and the partial ranking approach (Chen & Cheng 2009). AHP is the well-recognized method for aggregation of attributes weights. But in most cases it cannot take the relationship among the attributes into consideration. To solve the problem, a Bayesian-based AHP method was proposed by Altuzarra et al. (2007). Moreover, the experts express their standpoint in a less standard way like AHP method. And the preference orderings are much more fuzziness and imprecise. Thus TOPSIS (Xie et al., 2011), Entropy (Xia & Xu 2012), distance-based (Cook, 2006) method were employed.

As mentioned above, many powerful models have been proposed in group decision making. But when applying to the maritime accident response process, some issues have to be studied accordingly. There are multiple related departments in the process of decision making during a maritime accident. If the decision maker considers the problem from the perspective of his own department, it would give rise to inconsistency of execution. Thus all suggestions from other departments should be taken into consideration to obtain a highly uniformed SAR action sequence. Unfortunately, little research concerning about the multidivisional cooperation within group decision-making have been released. In order to address such problem, a multidivisional cooperation model was introduced in this paper.

The rest of this paper is organized as follows. In Section 2, the maritime accident group decision-making is described and some pre-assumptions are presented. Section 3 introduces a coordinative group decision-making model including the definition of cooperative network, cooperative matrix and cooperative coefficient. Section 4 presents algorithm of cooperation model. And an experimental investigation was studied by employing the cooperation decision-making model to address the maritime accident problem in Section 5. Finally, some concluding remarks are drawn in Section 6.

2 PROBLEM STATEMENT AND ASSUMPTIONS

Argumentation is a traditional but widely used method in group decision-making (Amgoud & Prade 2009). Through argumentation, the experts can easily express their preference orderings which enables others to easily get the key point. As information technology keeps on facilitating group decision-making in emergency response, the framework of group decision-making about emergent event based on network technology is proposed (Xie et al., 2011). Thus in order to apply the aforementioned framework to maritime accident case, the modeling and analysis of group decision-making plays a crucial role in emergency response based on network technology.

In term of maritime accident emergency response for inland transportation ship, several departments would be involved according to different accidental type. In the group decision making, each department should fully understand the opinions of other departments and reach a strategy in common as quick as possible. Three related departments that are closely related with maritime emergency response were taken into consideration in this paper. The role of each department and their basic principles in decision-making are shown as follows.

Ship owner department: it will try its best to damage control or risk migration by effective intervention methods. Thus normally they will not consider interests of other ships. But it should be noticed that the intervention methods may varies according to the navigational environment and the condition of ship.

Ships passing-by: ships passing-by concern more about the delay time, it cannot be tolerated to stay in a long time owing to the influence of this incident. As many ships would be blocked which will make the channel congested, the interests of ship passing by should be carefully considered.

Maritime safety administration (MSA): a department of governor in charge of waterway transportation safety and pollution prevention in China. In the perspective of MSA, the accidental ship should be properly disposed and the time of congestion would be as few as possible.

According to the aforementioned principles, the decision options of each department are shown in Table 1.

Table 1. Decision options for maritime accident

Option	Accidental ship	Ship passing-by	MSA
Option1	Anchoring in anchorage	Passing around the accidental spot	Traffic organization by Patrol Ships
Option2	Anchoring in channel	Navigate slowly	One-way traffic
Option3	Grounding Initiatively	Anchoring immediately	Restriction of navigation

In this study, the option of accidental ship may influence the options of ship passing by and MSA. For example, if the accidental ship try to anchor in the channel, it would be difficult to navigate freely for ship passing by and the MSA would have to choose the one-way traffic or closure of navigation most of the time. An important thing to be noticed is that the options of departments would be changed along with the decision option of related department.

The assumptions made in this study are as follows: (1) only one decision maker represent one involved department who is adequate for decision-making according to his knowledge. (2) All decision makers could understand the decision options proposed by other departments as well as make feasible determinations if the options are related. (3) The information obtained is sufficient for reasonable decision making. Thus the incomplete or imprecise information is negligible for this decision making model.

3 MODELING OF COOPERATIVE DECISION-MAKING

Chen et al. (2010) proposed an emergency decision model with multiple stages, multiple objectives, and multidivisional cooperation, this model is really well-organized and the definitions of cooperation decision-making model were adopted and introduced in the subsections of this paper.

3.1 *Cooperative network*

In the traditional MCDM models, the experts from different departments make comments and decisions on the predefined criteria which is proposed by a certain department, in that case the options or measures in each department was assumed independent from each other. In practice, the options may be related or even conflict. For example, in a ship accident, from the viewpoint of the crews, they would take effective and loss-least option like anchoring in the channel; but from the perspective of ship of passing by, they would like the channel to be unobstructed so that they could navigate without unencumbered. Thus the two departments may conflict in this problem and the MSA should balance the standpoints from both departments and make the best option which could be carried out by cooperative decision-making model.

As mentioned above, the emergency decision-making may involve many departments which may be independent from each other or not. The relationship between department j and department i can be denoted as $A(i,j)$, if the options of department j are independent from the options of department i, then $A(i,j)=1$, otherwise $A(i,j)=0$. Only in the case of $A(i,j)=1$ that each department should take the suggestions of other departments into consideration, through that the cooperative network was established and defined.

3.2 *Cooperative matrix*

The best option of each department is decided not only by the reliability (such as the possibility, expense and effectiveness) of the department itself, but also decided by the related department. Thus the cooperative matrix was proposed to consider the options made by other departments.

Cooperative matrix W is introduced to illustrate the importance of the other departments. And $W(i,j)$ is the cooperative coefficient of department i and department j. It could be easily obtained that the cooperative coefficient is 0 if the two departments are independent from each other. An essential theme to mention is that the cooperative matrix W is not always a symmetric matrix. This is very common in practice because the option of department i may influence department j but option of department j is independent from department i.

3.3 *Cooperative optimization model*

In the cooperative decision-making model, each department would take the other related departments options into consideration in the process of decision-making, from this perspective the cooperative decision option is a combination option of each department, and the goal of cooperative decision is to find a an optimization decision-making model which could balance the benefits and costs of the involving departments. The equation (1) is written as follows.

$$S_j(x_{iai}) = W(i,j)s_j^{self}(x_{iai}) + \sum_{t:A(t,j)=1} W(t,i)s_j^{other}(x_{iai} \mid x_{kak}) \quad (1)$$

The reliability of the combination option $S_j(C), j=1,2\cdots p$ under different criteria j is supposed to be Maximum. Thus the case can be addressed as an optimization problem with p object. As $S_j(C)$ is a N dimensional vector, thus the optimization model can be achieved by computing the maximum norm. And the multiple objectives cooperative decision-making model can be illustrated as equation (2).

$$\max \left\{ \|s_1(C)\|, \cdots \|s_j(C)\| \cdots, \|s_p(C)\| \right\} \quad (2)$$

$$s.t. S_j(x_{iai}) = W(i,i)s_j^{self}(x_{iai}) + \sum_{t:A(t,i)=1} W(t,i)s_j^{other}(x_{iai} \mid x_{kak}), \forall i \in \{1,2\cdots,N\}, j=\{1,2\cdots,p\}$$

$$W(i,j) \geq 0, \forall i \in \{1,2\cdots,N\}, j=\{1,2\cdots,N\}$$

$$0 \leq s_j^{self}(x_{iai}) \leq 1, \forall i \in \{1,2\cdots,N\}, j=\{1,2\cdots,p\}$$

$$0 \leq s_j^{other}(x_{iai} \mid x_{kak}) \leq 1, \forall i \in \{1,2\cdots,N\}, j=\{1,2\cdots,p\}, k=\{1,2\cdots,N\}, A(t,i)=1$$

where

$\| s_j(C) \|$ means the norm of vector $s_j(C)$.

N -Number of departments;

p - Number of evaluation criteria;

$R_i = \{x_{i1}, x_{i2} \cdots, x_{iki}\}$ - Alternatives of department i, where k_i means the number of alternatives of department i;

$C = \{x_{1a1}, x_{iai} \cdots, x_{NaN}\}, ai \leq ki$ - Combination option;

$s_j^{self} = (x_{iai})$ - Reliability of department i with the option j by itself;

$s_j^{other} = (x_{iai} | x_{tat})$ - Reliability of department t with the option j of department i when choosing x_{ta_t} ;

$s_j = (x_{iai})$ - Reliability of other related departments with the option j of department i ;

$s_j(C) = (s_j(x_{1a1}), \cdots, s_j(x_{iai}), \cdots s_j(x_{NaN}))$ - Reliability of all departments with the combination option c under criteria j ;

$S(C) = (s_1(C), \cdots, s_j(C), \cdots s_p(C))$ - Reliability of all departments with the combination option c ;

$W(i, j)$ - Cooperative coefficient of department i for department j ;

$A(i, j)$ - The value is 1 If the department i is related with department j otherwise is 0.

3.4 Simplification of cooperative model

In order to simplify equation (2) where weights of the alternatives were unknown or uncertainty, the cooperative optimization model can be addressed in a liner weighted way. Specifically, each weight of alternative can be supposed as $a(i)$, and this value can be decided either by the commander like MSA or by the information expressed by experts.

Thus the cooperative optimization model could be simplified as equation (3).

$$\max \sum_{j=1}^{p} \left\| a(j) s_j(C) \right\| \tag{3}$$

$s.t. S_j(x_{iai}) = W(i,i) s_j^{self}(x_{iai}) + \sum_{t:A(t,i)=1} W(t,i) s_j^{other}(x_{iai} | x_{kak}), \forall i \in \{1,2\cdots,N\}, j = \{1,2\cdots,p\}$

$W(i, j) \geq 0, \forall i \in \{1,2\cdots,N\}, j = \{1,2\cdots,N\}$

$0 \leq s_j^{self}(x_{iai}) \leq 1, \forall i \in \{1,2\cdots,N\}, j = \{1,2\cdots,p\}$

$0 \leq s_j^{self}(x_{iai}) \leq 1, \forall i \in \{1,2\cdots,N\}, j = \{1,2\cdots,p\}$

$0 \leq s_j^{other}(x_{iai} | x_{kak}) \leq 1, \forall i \in \{1,2\cdots,N\}, j = \{1,2\cdots,p\}, k = \{1,2\cdots,N\}, A(t,i) = 1$

4 ALGORITHM OF COOPERATION MODEL

4.1 Cooperative optimization model

As mentioned above, there are many effective methods to obtain the weights of alternatives Including the subjective method like AHP, and the aggregation method such as TOPSIS, Entropy. We introduce the subjective method in this paper, and for further research the other methods should be reviewed and an improved method should be proposed according to the obtained information in the process of decision-making.

Firstly, the experts will evaluate the weight of the department itself. And value of department A for option i under criteria j is defined as $s(A, i; k)$, where $0 \leq s(A, i; k) \leq 1$, moreover, the bigger the value, the importance of the alternative is.

While applied to the cooperative decision-making model, the related departments (if $A(i, j) = 1$) should make comments on the reliability of each alternative. The method is similar with the method

utilized by the department itself. But an important theme to notice is that the weight is related to different options, thus the department A chooses option A1 under the situation of the department B chooses option B2 under criteria can be defined as $s(B, B2, A, A1; k)$.

4.2 Computation of combination option under each criterion

The option of department i was defined as $C(i)$ when the combination option is C, thus the value of $C(i)$ under criteria k can be described as equation (4).

$$s(C(i), C; k) = \sum_{j: A(j,i)=1} W(j, i) s(i, C(i), j, C(j); k) + W(i, i) s(i, C(i); k) \tag{4}$$

The value of combination option is a vector of the value of each department. Suppose the number of departments is N, thus the value of the combination option under criteria k is shown in equation (5).

$$s(C; k) = s(C(1), C; k), \cdots, s(C(i), C; k) \cdots s(C(N), C; k) \tag{5}$$

4.3 Computation of combination option

The value of combination option is defined as equation (6).

$$s(C) = \sum_{k=1}^{p} a(k) s(C; k) \tag{6}$$

The value of combination option $s(C)$ is a vector in practice. Thus the optimization option can be obtained by the comparison of different vectors under different options.

4.4 Best option selection

In the process of cooperative decision making, the interest of each department should be taken into consideration in this cooperative decision-making model. Sometimes, one of department should take rigorous option while the combination is the best option. Thus a threshold value method should be predefined to exclude the rigorous option.

Firstly, each option should have a minimum value. If the value of any option for any department is smaller than the predefined minimum value 0.3, the corresponding option should be excluded.

Secondly, the best combination option is selected by normalized weighting method. Suppose the number of departments are N, thus each cooperative option is an N dimensional vector. If the vector elements are described as $s(1), s(2), \cdots, s(N)$, the combination value can be obtained.

$$s(C) = \sum_{k=1}^{C} b(i) s(i) \tag{7}$$

where $b(i)$ means the importance or the weight of department i, and the weight can be obtained by the commander of the emergency cooperative decision making.

5 COOPERATIVE DECISION-MAKING FOR MARITIME ACCIDENT

5.1 Scenario analysis

A ship accident was used as a case study in Jiangyin port in Yangzi River. The ship was not under control because of its rudder failure. This case was supposed to be an emergency event, and the ship would collide with other ships if any efficient methods were not adopted. The navigational environmental factors and the ship condition were shown as follows.

Ship condition: not under control for rudder failure, but the Jury Rudder was available. And main engines and anchors work well.

Ship Tonnage: 10, 000 DWT (Dead Weight Tonnage).

Channel: vessel Traffic Separation Scheme (TSS) was established, this channel is suitable for one-way traffic of 50, 000 DWT, and two-way traffic of 30, 000 DWT. But the depths of surrounding waterway area were good and suitable for ship under 5, 000 DWT.

Anchorages: the nearest anchorage is about 3 nautical miles (NM). Navigational structures: there are few navigational structures, but no obstructive navigational structures exited in the nearby waterway such as bridges. Shoals: there are many wharfs nearby; only one shoal was available for beach landing. Traffic flow: one-way traffic flow was about 70 vessels per hour including 60% of ship under 5000 DWT.

In order to achieve a cooperative decision-making option, three experts (decision makers) were invited to give expert knowledge on this scenario. Each expert represents one department in this cooperative decision-making case, and the background of each expert is as follows.

Chief mate: representative of the accidental ship (department A). He has been worked on an ocean-going ship as a chief mate for two years, and he has worked as a seaman more than ten years. Second mate: representative of the ship passing-by (department B). He has been worked on ocean-going ship as a second mate for three years, and he has worked as a seaman more than ten years. Staff in MSA: He has worked in Vessel Traffic System department for six years. And he is the reprehensive of MSA (department C).

5.2 Options and criteria

The three experts were invited to make judgment on the options listed in Table 3. Though some of the options are not suitable for this case study, all options were requested to be judged by each expert. This is required to make the whole model sensible and reasonable, but in practice especially in emergency event, the option that was unlikely to be taken up could be omitted according to the suggestions of experts.

The criteria utilized for judgment in this case study were cost (CT), effectiveness (EF) and possibility (PB) respectively. The assumption should be made that only the own interest would be taken into consideration since each of them represent one department. In order to simplify the decision-making process, the weight of each criterion is predefined. $W = (W_c, W_e, W_p) = (0.2, 0.4, 0.4)$.and the judgment of each option under different criteria is shown in Table 2 where comprehensive result (CR) was also given.

The table presents that the importance of each option under the abovementioned criteria. The bigger value means that the estimated cost of this option is smaller. While the bigger value effectiveness means the option is more effective. And the bigger value of possibility represent the option is more feasible. The judgments of these options are according to the following factors which ware proposed by experts. (1) Option A3 cost most because the ship would be damaged if the speed and heading of grounding is not well-controlled. Option A1 is more effective because the traffic flow would not be influenced after anchoring, While Option A2 and A3 could be easily adopted. (2) Option B1 owns the merits of less cost and more effectiveness because 60% ship could bypass the channel. Option B2 seems to be more possible because the influence of accidental ship. (3) Option C1 would be most useful to end the chaos situation.

Table 2. Judgment of each option under different criteria

option	CT	EF	PB	CR
option A1	0.7	0.7	0.5	0.62
option A2	0.5	0.6	0.8	**0.66**
option A3	0.4	0.6	0.8	0.64
option B1	0.7	0.8	0.6	**0.7**
option B2	0.3	0.8	0.7	0.66
option B3	0.2	0.5	0.5	0.44
option C1	0.5	0.6	0.7	**0.62**
option C2	0.4	0.5	0.4	0.44
option C3	0.5	0.4	0.4	0.42

5.3 Cooperative group decision-making (CGDM)

From the perspective of accidental ship, the best choice is anchoring in the channel. But this option would influence the ship passing-by even cause congestion. Thus the experts were requested to make judgment on the options of other departments. And the results were shown in Table 3.

Table 3 reveals that the result of option B3 and C3 was significantly lower than the other options. This is because if the ship passing-by stops sailing, the whole waterway would be congested. Moreover, the ship passing-by would also get loss during the stopover which could conclude from the judgment on B3 under different options of department A. similarly, the judgment on option C3 is also lower than others because closure of navigation will make the other ships congested.

Table 3. Judgment on options (OP) of other departments

OP	A1	A2	A3	B1	B2	B3	C1	C2	C3
B1	0.8	0.7	0.8	-	-	-	0.7	0.3	0.15
B2	0.6	0.6	0.4	-	-	-	0.2	0.3	0.15
B3	0.3	0.3	0.3	-	-	-	0.2	0.3	0.1
C1	0.8	0.8	0.4	0.6	0.5	0.15	-	-	-
C2	0.5	0.6	0.4	0.4	0.6	0.15	-	-	-
C3	0.2	0.2	0.2	0.2	0.3	0.25	-	-	-

The cooperative decision-making was then applied to model. In this study, the cooperative coefficient W is predefined as 0.5, otherwise if no cooperative is conducted, cooperative coefficient is 1.0. The result of computation was shown in Table 5. It should be mentioned that only the combination option (CO) and its three components (FC, SC, TC) under A1 was listed in Table 4.

Table 4. Result of computation

CO	FC	SC	TC
A1B1C1	0.71	0.65	0.66
A1B1C2	0.635	0.55	0.37
A1B1C3	0.56	0.45	**0.285**
A1B2C1	0.66	0.58	0.41
A1B2C2	0.585	0.63	0.37
A1B2C3	0.51	0.48	**0.285**
A1B3C1	0.585	**0.295**	0.41
A1B3C2	0.51	**0.295**	0.37
A1B3C3	0.435	0.345	**0.26**

It could easily obtained from Table 4 that some values was lower than 0.3. According to the algorithm proposed in section 4.4, these options should be excluded because they were not supposed to be recognized by other cooperative departments, and the value lower than a minimum value were in bold in Table 4. Thus 12 combination options were recognized by all departments which were shown in Table 5.

Table 5. Effective combination options

CO	FC	SC	TC
A1B1C1	0.71	0.65	0.66
A1B1C2	0.635	0.55	0.37
A1B2C1	0.66	0.58	0.41
A1B2C2	0.585	0.63	0.37
A2B1C1	0.705	0.65	0.66
A2B1C2	0.655	0.55	0.37
A2B2C1	0.68	0.58	0.41
A2B2C2	0.63	0.63	0.37
A3B1C1	0.62	0.65	0.66
A3B1C2	0.62	0.55	0.37
A3B2C1	0.52	0.58	0.41
A3B2C2	0.52	0.63	0.37

5.4 Cooperative group decision-making (CGDM)

The weight of each department was then utilized to computation of combination options. Since the accidental ship plays an essential role in triggering the actions of other departments, the weights of the three departments were supposed as $(W_A, W_B, W_C) = (0.5, 0.25, 0.25)$. The comparison between cooperative and uncooperative method was shown in Table 6.

As shown in Table 6, the result of the two methods varies from each other. The best choice of cooperative method is A1B1C1, while the uncooperative method is A2B1C1. The choice of A2B1C1 makes sense only if each department considers the interests of itself. But this choice ignores the interests of other departments, and the cooperative method could make a comprehensive choice. From the perspective of experts, the accidental ship could use Jury Rudder to control heading since the anchorage was not far away and the traffic flow was not density. Correspondingly, majority of ship passing-by could bypass the channel because the depths of surrounding waterway were suitable for majority of the ships. And MSA could make traffic organization to ensure the safety of the accidental ship.

Table 6. Comparison between cooperative and uncooperative method

option	CGDM	ranking	MCDM	ranking
A1B1C1	0.6825	1	0.63	5
A1B1C2	0.5475	8	0.585	11
A1B2C1	0.5775	5	0.62	6
A1B2C2	0.5425	9	0.575	12
A2B1C1	0.68	2	0.66	1
A2B1C2	0.5575	7	0.615	7
A2B2C1	0.5875	4	0.65	2
A2B2C2	0.565	6	0.605	8
A3B1C1	0.6375	3	0.65	2
A3B1C2	0.54	10	0.605	8
A3B2C1	0.5075	12	0.64	4
A3B2C2	0.51	11	0.595	10

6 CONCLUSIONS AND REMARKS

Compared with the traditional MCDM, the CGDM model could take the suggestions of other departments into consideration. Thus the combination option of this method seems to be more feasible and effective. And this methodology could be widely applied to emergency response of ship accident. In this study, the weights of different departments were predefined. An improved method should be proposed especially when the accidental ship is in traditional incident. And future work on this field should be conducted.

As mentioned in Section 5, the ship could anchor in the anchorage because few interference of ship passed. Actually the option would change according to the different navigational environment and ship condition. For example, if the ship is not under control owning to the machinery failure as well as rudder failure, the best option would be changed. Moreover, if the accidental ship should pass a bridge in this scenario, the option of anchoring in anchorage should be omitted for its high level risk. Thus a feasible decision-making model for accidental ship should be incorporated into this cooperative model in the future research.

ACKNOWLEDGEMENT

This Paper is supported by Research Fund for the Doctoral Program of Higher Education of China (Grant No.20130143120014).

REFERENCES

Altuzarra, A., Moreno-Jiménez, J.M. & Salvador, M. (2007). A Bayesian priorization procedure for AHP-group decision making. *European Journal of Operation Research* 182(1):367-382.

Amgoud, L. & Prade, H. (2009). Using arguments for making and explaining decisions. *Artificial Intelligence*173(3-4):413-436.

Cook, W.D. (2006). Distance-based and ad hoc consensus models in ordinal preference ranking. *European Journal of Operation Research* 172(2): 369-385.

Chen, X., Wang, Y., Wu, L.Y., Yan, G.Y. & Zhu, W. (2010). Emergency decision model multiple stages, multiple objectives, and multidivisional cooperation. *System Engineering-Theory and Practice* 30(11): 1977-1985.

Chen, Y.L. & Cheng, L.C. (2009). Mining maximum consensus sequences from group ranking data. *European Journal of Operation Research* 198(1): 241-251.

Dong, Y.C., Xua, Y.F. & Yu, S. (2009). Linguistic multi person decision making based on the use of multiple preference relations. *Fuzzy Sets and Systems* 160(5):603-623.

Fan, Z.P., Ma, J., Jiang, Y.P., Sun, Y.H. & Ma, L. (2006). A goal programming approach to group decision making based on multiplicative preference relations and fuzzy preference relations. *European Journal of Operation Research* 174(1):311-321.

Huang, Y.S., Chang, W.C., Li, W. H. & Lin, Z.L. (2013). Aggregation of utility-based individual preferences for group decision-making. *European Journal of Operation Research* 209(2):462-469.

Liu, F., Zhang, W.G. & Wang, Z.X. (2012). A goal programming model for incomplete interval multiplicative preference relations and its application in group decision-making. *European Journal of Operation Research* 218(3):747-754.

Mateos, A., Jiménez, A. & Ríos-Insua, S. (2006). Monte Carlo simulation techniques for group decision making with incomplete information. *European Journal of Operation Research* 174(3):1842-1864.

Ölçer, A.İ. & Ödabaşi, A.Y. (2005). A new fuzzy multiple attributive group decision making methodology and its application to propulsion/maneuvering system selection problem. *European Journal of Operation Research* 166(1):93-114.

Xia, M.M. & Xu, Z.S.(2012).Entropy/cross entropy-based group decision making under intuitionistic fuzzy environment. *Information Fusion* 13(1):31-47.

Xie, K.F., Chen, G., Wu, Q., Liu, Y. & Wang, P. (2011). Research on the group decision-making about emergency event based on network technology. *Information Technology and Management* 12(2): 137-147.

Xu, D. L., Yang, J.B. & Wang, Y.M. (2006). The evidential reasoning approach for multi-attribute decision analysis under interval uncertainty. *European Journal of Operation Research* 174(3):1914-1943.

Xu, Z.S. (2007). A method for multiple attribute decision making with incomplete weight information in linguistic setting. *Knowledge based Systems* 20(8):719-725.

Xu, Z.S. (2013). Group decision making model and approach based on interval preference orderings. *Computers & Industrial Engineering* 64(3):797-803.

Yeh, C.H. & Chang, Y.H. (2009).Modeling subjective evaluation for fuzzy group multi criteria decision making. *European Journal of Operation Research* 194(2): 464-473.

Yu, L., Wang S.Y & Lai, K.K. (2009). An intelligent-agent-based fuzzy group decision making model for financial multi criteria decision support: The case of credit scoring. *European Journal of Operation Research* 195(3):942-959.

Inland Shipping
Information, Communication and Environment – Marine Navigation and Safety of Sea Transportation – A. Weintrit & T. Neumann (eds.)

The Concept of Emergency Notification System for Inland Navigation

T. Perzyński, A. Lewiński & Z. Łukasik
University of Technology and Humanities in Radom, Poland

ABSTRACT: The paper presents solutions used in Polish inland navigation which raise safety in a water area, especially in water tourism. Solutions presented in the article allow limiting the dangerous events caused by weather phenomenon. As a result of the use of described solutions, a number of injured people can be decreased. In the article the conception of new system which allows faster reaction of emergency services (e.g. Aquatic Volunteer Emergency Corps) in dangerous situations is presented.
The conception of the system proposed in the paper can be very useful particularly in faster localization of capsized sailing yacht. The conception is based on telematic solutions (GPS, GSM). Introduced mathematical model of presented system related to theory of stochastic processes allows estimating the quantitative reliability and time criteria of safety.

1 INTRODUCTION

In recent years, we will notice a big change in water tourism in Poland. Investments in ports, marinas and all the water infrastructure is essential to attract tourists. The popularity of this type of tourism is not only a greater availability of water equipment and boat, but also a change in existing legislation, both for sailboats and motorboats [1],[2].

The current regulation regarding to water tourism allows sailing up to 7.5 m yacht length of the hull without documented skills (certificate of competency). In case of motor boats, the navigation is permitted without certificate of competency with an engine power up to 10kW. In the last amendment of the act about inland waterway there has been added one more law regulation (art. 37a § 4) with reference to the motorboat, which exempts the users of yachts with engine power up to 75 kW with a hull length up to 13m, where the maximum speed is structurally limited to 15km/h (Fig. 1.) from the requirement of the qualifying certificate, therefore the big popularity of cabin cruiser.

The changes in the construction of boats and their equipment as well as noticeable changes in the infrastructure still do not ensure a sufficient level of safety. Hence, the concept of the proposed system for tourist yachts, of which idea was born during sailing student camps. Changing weather conditions, even in the summer, violent gusts of wind, cause the

need for quick reaction in the event of a dangerous situation. More comfortable yachts cause the need to build boats with high board. The air mass in spite of folded sails may capsize boat. Currently there are no such solutions on the Polish lakes. Thus the authors proposed to build a system for enhanced safety which enables quick finding of people who need help. The proposed system is similar to those used in open waters - EPIRB (*Emergency Position Indicating Radio Beacon*). Thanks to the different form of sending a message (action based on the GSM network) it will not require registration at the Office of Civil Aviation [6].

2 CURRENT SOLUTIONS

The first system improving safety at Masurian Lakes is warning system before weather phenomena which has been activated in 2011. The people on lakes are informed about weather danger by light signals placed on the appropriate masts (Fig. 1). Without specifying the time and place of arrival of meteorological phenomena, sailors are informed by flash signal occurring at a frequency of 40 flashes per minute. It means that storm or high winds are expected. Imminent danger of storms and strong winds is indicated by 90 flashes per minute. In this case it is better to stay in harbour or quickly sail there. The principle operation of this system is presented on the Fig. 1.

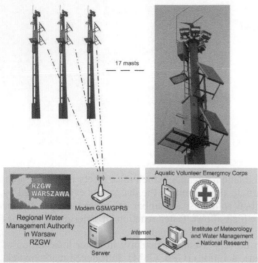

Figure 1. Simplified scheme of principle operation warning system before weather phenomena [7].

Information concerning the weather for Mazurian Lakes is transferred from the *Institute of Meteorology and Water Management – National Research* to *Regional Water management Authority in Warsaw*, where it is analysed. After receiving the information about the weather conditions, appropriate software interprets the message and if it is significant, it sends a signal to the masts and turns them on automatically. The light signals stop transmitting the light message after cancelling or expiration of the announcement. Currently the system consists of 17 masts and the led light which is visible from 8 – 9 km distance.

3 THE CONCEPT OF THE SYSTEM

The most common reason of capsizing the boat is unexpected and violent wind. The first step after capsizing is to check the status of the crew. The crew is often shocked. Next step is attempt to seek help, but very often sunken mobile phones disable emergency call. The pictures of capsized yacht are presented in the Fig. 2 (Mazurian Lakes – August 2007).

The concept of the proposed system (*Emergency Notification System – ENoS*) assumes full autonomous operation. In the situation like on the pictures above, the device automatically sends message (SOS) to the rescue service providing position. The device which sends the signal will be in a hermetic case with own power supply system. The algorithm of the system is shown on the Fig. 3.

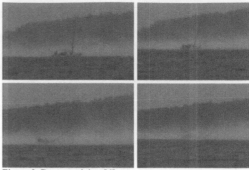

Figure 2. Boat capsizing [6].

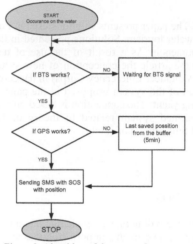

Figure 3. Algorithm of the system [own source]

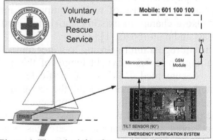

Figure 4. The principle of operation of the system [own source]

The proposed concept assumes that the accelerometer will be used as a tilt sensor. After crossing 90° tilt angle, system turns on and sends a message. In the case of absence of a current GPS signal from a satellite, system takes last saved position from the buffer (for example 5min before). The proposed system can also be equipped with additional functions such as belt tilt (inclinometers). In the event of bringing the boat to its initial position the center of buoyancy is changing and righting moment arises. The largest righting moment is at an angle of approximately 40°[4]. In this case system can signal the dangerous situation to the skipper. The principle of operation of the system is shown on Fig. 4.

4 MARKOV MODEL OF ENOS SYSTEM

Processes related to accidents on the waters can be assimilated to the group of stochastic processes, where a group of random variables is defined in the probability space. In order to make an analysis the proposed system suggested mathematical apparatus (the theory of stochastic processes) in the form of Markov model. Stochastic process X(t) is called the Markov process, if for any finite system $t_1 < t_2 < ... < t_n$ parameter value and for any real numbers $x_1, x_2, ...$, x_n:

$$P\{X(t_n) < x_n | X(t_{n-1}) = x_{n-1}, X(t_{n-2}) = x_{n-2}, ..., X(t_1) = x_1\} =$$
$$= P\{X(t_n) < x_n | X(t_{n-1}) = x_{n-1}\} \quad (1)$$

Conditional distribution $X(t_n)$ at a given values $X(t_1), ... , X(t_{n-1})$ depends only on $X(t_{n-1})$. Markov processes are described by the distribution of two dimensional (time and states of the process). The theory of stochastic processes is used for two reasons:
- safety analysis (probability of a critical event, the unavailability of the proposed system) of equipment based on typical probabilistic and exploitation characteristics,
- safety assessment of boat taking into consideration the standard procedures used in the occurrence of the event (including time parameters) with regard to the functioning of the proposed system under different environmental conditions and climate.

4.1 Markov model for GPS/GSM in ENoS system

Assuming exponential distribution of faults, stationary, homogenous and ergodic character of stochastic Markov process, the model of GPS/GSM in ENoS was built (Fig. 5).

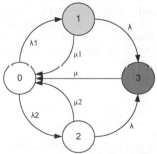

Figure 5. Markov model of GPS/GSM in ENoS system [own source]

where:
$\lambda - \lambda_2$ - failure rate ,
$\mu - \mu_2$ - intensity of service,
In the model from Fig. 5 we can distinguish the following states.
- 0 – state of safety. In the event of danger situation system is correctly working and sending the messages.
- 1 – no GSM signal.
- 2 – no GPS signal.
- 3 – no GPS and GSM signal.

For the proposed model can write equation:

$$\begin{cases} \dfrac{dP0}{dt} = (-\lambda_1 \cdot P0 + \mu_1 \cdot P1 - \lambda_2 \cdot P0 + \mu_2 \cdot P2 + \mu \cdot P3 \\[2mm] \dfrac{dP1}{dt} = \lambda_1 \cdot P0 - \mu_1 \cdot P1 - \lambda \cdot P1 \\[2mm] \dfrac{dP2}{dt} = \lambda_2 \cdot P0 - \mu_2 \cdot P1 - \lambda \cdot P2 \\[2mm] \dfrac{dP3}{dt} = \lambda \cdot P1 + \lambda \cdot P2 \end{cases} \quad (2)$$

In order to solve the system of computer-aided equations was used – *Mathematica* software. The window is presented on Fig. 6.

Figure 6. The window of calculation [own source]

The most dangerous state in the model is state number 3 and state 1 corresponds to GSM communication. Assuming the parameters of coefficients presented in the Fig. 5, as below:
- $\lambda_1 = 0,000114$,
- $\lambda_2 = 0,00114$,
- $\lambda = 0.0001$ h^{-1},
- $\mu = 1/t_1 = 0,1$ h^{-1},
- $\mu_1 = 1/t_1 = 2$ h^{-1},
- $\mu_1 = 1/t_1 = 20$ h^{-1},

the estimated safety value is:

$$B = 1 - (P_{nieb}) = 1 - (P_1 + P_3) = 0.999943 \quad (3)$$

4.2 Markov model of yacht safety with ENoS

The Fig. 7 presents model taking into consideration installed ENoS system on a yacht.

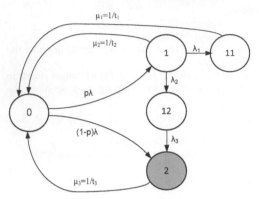

Figure 7. Markov model with ENoS on a yacht. [own source]

In the model we can distinguish:
- 0 – state of correct work,
- 1 – state of danger. Yacht equipped with EnoS,
- 2 – state of danger. Yacht not equipped with EnoS,
- 11 – no GPS signal,
- 12 – no GSM signal,
- p – the probability of equipment yacht in the system.

Assumptions:
- λ – capsized 25 yachts/year.
- λ_1 – no GPS - 1h/y.
- λ_2 – no GSM - 1h/y.
- λ_3 – intensity 60/1h
- $\mu_1 = 1/t_1$, t_1 –after 30 min. emergency service arrives (in spite of GPS defects, system sends the position from a buffer (5 min. before). The rescue services need additionally 10min. due to the movement of the yacht, relative to the last position).
- $\mu_2 = 1/t_2$, t_3 – after 20min. arrive the emergency service,
- $\mu_3 = 1/t_3$, t_3 – after 2h arrive the emergency service.

In the model presented on the Fig. 7 it was assumed, that the most dangerous state is number 2. The estimated probability P2 equals:

$$P2 = \frac{\lambda\lambda_3\mu_1 \cdot (\lambda_2 - p\lambda_2 + \lambda_2 + \mu_1 - p\mu_1)}{(\lambda_2\mu_1 \cdot (\lambda_1 + \lambda_2 + \mu_1)\mu_3 + \lambda \cdot (\lambda\lambda_3(\mu_1 - p\mu_1 + p\mu_1) + \mu_1 \cdot (\lambda_2 \cdot (\lambda_3 + p\mu_1) + \lambda_3(\mu_1 - p\mu_1 + p\mu_1))))} \quad (4)$$

According to the formula (3) the safety was analysed relation to the percentage equipment yachts in the ENoS system. The result of analysis is shown in Table 1.

Table 1.

No.	Percentage equipment yachts in the ENoS system	Safety: B=1-P2
1	10	0.99742
2	40	0.998279
3	60	0.998852
4	80	0.999425

The result presented in the Table 1 showing, that the more yachts equipped with ENoS the safety system is larger. Lack ENoS system causes, that 10.000 events with 10% participation in ENoS system, there is an average of 24 critical events, with 40% participation there is 17 critical events and with 80% participation there is only 6 critical events.

5 CONCLUSIONS

Every year during the sailing season boats capsize. This is related not only to the violent winds. Even a small wind can cause the boat capsizing which is associated with the lack of skippers' skills. The proposed system is at the stage of building and running a prototype. Small size and the cost of construction of such a system can bring a lot of interest and the implementation of a system for yachts. Currently there is no such solution dedicated to tourist yachts. Existing tracking devices are primarily dedicated to automotive vehicles. The main parameter influencing the safety of the system is characteristics of the GSM system, as the main component of the ENoS system. This applies to both the availability of the signal (the problem of coverage) but also the time parameters (time logging). The proposed Markov models reflect the typical situation of possession or lack of such a system on the boat as well as lack or loss of GPS and GSM. Due to the low frequency of occurrence of the described events on inland waters, assumed parameters for the analysis of the model are approximate. On the basis of analysis [5], in the model 5 assumed that the probability of simultaneous failure the BTS and the lack of GPS signal is at the level of the failure probability of a typical electronic system (λ = 1E-04). It must be emphasized after capsizing boat quick reaction of rescue services is the most significant. The crew may need immediate help and every minute of delay can lead to tragic situations. The analysis shows that the increase of installed systems on the boat could increase safety (search time is reduced). On the basis of available maps, the authors analyzed the availability of signal from the BTS (*Base Transceiver System*) in the Mazuria Lakes district [3]. The analysis allows to state that there is sufficient coverage of GSM for the analyzed district.

REFERENCES

[1] Dz. U. z 2003 r. Nr 212, poz. 2072.
[2] Dz.U. z 2006 r. Nr 123, poz. 857 - with changes.
[3] http://mapa.btsearch.pl/
[4] Kolaszewski A., Świdwiński P.: Żeglarz i sternik jachtowy; Almapress, Warszawa 2011 (in Polish).
[5] Lewiński A., Perzyński T.: The reliability and safety of railway control systems based on new information

technologies. Communications In Computer and Information Scienece 104. Springer 2010'. Transport Systems Telematics.

[6] Łukasik Z., Perzyński T.: Telematics systems to aid of safety in water inland turists. Communications in Computer and Information Science (395), Springer-Verlag Berlin Heidelberg 2013.

[7] Perzyński T.: Zastosowanie nowych technologii w zarządzaniu, logistyce i bezpieczeństwie w turystyce i rekreacji wodnej. LOGITYKA nr. 3/2014. (in Polish).

[8] Perzyński T.: Turystyka i rekreacja wodna – infrastruktura i bezpieczeństwo. Autobusy, technika, eksploatacja, systemy transportowe. Nr. 3/2013 (CD) (in Polish) .

[9] Perzyński T., Prokop-Perzyńska E.: Turystyka i rekreacja wodna – jachting w prawie i praktyce. Logistyka Nr 3/2012 (CD) (in Polish).

Ship Design Optimization Applied for Urban Regular Transport on Guadalquivir River (GuadaMAR)

A. Querol
Department of Naval Architecture, Universidad de Cádiz, Spain

R. Jiménez-Castañeda
Department of Engineering Electricity, Universidad de Cádiz, Spain

F. Piniella
Department of Maritime Studies, Universidad de Cádiz, Spain

ABSTRACT: GuadaMAR project, financed by European Regional Development Fund (ERDF) and by the Government of Andalusia (Spain), involves work on various objectives that are essentially aimed at the theoretical design of a Prototype vessel and maritime navigation system for the Guadalquivir River. Based in the port of Seville, this would serve as a transport system for passengers, under criteria of efficiency, technical operability, and economic and environmental sustainability. The aim of our paper, presented in this 11th International Conference "Transnav 2015" on marine navigation and safety of sea transportation in Gdynia, is to show a resume of the technical requirements of the ship, hull design and hydrodynamic behaviour prediction in terms of wave pattern and resistance, propulsion systems selection and dimension for both high speed version for the inter-urban line, using compressed natural gas and diesel as fuel, and slow speed version for the urban line using electric propulsion and the newest ion-lithium batteries technology and solar power supply. These technologies together, offer a sustainable navigation, without pollution during operation time in the use area of the electric ship. A review of the structure design, the analysis of safety equipment required and the study of the stability and equilibrium required by Spanish National Administration, are also presented.

1 INTRODUCTION

1.1 *Aim and scope*

Maritime and river transport and associated infrastructures that are not considered to be the responsibility of the State (i.e. the national Government) are among the various competences of the Regional Government, specifically the "Consejería de Fomento y Vivienda of the Junta de Andalucía". The "Agencia Pública de Puertos de Andalucía (APPA)" is an authority under public law, attached to the Consejería, which is concerned with the development and application of policy for ports and matters relating to the areas of transport of the Regional Government of Andalusia (Cruz, 1998).

The river Guadalquivir drains a basin of 57,000 km2 and forms a broad estuary of 1,800 km2. The Port of Seville is an integral part of the strategic and high-priority transport network of Europe, together with the E-60-02 Guadalquivir Eurovia navigable route. The lock is the point of connection between the E-60-02 Guadalquivir Eurovia and the port installations in Seville (Couser 1996, Alvarez 2001, Costa et al. 2009).

There are many cities where river or maritime means are successfully exploited for passenger transport. This is corroborated by the positive experience of Scandinavian, North American and other Spanish cities that have been the subject of study during the course of this project. Probably the best-known case is that of Venice with its famous waterbuses, the vaporettos, and other European cities like Hamburg where river boats on the Elbe are an established part of the urban transport system. In Spain, in the Bay of Santander, the capital has throughout its history been connected with the towns and villages located along its coast; closer to the geographical setting of the project reported here is the Bay of Cadiz. The case of the Bay of Cadiz is an example of how to take advantage of the natural platform represented by a location adjoining areas of protected waters where metropolitan passenger transport lines have long been established. The responsible body in Cadiz is a Transport Consortium which, since it was set up, has operated such services, as reflected in diverse documents that record its strategy of driving the provision of public transport and the basic lines on which it has acted.

1.2 The GuadaMAR project

The GuadaMAR project was proposed by researchers from several different departments of the University of Cadiz, as a multidisciplinary approach to the topic "The use of the river Guadalquivir as a passenger transport line, associated with the proposal of a prototype for a suitable and sustainable vessel". Bids for the execution of the project were invited by the Government of Andalusia, under its R&D+i Projects Service in the area of competence of the Regional Department (Consejería) of Public Works and Housing for the years 2012 to 2013. The project proposed was awarded a score of 77.8/100 by the various commissions of experts, and the budgetary allocation was €280,720.

The purpose underlying the GuadaMAR Project is based on the need to stimulate maritime transport systems in cities located on navigable rivers, as is the case of the metropolitan nucleus of Seville on the River Guadalquivir. Three specific objectives were set for this study: to determine the needs and alternatives; to develop theoretically the prototype vessel; and to study the direct economic impact.

In an initial phase, a study has been made of the regulations applicable to the maritime transport of passengers on the lines foreseen; a Navigation Study has been prepared on aspects related to the bathymetry, draughts, effects of current and wind, velocities, beaconing, interferences in the routes or headings, etc; and also a summary of the models already applied in other metropolitan zones with comparable expanses of water (river or bay) (Piniella et al. 2014).

As working hypotheses, two possible lines of navigation are put forward: one is an external Interurban route along the course of the river that could link the nearby population centres (Coria, Gelves) to the proximities of the intermodal point of the San Juan de Aznalfarache Metro Station, which we have designated Line 2; the other is an urban route, Line 1, that in successive phases of expansion of the project, could link intermodal stations on one or other riverbank, to facilitate urban mobility, especially in zones close to the course of the river (Triana, the City Centre, San Jerónimo, the Recinto ferial,...) where access would be provided to the Metro, to buses and to the Sevici system.

1.3 Requirements for Urban and Interurban Transport Lines

A resume of the main requirements stablished by the transport lines are shown in Table 1, in terms of passenger and bicycles capacity, dimensional restriction for water depth or height of bridges, maximum time permitted for embarking, mooring and disembarking operations, and service necessary speed for each line. Some of these requirements were common for both lines and some other were specifically stablished for each line. Anyhow, it was also decided that hull, external configuration and capacity of prototypes vessels would be the same for both lines.

Table 1. Requirements stablished by transport lines

Requirements	Value
Number of passengers	110
Number of bicycles	6
Time for embarking-disembarking (minutes)	2
Time for mooring operation (minutes)	2
Maximum speed for urban line (knots)	7
Maximum speed for interurban line (knots)	20
Maximum draft (meters)	3
Maximum air draft (meters)	6
Propulsion systems for each line	Different
Ship platform for each line (hull and G.A)	Equal

2 DESIGN REQUIREMENTS: RIVER BOATS VERSUS BUSES AND TRAMS.

In order to understand the design requirements of our prototype vessel we must bear in mind that the river transport system within a city has to compete with other means of public transport such buses or trams, in terms of speed and reliability of the time schedules of the services. One of the advantages of a river transport system could be the absence of traffic jams in the river, but on the other hand, boats may have more complicated operations for mooring and passenger embarking and disembarking than their competitors buses and trains, and therefore, they need to be done in the safest and fastest possible way, reducing also the minimum number of crew or personnel involved in these operations.

Other aspect that can make less competitive the river transport system is the limitation in speed that some maritime authorities may stablishes, not only due to manoeuvrability requirements, but also for reducing the effect of boats wave pattern in the banks of the rivers, in infrastructures and other sport nautical activities like canoeing or fishing. The optimum river boat should be designed to reduce the wave pattern as possible in order to be exceptionally be allowed for increasing its operational speed limit and therefore its competitiveness with land transport systems.

Bearing in mind that Society demands that transport systems should be increasingly respectful of the natural environment, efficient and socially sustainable, when trying to stablish a regular river public transport system is necessary to assure that this is not more contaminant than similar land transport system with buses or trams (Moral 1990).

3 DESIGN PROCESS

The methodology used for optimizing our river boat is based on the well-known design spiral, focused to the requirements stablished by the transport lines and those consequential of the competence with other transport system such buses or trams.

3.1 *Main dimensions and general arrangement*

These items have been defined in order to achieve both requirements: passenger and bicycles capacity and speed of embarking and disembarking operations. During this design phase the routes for passenger with and without bicycles, the number and positions of doors, seats and bicycles storage area were stablished. Different time simulations were made using alternatives combination of all these elements in order to find the fastest arrangement. Figure 1 shows the final general arrangement and routes for passengers.

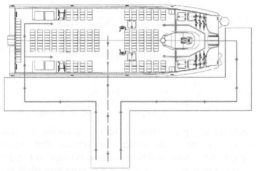

Figure 1. Passenger routing at boat stops

3.2 *Mooring system*

In order to achieve the requirement of quickness in the mooring operations and also minimizing the personnel involved, the design of mooring floating dock was carried out in conjunction with the general arrangement of the prototype boat. At figure 2 is showed an innovating mooring system that was developed to assure the boat stays in the correct position of mooring dock without the need of using ropes, by mean of a special asymmetrical fore fender design which make the stern rotates while pushing dead slow ahead with inner propeller.

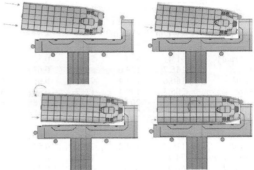

Figure 2. Mooring system without lines.

3.3 *Hull Design*

Together with the requirements of resistance reduction, stability and manoeuvrability, which normally are involved in any hull design, we had to carry out a complex CFD analysis to minimize the wave pattern at 7 knots, which was stablished as the necessary operating speed for the urban line. Bearing in mind that the river channel though Seville is use for international high level canoeing competitions, and that local Maritime Administration has limited the speed of any boat or vessel up to 5 knots, we had to develop a hull with equivalent wave pattern at 7 knots. In Figure 3 some results of wave making analysis is shown. A catamaran hull with high length—breath (L/B) and breadth-draft (B/) ratios was finally selected as the best option from a reduction wave pattern point of view (Tuck 1987, 2003).

3.4 *Propulsion system for high speed inter-urban line*

The propulsion system originally decided for the inter-urban line was a conventional diesel propulsion system with two shaft lines and fixed pitch propellers connected with two 715 BHP diesel marine engines, though both transmission gears with a reduction factor of 4:1. A special consideration has been taken to allow the vessel to operate also with Compress Natural Gas (CNG) as fuel for diesel engines, in order to reduce the pollutions emissions at the river area. Even thou the final authorization for using CNG as fuel for a river passenger ship is to be given by the Administration, a special exterior area at upper deck for storage the CNG together with a possible external routing of fuel pipes avoiding passing through the passenger area, it was taken into account during the early design phases.

Table 2. Electric power balance for each propeller group

	GlobHor kWh/m2	GlobEff kWh/m2	E Avail kWh	E Unused kWh	E Miss kWh	E user kWh	E Load kWh	Solar Frac
January	78.0	73.0	485	0.531	7339	411	7750	0.053
February	91.4	86.6	568	0.523	6671	329	7000	0.047
March	145.7	139.6	895	0.639	6931	819	7750	0.106
April	160.6	154.4	976	0.084	6840	660	7500	0.088
May	205.3	198.4	1217	0.587	6770	980	7750	0.126
June	215.9	209.1	1234	0.788	6416	1084	7500	0.145
July	235.1	228.3	1327	1.428	6649	1101	7750	0.142
August	201.6	194.8	1148	0.000	6897	853	7750	0.110
September	161.2	155.1	946	0.604	6726	774	7500	0.103
October	119.8	114.2	723	0.536	7155	595	7750	0.077
November	74.9	70.5	458	0.523	7165	335	7500	0.045
December	66.9	62.3	411	0.335	7412	278	7750	0.036
Year	1756.3	1686.3	10388	6577	83032	8218	91250	0.090

GlobHor: Global Horizontal solar radiation
GlobEff: Effective global radiation (over PV modules including IAM and shadows effects)
E Avail: Total solar energy available
E Unused: Energy loss (full charge battery)
E Miss: Energy supplied in the night charging process from grid power charger
E user: Energy supply to ship from PV on board installations
E Load: Total energy requirement for full electric ship version
Solar Frac: Solar fraction. Ratio Euser/Eload

Table 3. Results of pollution emissions study

Contaminant	Period	Legal Limit (µg/m3)	Winter Scenario 1 (µg/m3)	Winter Scenario 2 (µg/m3)	Summer Scenario 1 (µg/m3)	Summer Scenario 2 (µg/m3)
SO_2	1 h	350	48.1	55.0	43.2	70.1
SO_2	1 d	125	0.9	0.9	0.9	0.9
NO_2	1 h	200	34.3	20.2	21.5	27.5
CO	8 h	10 mg/m3	10.2	2.57	10.2	3.51
PM_{10}	1 d	50	0.9	1.0E-04	0.9	1.0E-04
$PM_{2.5}$	1 d	25 / 20	4.8	1.8E-04	11.0	6.0E-04

Figure 3. CFD wave pattern simulation.

3.5 Propulsion system for low speed urban line

With a maximum service speed of 7 knots, an electric propulsion system was decided as the most efficient and less contaminant possible alternative. The optimum configuration of propulsion plant was stablished by two shaft lines direct connected with two electric motors (2x 22kW 400V 50 Hz). Electric power is supply by two banks of 128 ion-lithium batteries each (HE 24098 type), giving a storage capacity of 2 x 792 Ah at 400VDC, and an operating autonomy of 12 hours. Charging of batteries are carried out during day time by means of 52 solar panel of 320 W each, located at deck, together with shore power current during no operating period at night. Table 2 shows the electric power balance of the system obtained from radiations simulations at Seville, solar panel configuration at ship and Guadalquivir geographic course at urban line.

A weight analysis was also carried out to assure that the electric propulsion system was similar to the conventional propulsion system used for high speed, including fuel, in order to assure the same equilibrium and stability characteristics for both propulsion system installed in the common platform hull (Falk et al. 2006, Fraile 2011).

3.6 Pollution emissions

A simulation of atmospheric contamination in Seville area was carried out, using a Gaussian stationary dispersion model and based in modelling the NOx, SPM y SOX diary concentrations, and the influence of pollution emitted from our prototype vessels in both lines at Guadalquivir River. This study was made for different weather conditions and scenarios. In Table 3, a resume of results is presented and it can be observed that in the worst possible scenarios the pollution emissions from the fleet are very lower than legally stablished limits.

3.7 Naval architecture

The structure was designed bearing in mind the necessity of a light hull which reduces the fuel consumption and therefore pollution emissions, but keeping the safely structural resistance necessary for the operational working solicitations. In figure 4, the amidships section of the structure in composites can be observed. The minimum scantlings of structure were obtained using the Special Service Craft rules from Lloyd's Register.

Lightship estimation was done for both different propulsion systems of prototype vessel: diesel-CNG for high speed inter-urban line and electric for slow speed urban line. Equilibrium and stability assessment was also carried out for both propulsive versions.

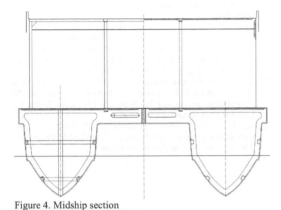

Figure 4. Midship section

3.8 *Cost estimation.*

The budget of both propulsive version was estimated and compared with similar vessels used by Spanish Public Transport Consortium at Cadiz Bay. The maintenance cost was also analysed and compared between the different propulsive systems, taking into account the lifetime of batteries given by the manufacturer and also the maintenance program given by the combustion engine manufacturer. As a conclusion was stablished that the cost of purchase of the high speed conventional version was similar to the low speed electric version, with a minimum approximated value of 1,822,000 € and a maximum value of 2,275,000 €.

4 CONCLUSIONS

Ship design optimization is not just a matter of good naval architecture; it is also a matter of understanding the different facts around any new transport system implementation. The optimized vessel will be the one which is dimensioned to get the passenger capacity just needed, able to compete with some other means of transport at land with the same or even better conditions of safety, timing reliability and environmental protection. The optimized prototype vessel, which main particulars are listed in Table 4, has been found by GuadaMAR research project as the optimum for a regular

transport system on Guadalquivir River in Seville metropolitan area, bearing in mind that could not be the optimum for a different river, city or country with different design requirements. In Figure 5 some external general view of the prototype can be observed.

Table 4. Main principals of optimized vessels

Main Principles (both versions)	
Length overall	21.25 m
Length between perpendiculars	19.21 m
Waterline Length	19.71 m
Breadth overall	7.38 m
Waterline breadth	2 x 2.18 m
Depth	2.62 m
Maximum Draft	1.29 m
Maximum Displacement	50 t
Passengers	110
Crew	De 1 a 3
Bicycles	6
Classification for Spanish Adm.	Passenger Ship- "K" Class
Structure	GRP
Type of hull	Catamaran
Propulsion system	2x shaft line & fixed pitch props.
Bow thruster	2 x 20kW
Solar electric panels area	82 m2
Maximum solar power	52 x 320 W
Minimum budget	1,822,000 € (VAT inc.)
Maximum budget	2,275,000 € (VAT inc.)

High Speed (inter-urban)	
Maximum speed	20 knots
Power installed	2 x 715 BHP
Type of main engines	2x Marine Diesel wet exhaust system
Reduction gears (2x)	Ratio 4:1
Auxiliary engine	Diesel 64.5 kVa @ 1500 rpm
Emergency generator	Diesel de 9cv (type HATZ B-40)
Fuel Capacity	2 x 2100 litres
Autonomy	2 days x 12 hours operation

Low Speed (urban)	
Maximum speed	7 knots
Power installed	2x22kW
Type of main engines	2x electrical 22kW 400V 50 Hz
Number of batteries	256
Type of batteries	Ion Lithium HE 240
Electric storage capacity	2 x 792 Ah a 400VDC
Autonomy	1 day x 12 hours operation

DISCLAIMER

The authors would like to thank the ERDF of European Union for financial support via project "GuadaMAR" of the "Programa Operativo FEDER de Andalucía 2007-2013". We also thank all Public Works Agency and Regional Ministry of Public Works and Housing of the Regional Government of Andalusia staff and researchers for their dedication and professionalism.

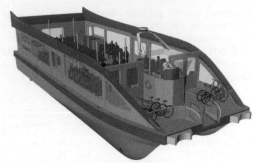

Figure 5: Exterior without deck 3D model

REFERENCES

Alvarez, O., Tejedor, B. & Vidal, J. 2001. *La dinámica de marea en el estuario del Guadalquivir: un caso peculiar de resonancia antrópica.* Cádiz

Costa, S., Gutiérrez Mas, J.M. & Morales, J.A. 2009. Establecimiento del régimen de flujo en estuario del Guadalquivir, mediante el análisis de formas de fondo con sonda multihaz. *Revista de la Sociedad Geológica de España* 22(1-2):23-42

Couser, P.R. 1996. *An Investigation into the Performance of High-Speed Catamarans in Calm Water and Waves,* PhD Thesis, Department of Ship Science: University of Southampton, UK.

Cruz Villalón, J. 1998. La intervención del hombre en la ría y marismas del Guadalquivir. *ERIA*:109-123

Falk, A. Dürschner, C. & Karl-Heinz, R. 2006. *Fotovoltaica para profesionales,* Progensa: Madrid.

Fraile Mora, J. 2011. *Máquinas Eléctricas,* McGraw-Hill

Moral (del) Ituarte, A, 1990. La pugna por el agua en el valle del Guadalquivir. *Revista de Obras Públicas,* May-1990:13-33

Piniella, F., Querol, A. & Rasero, J.C. 2014. Maritime passenger transport as an urban and interurban alternative on the river Guadalquivir: GuadaMAR In Martinez, X (ed.), *Maritime Transport VI – 6th International Conference on Maritime Transport*: Barcelona. 443-452

Tuck, E.O. 2003. Computation and Minimisation of Ship Waves, ICIAM03, Sydney, Chapter 17 in SIAM *Proceedings in Applied Mathematics* 116. Edit. James M. Hill and Ross Moore: SIAM.

Tuck, E.O. 1987. *Wave Resistance of Thinships and Catamarans,* Report T8701, Dept. of Applied Mathematics. The University of Adelaide, Australia.

Inland Shipping
Information, Communication and Environment – Marine Navigation and Safety of Sea Transportation – A. Weintrit & T. Neumann (eds.)

Ship Emission Study Under Traffic Control in Inland Waterway Network Based on Traffic Simulation Data

X. Chen & J. Mou
School of Navigation, Wuhan University of Technology, Wuhan, China

L. Chen
Delft University of Technology, Delft, The Netherlands

X. Yue
Wuhan Maritime Professional School, Wuhan, China

ABSTRACT: The inland shipping plays a more and more important role in transportation. As it is growing rapidly, the impact from ship emission becomes bigger than ever before. The emission from ships can be estimated by the vessel activity data. However, the quantity and quality of these data is quite defective in China inland shipping. It is challenging to conduct emission study in inland waterway. Simulation modeling is a proven method that provides insight into and truly reproduces the complicated traffic situations. It provides a possibility to estimate the emission based on traffic simulation. As a case study, a city-wide inland waterway network in the North region of Zhejiang Province, China, is demonstrated. The traffic flow is simulated via SIVAK software. In the simulation, by shutting down one route or controlling the traffic lane number in the waterway, four kinds of traffic control scenarios are applied. Based on the simulated traffic flow data, by the power-based bottom up approach, emission inventory for the waterway network per week is estimated for four kinds of scenarios. According to the emission inventory and simulation traffic data, the traffic characteristic and emission efficiency index are calculated and analyzed, such as EEOI, CO_2 efficiency. Meanwhile, considering the traffic congestion, a CO_2 density liner regression model and a CO_2 efficiency quadric regression model are performed. Through the quantification of ship emission, analysis of emissions with traffic congestion and comparative study on emission in different kinds of traffic control scenarios, strategies to reduce emission and balance the traffic service are presented.

1 INTRODUCTION

Global warming is an increasingly serious issue for earth's ecosystem, climate change and human living. These days, the frequency and intensity of extreme weather and climate events have assumed significant change. The growing vessel number and emission are part of the reason. International shipping is currently estimated to have emitted 870 million tons of CO_2 in 2007, and the quantity is in a rapid growth (Buhang et al. 2009). Also, the emission from shipping has stimulated wide-ranged study and discussion. Winther et al. (2014) presents a detailed BC, NOx and SO2 emission inventory for ships in the Arctic in 2012 based on satellite AIS data and predicts a projection for2020, 2030 and 2050. Corbett et al. (2010) produced a 5 km×5 km Arctic emissions inventories under existing and future scenarios considering traffic growth, diversion routes and possible control measures. IMO performed a study on shipping emission from 1990-2007 and discussed reduction by legislation, technology and operation, also forecasted scenarios

for future emissions (Buhang et al., 2009). Using AIS data for vessel speed profiles, Yau et al. (2012) developed a detailed maritime emission inventory for ocean-going vessels in Hong Kong with the year 2007. Fu et al. (2013) obtains emission data of inland ships on the Grand Canal of China by on-board emission test for decisive emission factor for local ship. Song et al. (2014) estimated both the in-port ship emissions inventory and the social cost of the emission impact on Yangshan port in Shanghai and propose eco-efficiency concept. Yang et al. (2007) developed an air pollutant emission inventory for marine vessels in the Shanghai Pore in 2003 under cruising and maneuvering conditions based on two categories of vessel. Fu et al. (2012) made a detailed investigation about the Shanghai outer port, inner waterway. Combining the AIS data, present an emission inventory based on 9 vessel types and 4 operation modes. Zheng et al. (2009) develop a spatial Pearl River Delta (PRD) regional emission inventory for the year 2006 with the use of domestic emission factors and activity data.

Some scholars try to perform study on cutting down the ship emission by operational measures for the reason that technical method has along payback periods with new investment. Lack et al. (2012) evaluated current insights regarding the effect of ship speed, fuel quality and exhaust gas scrubbing on BC emissions from ships. Xing et al. (2013) studies on the energy consumption and green house gas emissions of inland shipping, and compares with seagoing vessel. Acomi & Acomi (2014) develop a software solution for on board measures over voyage energy efficiency. Ni & Zhao (2010) studied the factors affecting the EEOI and evaluate the measures improving energy efficiency.

In many regions, inland waterways are characterized with network. And the traffic in complex channel network is quite different from open waters. Some congestion indexes are proposed to judge the condition of waterway or network. Certainly, traffic flow characteristic has a great connection with shipping emission. For example, in different degree of traffic congestion, the emission can be quite various. However, the shipping activity data in inland waterway is quite scarce. It is hardly precise to estimate the emission. But via computer simulation and detailed investigation, the traffic flow is reproduced and the simulation data is useful for emission study. Meanwhile, in inland waterway network, administration will adopt different traffic control scenarios for better service or respond to emergencies. However, the function of emission reduction is always ignored.

In this paper, the simulation investigates different scenarios, e.g., normal (two-lane shipping) and traffic control (one-lane shipping, suspended and hybrid mode). Based on the simulation output of traffic flow data, emission inventories of 4 kinds of traffic scenarios for a city-wide inland waterway network have been developed in the North region of Zhejiang Province, China. According to the emission inventory and traffic data in simulation, the traffic characteristic and emission efficiency index are calculated and analyzed, such as EEOI, CO_2 efficiency. Meanwhile, considering the traffic characteristic, emission features, CO_2 density and CO_2 efficiency are applied with regression analysis correlated with traffic volume and PWT (percentage of waiting time) respectively. Through the quantification of ship emission, analysis of emissions with traffic congestion and comparative study on emission in different kinds of traffic control scenarios, strategies to reduce emission and balance the traffic service are presented, which will provide deeper insight of emission for inland shipping and support the administration to make decision and manage the water region.

2 VESSELS TRAFFIC SIMULATION OF INLAND WATERWAY NETWORK

Simulation modeling is a useful tool to reproduce the traffic scenario for emission estimation as the real vessel activity data is defective in inland shipping. And the detailed spatial traffic situation can be obtained and it is good for better analysis and emission estimation. The traffic simulation can be referred to Chen et al. (2013), but the data selection and analysis are different for various research purpose. Because of the limited space, here we only show the main point useful for emission estimation, other detailed steps and parameters can be referred to Chen et al. (2013).

2.1 SIVAK

The simulation model SIVAK (SImulatie VAarwegenen Kunstwerken , Simulation model for waterways and civil works) is made by Rijkswaterstaat (Public Works Agency) in the Netherlands to analyze traffic flows of ships and road traffic at bridges, locks, narrowing and waterway sections(De Gans, 2010).

2.2 Research area

The study area is in Huzhou city, ranging from Latitude 120º6' E to 120º42'E, and Longitude 30º42' N to 30º54' N (Fig.1). Changhushen and Hujiashen are two main Fairways in this region. Huzhou Lock and Wushenmen Lock are two deserted locks located in the two fairways, but presently function as stations to observe the traffic flow.

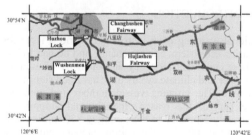

Figure 1. Research area

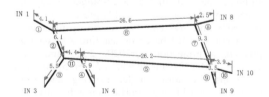

Figure 2. Topology of the inland waterway network

To simplify the model, the waterway network is converted to the topological structure in Figure 2.

There are 11 waterway sections. Waterway section ⑦ is a Class IV fairway of China with width of 40 meters and depth of 2.5meters. The others sections are Class III fairway of China with width of 45 meters and depth of 3.2 meters.

2.3 *Traffic control scenarios and volume*

Investigation shows that the traffic volume of this area in 2011 is 9939 ships/week. Usually, the N0 scenario is adopted in the waterway network. Also, the time period of the simulation is 1 week. 4 kinds of traffic scenarios are shown as below:
- Normal/N0: all the waterway sections are two-lane shipping;
- Control 1/C1: waterway section ⑥ is one-lane shipping;
- Control 2/C2: both waterway section ⑤ and ⑥ are one-lane shipping, section ⑤ for upstream while section ⑥ for downstream;
- Control 3/C3: waterway section ⑥ is closed.

2.4 *Vessel characterization*

The emission is quite different from vessel types. According to our investigation, over 90% is cargo vessel (Chen. 2014). To simply the traffic, the vessel in simulation is all set as cargo vessel. For studying the relation between traffic congestion and CO_2 efficiency, and the effect of ship size on traffic congestion and power difference cannot be ignored, the ship class is identical as the reality (table 1). And the distribution of ship class is shown in figure 3. The power per ship type and other characteristic are referred to the principal dimensions series of standard ship type in the Grand Canal.

Table 1. Ship classes

Ship class	DWT	Width(m)	Length(m)	Percentage
WN 1	21-42	3.4-4.41	15.95-20	0.06%
WN 2	38-98	3.8-5.4	20.1-25	1.16%
WN 3	60-240	4.6-6.2	25.1-30	4.34%
WN 4	80-530	4.98-7.18	30.03-35	15.51%
WN 5	126-550	5.5-8.48	35.01-40	27.33%
WN 6	183-830	5.6-9.97	40.04-45	39.79%
WN 7	296-800	7.6-10	45.1-49.9	8.71%
WN 8	450-1010	6.6-11.85	50.1-55	2.88%
WN 9	830-1450	10.1-11.8	55.1-60	0.22%

2.5 *Simulation outputs*

SIVAK provides many outputs. In this paper, the outcome of vessel activity is counted every two connective node. As there are 11 waterways in the network, traffic flow in 22 directions is counted for one scenario. The statistics is shown in table 2 (only showing the data for N0 scenarios and detailed data of other 3 scenarios are not shown). For this research, the following outputs are used:
- Passing time: the time it takes for the ship from passing the entrance to the destination.
- Waiting time (delay): including the time a ship is waiting at a node and the time ship waiting to enter the network after generating.
- PWT (percentage of waiting time): the percentage of average waiting time in passing time.
- Navigation duration: the time when the vessel is navigating in the waterway.
- Number of ships per ship type sailing out and en route.
- Average speed: the vessel sailing speed in the waterway.

Table 2. Traffic flow statistic for N0 Scenario (Other scenarios are not shown)

waterway section	Direction	length (km)	traffic volume (ships/week)	average delay (h)	passing time (h)	nav duration (h)	average speed (km/h)	PWT (%)
1	east	4.1	4011	0.062	0.226	0.164	25.162	27.35%
1	west	4.1	1912	0.094	0.309	0.215	19.147	30.42%
2	south	6.1	1977	0.117	0.384	0.267	23.001	30.45%
2	north	6.1	1362	0.104	0.423	0.319	19.14	24.59%
3	east	5.5	1451	0.134	0.416	0.282	19.54	32.20%
3	west	5.5	1929	0.088	0.365	0.276	19.994	24.25%
4	north	5.9	1000	0.067	0.363	0.295	19.996	18.58%
4	south	5.9	985	0.122	0.389	0.267	22.219	31.37%
5	east	26.2	1570	3.366	4.707	1.341	19.737	71.51%
5	west	26.2	1417	1.363	2.756	1.393	18.868	49.47%
6	east	26.6	2772	1.433	2.649	1.216	22.181	54.10%
6	west	26.6	1288	4.935	6.362	1.427	18.768	77.57%
7	north	9.3	371	0.695	1.198	0.504	18.534	57.97%
7	south	9.3	1772	0.128	0.609	0.481	19.351	20.96%
8	west	2.5	2421	0.022	0.151	0.129	19.412	14.73%
8	east	2.5	2504	0.036	0.151	0.116	21.989	23.70%
9	north	3.5	659	0.023	0.194	0.171	20.48	11.71%
9	south	3.5	910	0.018	0.176	0.158	22.271	10.33%
10	west	3.9	397	0.058	0.251	0.193	20.2	23.15%
10	east	3.9	1700	0.012	0.201	0.188	20.818	6.09%
11	east	4.4	2305	0.117	0.318	0.201	21.936	36.68%
11	west	4.4	2168	0.079	0.311	0.232	19.011	25.33%

3 EMISSIONS ESTIMATION

Simulation traffic flow data is used as ship activity data to produce emission inventories for every scenario lasting for one week. This inventory methodology employs a bottom-up approach (based on vessel activity data) to estimate the shipping emission. Emission is calculated for each direction in every waterway section

3.1 *Emission factor and fuel correction factor*

Generally speaking, there are three sources that produce emissions from ships: main engine power, auxiliary power and the boiler. Considering the situation about inland vessel described in Fu et al. (2012). We only consider the main engine power and auxiliary power in the emission calculation. Also, as the vessels in this region didn't call at any ports, so the vessels only have two statuses, one is navigating and the other is waiting. We assumed that when the ship is navigating, the main and auxiliary engines are both working. When the ship is waiting, the main engine will stop and leaving the auxiliary engine working alone.

Emission factors vary from engine type, fuel type and engine production year. Also, the emission factors reflect the advance of technology and the effect of some kind of law or regulation. In this study, because of defective data, only fuel type and engine type are considered. The emission factor and fuel correction factor used in this paper are referred from Shanghai emission inventory (Fu et al. 2012) and report of Port of Los Angeles inventory of air emission (Agrawal et al. 2012). The EF (emission factor) is expressed in terms of g/kW-hr, and the FCF (fuel correction factor) is dimensionless. Considering the ship type, engine type and fuel type, specific EF and FCF are chosen. (Show in table 3).

Table 3. EFs and FCFs

	PM10	PM2.5	DPM	NO_x	SO_x
EF(g/kW-hr)	1.5	1.2	1.5	13	11.5
FCF	0.19	0.19	1	0.94	1

	CO	HC	CO_2	N_2O	CH_4
EF(g/kW-hr)	1.1	0.5	683	0.031	0.01
FCF	1	1	1	0.94	1

3.2 *Calculation method*

In every waterway section, and for combination of ship class, main engine power, auxiliary engine power, load factor, navigation duration, by the basic concept that work equals to power multiply duration, the work produced by the ship is calculated as Eq. (1), load factor is calculated as Eq. (2):

$$W_{i,j} = \left[\left(ME_j + AE_j \right) \times LF \times NT_{i,j} + AE_j \times WT_{i,j} \right] \times n_{i,j} \quad (1)$$

$$LF = \left(S_a / S_m \right)^3 \quad (2)$$

where W is work in kW-hr, i is waterway section and direction, j is ship class, ME is main engine power in kW, AE is auxiliary engine power in kW, LF is load factor of main engine related with ship speed in dimensionless, NT is navigation duration in h, WT is waiting time in h. S_a is actual speed of ship sailing in waterway in km/h, S_m is maximum speed in km/h.

With basis of work produced by the vessel, the emissions are calculated by multiplying emission factors and fuel correction factors as Eq. (3):

$$E_s = \sum_{i,j} W_{i,j} \times EF_s \times FCF_s \quad (3)$$

where E is emission in waterway network per week in g, s is pollutant gas type, EF is emission factor in g/kW-hr, FCF is fuel correction factor in dimensionless.

3.3 *Emission result*

Simulation activity data, vessel characterization, operation models, EFs and FCFs were combined to produce the emissions inventory for four kinds of traffic scenarios (shown in table 4). Usually, the N0 scenario is adopted in the waterway network. The result shows that the emission in C2 is the lowest and C3 is the highest. That is for the reason that in C2 scenario, the traffic flow is simpler, fewer conflicts will be occurred in the node, so the average delay and the waiting time of C2 is in a better value (shown in figure 3). Although the average speed of C2 is the lowest, but the difference is very small. In addition, the value difference between PWT is big and means greater.

Emissions of every waterway section and direction are calculated. The emission result and traffic flow statistics of every route will be useful for studying the relation between traffic flow and emission. The values in waterway section ⑤⑥ are higher. Longer distance is the main reason as the passing vessels are not more than other waterway.

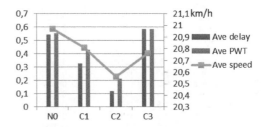

Figure 3. Average value of four scenarios

Table 4. Emission inventory for four traffic scenarios

1 week traffic scenario	PM_{10} kg	$PM_{2.5}$ kg	DPM kg	NO_x kg	SO_x kg	CO kg	HC kg	CO_2 kg	N_2O kg	CH_4 kg
N0	212.44	169.95	1118.08	9108.67	8571.98	819.93	372.69	509101.33	21.72	7.45
C1	199.53	159.62	1050.15	8555.23	8051.16	770.11	350.05	478168.80	20.40	7.00
C2	162.39	129.91	854.66	6962.65	6552.41	626.75	284.89	389156.15	16.60	5.70
C3	244.06	195.25	1284.51	10464.45	9847.89	941.97	428.17	584878.77	24.95	8.56

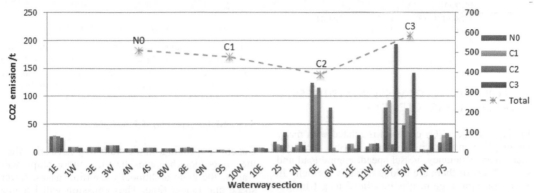

Figure 4. CO_2 emission result of every waterway section in four traffic scenarios

As there is one ship type and engine type in our study, the value of each emitted gases are proportional, so we only choose CO_2 on behalf of ship emission for the analysis and relation study. Figure 4 illustrates the CO_2 emission difference in every waterway section between difference scenarios. Waterway sections 1, 2, 3, 4, 8, 9, 10 all are connected to end node of the network, or we can call them the outer waterway, as they are all the entrance or exist of the network. Waterway 2, 6, 11, 5, 7 are the inner ring of the network; we can call them inner ring waterway. The emission of the outer waterway is almost the same while in inner ring waterway is quite different. Because the traffic control change the condition of the inner ring waterways and these waterways affect the traffic directly, which leads to emission change. C2 scenario like a one-way circulation in the inner ring waterway which simplifies the traffic flow while C3 scenario shutting down one section. Many vessels have to detour this blocked section making the passing vessels of C3 is the highest, 44522, comparing to N0 36881, C1 39609, C2 38726. That implies the travelling distance of C3 is the longest and vessels pass more nodes as the traffic volume are the same.

3.4 Emission validation

Although the methodology of our estimation is widely used around the world and the parameters fits our study, but it is essential to validate our emission result. The emission in one ship can be easily measured by device while it is challenging to test

thousands ships in the waterway network. In addition, because the traffic volume, research area and time period are quite different from other research, it is hard to compare our emission result with them. Thus, we calculate EEOI to validate our emission to solve the problems we mention above. EEOI (Energy Efficiency Operational Indicator) is defined as the ratio of mass of CO_2 emitted per unit of transport work. The basic expression is shown as eq. (4). When the EEOI is similar to other research, we think the emission estimation is reasonable.

Before getting the EEOI, we should determine the SFOC (specific fuel oil consumption) and C_f, the values are referred to Second IMO GHG study (Buhang et al. 2009), standard of fuel oil consumption for transportation ships-part 2: calculation method for inland ships (GB/T 7187.2-2010 in Chinese), Xing et al. (2013), MEPC.1 - Circ.684 (Guidelines for voluntary use of the ship energy efficiency operational indicator).

$$EEOI = \frac{\sum_j FC_j \times C_{Fj}}{m_{cargo} \times D} \tag{4}$$

where j is the fuel type; FC is the mass of consumed fuel; C is the fuel mass to CO_2 mass conversion factor

Table 5 shows the EEOI in our study comparing to other research. The EEOI is calculated based on numbers of vessels passing waterway section instead of single vessel, while MEPC 684 and Xing et al. studied the single vessel. The comparison shows that the number is in same level and because of some different condition, small difference exists which can be accepted.

Table 5. Comparison of EEOI

EEOI g/(t*n mile)	Our study Average ofcase waterway section	MEPC 684	Second IMO GHG study general/ bulk	Xing et al. study(2013) Inland Container ship
N0	5.78-12.8	13.5	14-27	84-419
C1	5.44-15.2		5.5-46.3	
C2	4.98-12.1			
C3	5.67-11.1			

4 REGRESSING CO2 EMISSION WITH VESSEL TRAFFIC CHARACTERISTIC

In single vessel, CO_2 emission is related with many factors, like engine type, engine built year, fuel type, machinery efficiency, vessel speed, engine load and so on. Most of them are belong to hardware field. In a macroscopic region, it is beneficial to get insight into the relationship between emission and some kinds of traffic characteristic. Consequently, the discovery can support the administration to manage the complex waterway network for lower carbon.

4.1 *CO2 density and CO2 efficiency*

CO_2 density means the quantity of CO_2 per kilometer in the waterway for 1week. By combination of CO_2 emission, waterway length, the CO_2 density is counted for each direction in every waterway section in four traffic scenarios.

CO_2emission efficiency of transport can be expressed as CO_2/tone*kilometer, it is defined as eq.5 (Buhang et al., 2009). This index can be applied in all transport sectors, such as shipping, rail, road and aviation. Usually, it is used for comparison between different transport mode, also performed in the Energy Efficiency Design Index (EEDI) and Energy Efficiency Operational Indicator (EEOI). Here, after a series of experiments, we try to find out some relation with traffic characteristic.

$$CO_2 \text{efficiency} = \frac{CO_2}{\text{tonne*kilometer}} \qquad (5)$$

where CO_2 is total CO_2 emitted from the vessel within the period, tonne*kilometer is total actual number of tone-kilometers of work done within the same period.

4.2 *PWT*

The definition of PWT (percentage of waiting time) is the same as Chen et al. (2013) (show in eq. 5). But differently, we calculate for each waterway instead of the whole network and there is only one kind of traffic volume (2011, 100%, 9939 ships per week) in this case.

$$PWT = \frac{\text{waiting time}}{\text{passing time}} \qquad (6)$$

4.3 *Correlation coefficient and regression analysis*

After selecting and calculating the index above, we can calculate correlation coefficient to test the relation between them. If the coefficient is higher than 0.70, then we can perform regressing analysis.

5 RESULT AND DISCUSSION

Generally speaking, in the waterway network, when the congestion becomes heavier, more ships will wait at the node additionally emitting gases, lower speed and longer time. Thus emission will become higher. The correlation coefficient test result is shown as table 6 indicating that traffic volume has a strong connection with CO_2 density and the same situation for PWT and CO_2 efficiency.

Table 6. Correlation coefficient

	CO_2 volume	CO_2 density	CO_2 efficiency	PWT
traffic volume	0.447	**0.892**	0.006	0.018
traffic density	-0.237	0.524	-0.133	-0.259
PWT	0.496	0.285	**0.744**	/
CO_2 volume	/	0.578	0.455	/

The regression analysis is performed based on the emission and traffic characteristic of waterway sections. There are 11 sections in the network, and we have 4 scenarios. After eliminating some abnormal data, we have 82 units of statistic.

The regression study shows that the CO_2 density has a great match with traffic volume (see fig.5 and fig. 6), in spite of waterway length. Liner regression model fits well for their regression, where the regression model is as eq.6 and the R^2 is 0.796, the variable x (traffic volume) can explain the CO_2 density in 79.6%.

$$y = 0.001x - 0.115 \qquad (7)$$

here y represents CO_2 density, x represents traffic volume.

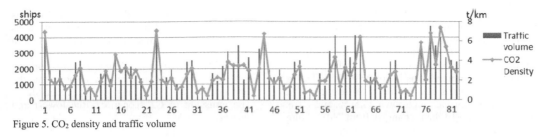

Figure 5. CO_2 density and traffic volume

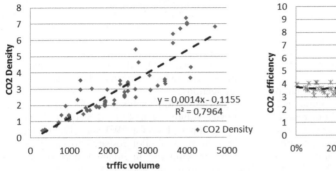
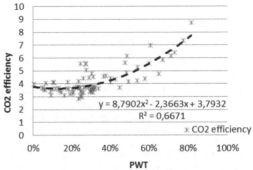

Figure 6. Regression of emission index with traffic characteristic

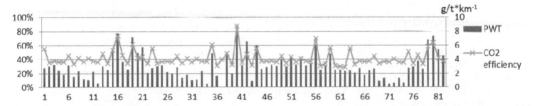

Figure 7. CO_2 efficiency and PWT

Table 7. Regression Parameter Test

	Multiple R	R Square	Adjusted R Square	Standard Error	Significance F	X Variable 1 P-value	X^2 Variable P-value
Regression 1	0.8924	0.7963	0.7938	0.7552	2.24E-29	2.24E-29	\
Regression 2	0.8168	0.6671	0.6587	0.6245	1.35E-19	0.0876123	1.545E-06

As the defective data of inland shipping, it cannot be verified by the reality data. But the regression parameters test is performed and the result shows in table 7. Where the Significance F, 2.24E-29, much less than 0.05, X variable 1 P-value, 2.24E-29, less than 0.0001, means the confidence is 99.99%.

In addition, CO_2 efficiency shows a low connection with traffic volume, traffic density which is beyond expectation. Form Fig. 7, CO_2 efficiency and PWT shows a connection and after calculation, the correlation coefficient is 0.744, also means a strong connection between them. The quadratic regression is performed for CO_2 efficiency and PWT (show in eq.8) and the parameter test is shown in table 7. In the test, multiple R is 0.8168 showing they are high positive correlation. R^2 is 0.6671, which means the independent variable (PWT) can

explain dependent variable (CO_2 efficiency) in 66.71%. Standard error, 0.6245 is quite low. Significance F, 1.35E-19 is much less than significance level 0.05, showing the regression effect is remarkable. And, the x P-value is 0.0876, much bigger than the x^2 P-value, 1.545E-06, which means the dependent variable has bigger correlation with x^2,

$$y = 8.790x^2 - 2.366x + 3.793 \qquad (8)$$

here y represents CO_2 efficiency, x represents PWT, it should be noted that the value of CO_2 efficiency is bigger, means the efficiency is lower due to its definition.

This study uses a kind of new activity data to perform emission inventory calculation. And the emission result is validated by comparing emission index with other study. In addition, presenting three other traffic scenarios for emission comparison study provide a meaningful reference for reality to cut down the carbon emission. C2 traffic control strategy is recommended as its lower PWT, reducing by 51.01% (means higher traffic efficiency) and lower carbon emission (cut down 23.56%, and C1 is -6.08%, C3 is +14.88% changed from N0). Using the statistics of traffic flow and emission result of a number of waterway sections, a liner regression model for CO_2 density and a quadric regression model for CO_2 efficiency are established. By these models, preliminary relation is built up for emission and macroscopic traffic flow. Rough prediction could be made by these models based on assumption of traffic change.

However, even though the traffic simulation is well verified including arrival pattern, output etc. After all, simulation has a difference from reality, for example, in the simulation the node is only allowed one ship to pass and the cargo load situation of vessel is not clear here. Additionally, some model like generic ship design model (SHIP-DESMO) to predict design power and the propulsive power at certain speed is not applicable in this case which makes hard for power determination, hence the power parameter is less certain here, only referring to the principal dimensions series of standard ship type in the Grand Canal.

Anyhow, this is a daring attempt to perform study on the emission in inland waterway network under traffic control and explore correlation between emission feature and traffic characteristic.

6 CONCLUSION

In this paper, a part of an inland waterway network in the North part of Zhejiang Province, China, is studied. The traffic flow in this network is simulated via SIVAK, applying four scenarios, normal (N0, two-lane shipping) and traffic control (C1, waterway section ⑥ is one-lane shipping; C2, both waterway section ⑤ and ⑥ are one-lane shipping; C3, waterway section ⑥ is closed) and the 2011 traffic volume (9939 ships per week). Based on the simulation output, a power-based bottom up approach for emission calculation is performed. Emission is verified by emission index comparison. And the vessel traffic statistics and ship emission are applied for each direction in every waterway. Based on the spatial detailed study, analysis of correlation between some emission features and traffic congestion characteristic and two regressions are carried out. The study shows:
- The traffic simulation output data contains detail information and can be used as ship activity data for emission estimation.

- The emission of CO_2 in the waterway network is 509.1t/week for N0 traffic scenario (the same as reality) and other emission result can be seen in the inventory.
- As different research condition, the emission can be validated by emission index and shows that the emission result is a little bit low but acceptable.
- The emission in C2 scenario is the lowest and changed -23.56% from N0, also the traffic efficiency is highest where average PWT is 27.18% (changed -51% from N0), so C2 plan is a good reference for traffic control in waterway network for lower emission and higher traffic efficiency.
- The correlation coefficient is high between traffic volume and CO_2 density, PWT and CO_2 efficiency, which means they have strong connections.
- Liner regression analysis can be performed for traffic volume and CO_2 density and quadric regression analysis for PWT and CO_2 efficiency.

This study is a pilot research to calculate and analyze the emission in inland waterway network based on traffic simulation data for different kind of traffic controls. Also, the regression analysis builds up a preliminary relation for traffic characteristic and emission. This research will provides deeper insight of emission for inland shipping and support the administration to make decision and manage the water region.

ACKNOWLEDGMENT

The work presented in this paper is financially supported by Self-determined and Innovative Research Funds of WUT (Grant no. 2014ZY141), NSFC and NOW (Sino-Dutch Cooperation, Grant no. 51061130548) and Zhejiang Provincial Department of Communication in China (Grant no. 2010W11).

REFERENCES

Acomi, N., & Acomi, O. C. 2014. Improving the Voyage Energy Efficiency by Using EEOI. *Procedia-Social and Behavioral Sciences, 138*, 531-536.

Agrawal, A., Aldrete, G., Anderson, B., et al. 2012. Port of Los Angeles Air Emissions Inventory. Starcrest consulting group, LLC, Los Angeles, USA.

Buhaug,Ø., Corbett, J. J., Endresen Ø., et al. 2009. Second imoghg study 2009. International Maritime Organization (IMO), London, UK, 24.

Chen, L., Mou, J., Ligteringen, H. 2013. Simulation of Traffic Capacity of Inland Waterway Network. IWNTM13: International Workshop on Nautical Traffic Models 2013, Delft, The Netherlands, July 5-7, 2013. Delft University of Technology.

Chen, L. 2014.Study on Traffic Capacity of Waterway Network. Wuhan University of Technology.

Corbett, J. J., Lack, D. A., Winebrake, J. J., et al. 2010. Arctic shipping emissions inventories and future scenarios. Atmospheric Chemistry and Physics, 10(19), 9689-9704.

Fu, M., Ding, Y., Ge, Y., et al. 2013. Real-world emissions of inland ships on the Grand Canal, China. Atmospheric Environment, 81, 222-229.

Fu, Q. Y., Shen, Y., Zhang, J., 2012. On the ship pollutant emission inventory in Shanghai port. Journal of Safety and Environment 12(5):58-63.

Lack, D. A., Corbett, J. J. 2012. Black carbon from ships: a review of the effects of ship speed, fuel quality and exhaust gas scrubbing. Atmospheric Chemistry and Physics, 12(9), 3985-4000.

Ni, J. K., Zhao, Y. F., 2010. Study on energy efficiency operational indicator of container ships. Ship & Ocean Engineering 39(5): 140-143.

O. De Gans, 2010, SIVAK II Handleiding. Rijkswaterstaat, the Netherlands.(In Dutch)

Song, S. (2014). Ship emissions inventory, social cost and eco-efficiency in Shanghai Yangshan port. Atmospheric Environment, 82, 288-297.

Sun, X., Yan, X., Wu, B., et al. 2013, Analysis of the operational energy efficiency for inland river ships. Transportation Research Part D: Transport and Environment,22, 34-39.

Winther, M., Christensen, J. H., Plejdrup, M. S., et al. 2014. Emission inventories for ships in the arctic based on satellite sampled AIS data. Atmospheric Environment, 91, 1-14.

Yang, D. Q., Kwan, S. H., Lu, T., Fu, Q. Y., Cheng, J. M., Streets, D. G., ... & Li, J. J. 2007. An emission inventory of marine vessels in Shanghai in 2003. *Environmental science & technology*, *41*(15), 5183-5190.

Yau, P. S., Lee, S. C., Corbett, J. J., et al. 2012. Estimation of exhaust emission from ocean-going vessels in Hong Kong. Science of the Total Environment, 431, 299-306.

Zheng, J., Zhang, L., Che, W., Zheng, Z., & Yin, S. 2009. A highly resolved temporal and spatial air pollutant emission inventory for the Pearl River Delta region, China and its uncertainty assessment. *Atmospheric Environment*,*43*(32), 5112-5122.

Inland Shipping
Information, Communication and Environment – Marine Navigation and Safety of Sea Transportation – A. Weintrit & T. Neumann (eds.)

The Using of Risk to Determination of Safety Navigation in Inland Waters

W. Galor

Maritime University of Szczecin, Poland

ABSTRACT: Many kind of ships are navigated in inland waters. There are sea, river- sea and inland ships. The navigation in inland waters is hard by small relation between the ship size and dimensions of water area. The ships can came natural objects (coast, water bottom) and artificial obstructers (port structures, locks, bridges etc.). The determination of safety of ship's movement can be identified by the risk as a combination of probability of accident and its results (losses). The paper presents two methods of measure of safety. The first one is based on determination of overall risk as a sum of independent components connected with different possibilities of potential accidents. There are contain under keel clearance, distance to obstructers, closed distance of ships approach, air drought and energy of berthing. The second method is based on comparing of assessed the overall risk to an absolute value, constant value or relative change of risk.

1 INTRODUCTION

Process of ships movement in water area should be safely. Its estimation is executed by means of notions of safety navigation. It may be qualified (Galor W. 2001) as set of states of technical, organizational, operating and exploitation conditions and set of recommendations, rules and procedures, which when used and during leaderships of ship navigation minimize possibility of events, whose consequence may be loss of life or health, material losses in consequence of damages, or losses of ship, load, port structures or pollution of environment. That events are is calling as navigational accidents. Very often, the ships move on waterways (natural or artificial) inside of land for many kilometres (Weintrit A. 2010). The manoeuvring of ships on each water area is connected with the risk of accident, which is unwanted event in results of this can appear the losses. There is mainly caused by unwitting contact of ship's hull with other objects being on this water area. The safety of ship's movement can be identified as admissible risk, which in turn can be determined as:

$$R_{adm} = P \, C_{min} \tag{1}$$

where R_{adm} - admissible risk; P_A = probability of accident; and C_{min} = acceptable losses level.

As a result, a navigational accident may occur as an unwanted event, ending in negative outcome, such as:
- loss of human health or life,
- damage or damage of the ship and cargo,
- environment pollution,
- damage of port's structure;
- loss of potential profits due to the port blockage or its parts,
- coast of salvage operation,
- other losses.

The inland waterways are restricted areas those where ship motion is limited by area and ships traffic parameters. Restricted areas can be said to have the following features:
- restriction of at least one of the three dimensions characterizing the distance from the ship to other objects (depth, width and length of the area),
- the ship has no choice of a waterway,
- restricted ship manoeuvring,
- necessity of complying with safety regulations set for local conditions and other regulations.

In a few cases especially for ports situated inside of land there are the waterways and canals with great lengths of hundreds kilometres that leading mainly by natural water region and rivers. Thus the navigation on such waterways is different than on approaching waterways and coastal water areas. The realization of navigation on limited water areas is consisted on (Galor W. 2001):

- planning of safety manoeuvre,
- ship's positioning with required accuracy on given area,
- steering of ship to obtain the safety planned of manoeuvre,
- avid of collision with other ships.

Approach channels to port, port water area and inland waterways are characterised by occurrence of hydrotechnical construction. These constructions are result of activity of man and embrace aquatic or under-water structure which together with installations, builder's devices connected of its technical devices and other advisable necessary equipment to realisations of its intended function to state whole of technical using. From sight of view limitation of movement in water area, ports structures envelop following component:

- objects arising in result of executing of dredged works such port and shipyard water, especially and basin, sea and lagoon fairways, channels, turning basins, passing area,
- channels,
- wharfs determining of water area coast and largely making possible berthing to them and mooring of ships,
- constructions of coast protection such breakwaters, under-water thresholds, strengthening of bottom, scarp of fairways deepened,
- constructions of fixed navigational marks such lighthouses, situated on shore of sea water area and aquatic, light lines and navigational marks, navigational dolphins,
- locks,
- port structures, situated in area of sea harbour in particular breakwaters, breakers of waves, wharfs trans-shipment and berthing and other,
- structures connected with communication, in particular road – bridges, railway, submarine tunnels,
- structures connected with exploitation of sea bottom (drilling towers, platforms, submarine pipelines).

2 THE CRITERION OF SHIPS SAFETY ASSESSMENT

The restricted area is characterized by a great number of factors being present at the sometime. It caused that possibility of navigational accident in these areas is more than in other ones. It means the navigation safety is lower in restricted areas. The assessing of navigation safety requires the application of proper criteria, measures and factors. The criteria make it possible to estimate the probability of navigational accident for certain conditions. The ship during process of navigation has to implement the following safety shipping conditions:

- keeping the under keel clearance,
- keeping the proper distance to navigational obstruction,
- keeping the proper air drought,
- avoid of collision with other floating craft,

To assessment and analyse the safety, especially in the quantitative manes, the necessary to select values that can by treaded as a safety measures. It permits to determine the safety level by admissible risk (Galor W. 2003):

$$R_a = P_A [d(t)_{max} < d_{min} (0 < t < T_p)] \text{ for } c < C_{min} \quad (2)$$

where $d(t)_{max}$ - distance of craft hull to other objects during manoeuvring; C_{min} - least admissible distance of craft hull to other objects; T_p - time of ships manoeuvring; c - losses as result of collision with object; and C_{min} - the acceptable level of losses.

Because the losses can be result different events, the following criterion of safety assessment can be used:

1. Safety under keel clearance (SUKC)
2. Safety distance to structure (SDS)
3. Safety distance of approach (SDA)
4. Safety air drought (SAD)
5. Safety of berthing (SOB)

Thus, there are many categories of risk due to ship movement in water area. In each case the accident rate (probability) is determined for each of the accident categories. The overall risk of ship movement in water area in then the sum of these single, independents risks:

$$R_o = R_g + R_n + R_c + R_{ad} + R_b \quad (3)$$

where R_o - overall risk of ship movement in water area; R_g - risk of grounding; R_n - risk of collision with navigations obstructers; R_c - risk of collision with other ships; R_{ad} - risk of impact the object over the ship; and R_b - risk of damage of ship's hull and port structures during berthing.

3 OVERALL RISK OF SHIP MOVEMENT IN WATER AREA

3.1 *Safety under keel clearance (SUKC)*

The under keel clearance is a vertical distance between the deepest underwater point of the ship's hull and the water area bottom or ground. That clearance should be sufficient to allow ship's floatability in most unfavourable hydrological and meteorological conditions. Consequently:

$$H \geq T + R_B \quad (4)$$

where H – depth; T - ship's draft; and R_B - safe under keel clearance (UKC).

The safe under keel clearance should enable the ship to manoeuvre within an area so that no damage to the hull occurs that might happen due to the hull impact on the ground. A risk of an accident exists when the under keel clearance is insufficient. When determining the optimized UKC we have to reconcile contradictory interests of maritime administration and port authorities. The former is responsible for the safety of navigation, so it wants UKC to be as large as possible. The latter, wishes to handle ships as large as possible, therefore they prefer to accept ships drawing to the maximum, in other words, with the minimized UKC. The maximum UKC requirement entails restricted use of the capacity of some ships, which is ineffective in terms of costs for ports and ship operators. In the extreme cases, certain ships will resign from the services of a given port. Therefore, the UKC optimization in some ports will be of advantage. It is possible if the right methods are applied. Their analysis leads to a conclusion that the best applicable methods for UKC optimization are the coefficient method and the method of components sum.

In the coefficient method one has to define the value R_{min} as part of the ship's drought (Mazurkicwicz, 2008):

$$R_{min} = \eta \ T_c \qquad (5)$$

where η - coefficient; and T_c - deepest draught of the hull.

The other method consists in the determination of R_{min} as the algebraic sum of component reserves which accounts for errors of each component determination:

$$R_{min} = \sum R_i + \delta_r \qquad (6)$$

where R_i - depth component reserves; and δ_r - sum of errors of components determination.

The UKC is assumed to have the static and dynamic component. This is due to the dynamic changes of particular reserves. The static component encompasses corrections that change little in time. This refers to a ship lying in calm waters, not proceeding. The dynamic component includes the reserve for ship's squatting in motion and the wave impact. One should emphasize that with this division the dynamic component should also account for the reserve for ship's heel while altering course (turning).

3.2 Safety distance to structure (SDS)

The accessible port water area (for given depth) warrants safety manoeuvring for fulfil condition:

$$\omega \in \Omega \qquad (7)$$

where: ω - requisite area of ship's manoeuvring; and Ω - accessible water area.

Ships contact with structure can be intentional or not. Intentional contact steps out when ship berthing to wharf. During this contact energy dependent from virtual ship masses and its perpendicular component speed to the wharf is emitted. In result of ship pressure on wharf comes into being reaction force. Both emitted energy during berthing and bulk reaction force cannot exceed admissible value, definite by reliability of ship and wharfs. These values can be decreased by means of fenders, being usually of wharf equipment. Ship should manoeuvre in such kind to not exceed of admissible energy of fender-structure system. Unintentional contact can cause navigational accident. Process of ship movement in limited water area relies by suitable manoeuvring. During of ship manoeuvring it can happen the navigational accident. Same events can occurred strike in structures, when depth of water area is greater than draught ship. There are usually structures like wharf, breakwater, etc., and also floated objects moored to structure.

3.3 Safety distance of approach (SDA)

Where:

The fundamental measure of ships passing is distance to closest point of approach (DCPA). Its value should be safety, it means:

$$DCPA \geq DCPA_{min} \qquad (8)$$

where $DCPA$ - distance to closest point of approach; and $DCPA_{min}$ -acceptable distance to closest point of approach.

The accident can happen, when above condition will not be performance. Knowing the number of entries of ships in a year (annual intensity of traffic), one can determine the probability of ships collision for one ship transit:

$$p_A = \lambda / I_R \cdot t \qquad (9)$$

where p_A - probability of ships collision in one transit; λ - accident frequency; I_R - annual traffic intensity and t - given period.

3.4 Safety air drought (SAD)

Air drought is distance over ship, when manoeuvre under construction. They mainly consist:
– bridges (road, railway) over waterway,
– high voltage lines,
– pipelines over waterway,
The condition of safety ship movement is following:

$$H_S < H_C \qquad (10)$$

where H_S - the height of highest point of ship; and H_C - the height of lowest point of construction over waterway.

In many cases, the sea-river ship's superstructure is regulated. It permits to decrease of ships height. Also other elements of ship's construction can be disassembled – for instance masts of radar antenna, radio etc.

3.5 Safety of berthing (SOB)

Condition of the safety of the manoeuvre while berthing the ship to the quay can be as follows :

$$E(t) \leq E_k^{berth} \text{ or } E(t) \leq E_k^{ship} \qquad (11)$$

where $E(t)$ - maximum kinetic energy of the ships impact absorbed by the system berth - fender – ship; E_k^{berth} - admissible kinetic energy absorbed by the system berth – fender; E_k^{ship} admissible kinetic energy, near which the formed strengths of the reaction of the system berth – fender do not cause the durable deformation of the ship's hull yet .

Factors which have the influence on the size of the maximum kinetic energy of the ship's impact against the berth construction are as follows:
– ship manoeuvrability (kind and the power of the propulsion, thrusters),
– hydro meteorological conditions (wind, current),
– tugs service (the number of tugboats, their power),
– the manoeuvring tactics (captain's or pilot's skill).

4 THE MODES OF APPLYING RISK

As it has already been mentioned, a picture of the situation closer to completeness is presented by navigational risk. Knowing the probability of the accident, its effects can be taken into account in the calculations. When applying the risk, three modes can be distinguished (Galor, 2010):
– the value of absolute risk;
– risk constant;
– relative increase of risk.

In the first method the determined risk value refers to the assumed value limit. Sometimes this value is specified in regulations.

$$R_N \leq R_p \qquad (12)$$

where R_N – determined risk value; and R_p – assumed risk value limit.

The next method of constant risk is a development of the first. It is applied in situations where the effects of the accident are likely to change. This can take place when a ship of a specified size manoeuvring on a certain water area is carrying in turn cargoes of various hazard degrees (various effects in case of accident, e.g. coal, and the other time crude oil products). Then, to maintain a steady level of safety, the risks must be equal to each other.

$$R_{N1} = R_{N2} = P_{A1} \cdot C_1 = P_{A2} \cdot C_2 \qquad (13)$$

where R_{N1}, R_{N2} - navigational risk; P_{A1}, C_1 - probability of an accident and consequences of the accident with cargo of low hazard degree; and P_{A2}, C_2 - probability and consequences of an accident with cargo of higher hazard degree.

If the effects of an accident change, then, in order to maintain the assumed risk level the value of accident probability is changed, too.

$$P_{A2} = P_{A1} \cdot C_1 / C_2 \qquad (14)$$

If the effects of an accident increase, the probability of an accident has to diminish. In the case of the criterion of required manoeuvring area size, its size should be increased. If it is impossible because of local conditions, other measures must be undertaken, lowering the permissible boundary of hydro meteorological conditions speed and direction of wind and current, or increasing the tug service. In special cases, when the accident effects can be so enormous, extraordinary measures are undertaken. Let me mention an example of the transport of explosives brought out in the port water area and taking them to a maritime firing areas. Manifold measures are undertaken then, i.e. certain units are prohibited to move, people from the coastal area are evacuated etc.

In certain cases there may appear plans of increasing the size of operated ships. Then, the experience of many years' observations concerning the operation of ships on a given area will not permit to directly support a decision to allow the traffic of larger units. Simulation research which could solve the problem is not always possible to be carried out at a particular time. And yet, the maritime administration should make a decision based on justified consideration, as it happens in many cases, that the decision is made intuitively and the one who makes it is led by irrational factors, as for instance pressure on the part of the port authorities. Therefore, in such cases a method is suggested to determine the relative risk increase (Galor, 2010):

$$\rho_R = \Delta R / R_P = (R_U - R_p) / R_p \qquad (15)$$

where ρ_R – relative risk increase; ΔR – unconditional risk increase; R_P - navigational risk before the introduction of changes; and R_U - navigational risk after the introduction of changes.

Taking into consideration that the risk is a combination of accident probability and its consequences, the following can be written:

$$\rho_R = \rho P_A + \rho S \qquad (16)$$

where ρP_A – relative increase of accident probability; and ρS – relative increase of accident consequences.

5 CONCLUSION

The ships move on waterways (natural and artificial) inside of land in many cases for hundreds kilometres. The ship (many types-sea, inland or sea-river) can came natural objects (coast, water bottom) and artificial (water port structures-locks, bridges etc.) obstructers. Also many other ships can encounter. It caused that the navigation in inland waters is harder than on open seas. The criterions of safety determination of ship movement need more precisely of qualify. The risk can be used as measure of safety. They are a result of unwanted contact with objects on inland water area. The presented above consideration can permit to analysis of safety sea-river ships in inland shipping. Two methods were presented to measure of navigation safety. The first method is based on overall risk as a sum of component based on different potential of accident. The second method contain the comprising three modes of applying risk.

REFERENCES

Galor W. (1999): Determination of navigational risk due to planned increased ship size in port channels. *Proceedings of the "1th International Congress on Maritime Technological Innovations and Research"*. Barcelona, 1999.

Galor W. (2001): The methods of ship's manoeuvring risk assessment in restricted waters. *Proceedings of the 14th International Conference on Hydrodynamics in Ship Design*. Szczecin-Międzyzdroje, 2001.

Galor W. (2003): The modelling of ships movement in limited waters to determine of navigational risk. *Proceedings of 9-th International Conference IEEE "Methods and models in automation and robotics"*. Międzyzdroje, 2003.

Galor W. (2004): Kryteria bezpieczeństwa ruchu statku po akwenie portowym. *Zeszyty Naukowe Akademii Morskie w Szczecinie nr 3(75)*. Szczecin, 2004.

Galor W., (2010), The model of risk determination in sea- river navigation. *Journal of KONBIN 2,3 (14,15)*.Warsaw, 2010.

Mazurkiewicz B. (2008*): Morskie budowle hydrotechniczne. Zalecenia do projektowania i wykonywania. Wyd. FPPOiGM*. Gdańsk, 2008.

Weintrit A. (2010), Nawigacyjno-hydrograficzne aspekty żeglugi morsko-rzecznej w Polsce, (red). Akademia Morska w Gdyni, 2010.

Inland Water Transport and its Impact on Seaports and Seaport Cities Development

A.S. Grzelakowski
Gdynia Maritime University, Poland

ABSTRACT: Inland water transport has played a significant role in the development of port cities and seaports all over the world for ages. As a dominant mode of inland transport until the rail and road have begun to prevail in the European transport systems, it has at first created and then shaped seaport hinterland in the majority of the European ports. However, its impact on these organisms, i.e. seaports and port cities has changed in many European countries in recent decades, due to the fact that it was getting weaker (logistics time constraints) and less competitive on the transport market. Consequently, in many European countries it has lost its impact on seaports hinterland and has been replaced by other less environmentally friendly modes of transport. This tendency is, however, not observed in the North Sea mega-ports as well as in many world biggest ports in Asia and North America, where inland shipping still maintains significant share in the port total turnover modal split. The reasons for such situation are analyzed in this paper.

1 INTRODUCTION

The main objective of the article is to analyze the development of inland water transport in the EU in recent years and indicate its active role as a still important factor influencing the prosperity of seaport cities as well as the creation and expansion of mega-ports in the North Sea Region. Admittedly, this mode of transportation has lost its absolutely dominant position in the European transport systems since the mid 30s in the 20th century in favor of rail transport and later on, partially, even road transport; however, in some EU countries, especially those situated along the main European inland waterways, it still holds significant market shares.

Nowadays, however, on the open highly competitive EU transport market, under the pressure of growing logistics constraints, its position against other modes of transportation is getting weaker. Its traditional competitive advantages such as low costs per ton and low freight rates as well as ability to transfer large quantities of bulk cargos on long distances have not been attractive enough for shippers and logistics supply chain operators who require transport services of the highest quality in terms of costs and time. Fortunately, in such case the ongoing pressure of transport market forces is mitigated by the gradually imposed EU public regulatory measures which, undoubtedly, will positively affect this mode of transport in the long run. The EU sustainable transport policy launched two decades ago, oriented on the external costs reduction, strongly supports this environmentally friendly mode of transport, favoring its development in many ways. In the EU strategy of combined transport development along with TEN-T projects of priority transport axles including motorways of the sea, short sea shipping program as well as the concept of boosting co-modality, inland water transport besides rail transport is regarded as the key instrument facilitating the implementation of the main EU transport policy goals established in the first (1992) the second (2001) and the third (2011) EC White Paper on the EU transport policy. [19]

Supported by the EU transport policy measures the development of inland water transport within the EU transport system ought to strengthen directly and indirectly the EU seaport sector and especially those ports where their role in servicing their hinterland is still considerable, i.e. those belonging to the core network of TEN-T . Hence, analyzing the relations between the seaports and inland water transport development, from the European North Sea perspective, in recent years might, on the one hand, smooth over better recognition of its potential as a driving force for gaining new market shares on port hinterland and, on the other, draw the line for better planning of these relations in the framework of the

European transport system in the next decades and cooperation between these links of the transport chain (co-modality and intermodality).[5]

2 EU NORTH SEA MEGA-PORTS ON THE GLOBAL AND EUROPEAN SCALE

In the contemporary global economy international trade determines the worldwide flow of goods. The absolutely dominant position belongs to international seaborne trade (9,9 billion tons in 2013). Maritime transport share in servicing the global trade, nowadays, accounts for more than 90 percent (in ton-miles) and 82 percent in terms of volume, and it is still growing. [11]

Increases in the volume of commodities carried out by the maritime transport sector have been induced by growing globalization accompanied by the international economic integration. However, such tendencies facilitating free unconstrained flow of capital, goods and services among the countries, strongly stimulate competition on a worldwide scale. As a result, cooperation in the form of vertical and horizontal concentration has been intensified, and relatively low transport fees, especially in the maritime sector, are maintained. With such low freight rates the question of where in the world the goods are manufactured has become subsidiary. [12]

Consequently, as a result of developing asymmetric trade pattern in the world – spatial concentration of main production and consumption center, more than half of the largest ports (mega-ports) are situated in Asia. Chinese ports in particular are remarkable (see tab. 1).

Table 1. Top twenty world ports in 2012 (in 1000 tons)

Rank	Port	Country	Measure	Tons (000s)
1	Ningbo-Zhoushan	China	MT	744,000
2	Shanghai	China	MT	644,659
3	Singapore	Singapore	FT	538,012
4	Tianjin	China	MT	477,000
5	Rotterdam	Netherlands	MT	441,527
6	Guangzhou	China	MT	438,000
7	Qingdao	China	MT	407,340
8	Dalian	China	MT	303,000
9	Busan	South Korea	RT	298,689
10	Port Hedland	Australia	MT	288,443
11	Hong Kong	Hong Kong, China	MT	269,282
12	Qinhuangdao	China	MT	233,235
13	South Louisiana	United States	MT	228,677
14	Houston	United States	MT	216,082
15	Nagoya	Japan	FT	202,556
16	Shenzhen	China	MT	196,458
17	Port Kelang	Malaysia	MT	195,856
18	Antwerp	Belgium	MT	184,136
19	Dampier	Australia	MT	180,366
20	Ulsan	South Korea	RT	174,117

Source: IAPH Statistics on world ports. Tokyo 2013 [7]
MT=Metric Tons; FT=Freight Tons; RT = Revenue Tons

On the list of top twenty world seaports there are nine Chinese ports altogether and among them two which handle more than 640 million tons a year (2012). [7, 11] What is characteristic for both of them (Ningbo & Zhoushan and Shanghai is that they are situated at the estuaries of great rivers and well connected by inland waterways with the main Chinese manufacturing and consumption centers.

It is worth to mention that the world largest ports in terms of total volume handled also belong to the group of the biggest container ports. The dominance of the Chinese ports is in this area particularly visible (see tab. 2). Among the ten largest world container ports are seven Chinese seaports. Frontrunner on this list is Shanghai, strongly supported by inland water transport, connecting this mega-port with its large hinterland.

Table 2. The world largest container ports in Asia and Europe in 2013 in terms of handled volume

Rank	Port, Country	Volume 2013 (Million TEUs)
1	Shanghai, China	33.62
2	Singapore, Singapore	32.6
3	Shenzhen, China	23.28
4	Hong Kong, China	22.35
5	Busan, South Korea	17.69
6	Ningbo-Zhoushan, China	17.33
7	Qingdao, China	15.52
8	Guangzhou Harbor, China	15.31
9	Jebel Ali, Dubai, United Arab Emirates	13.64
10	Tianjin, China	13.01
11	Rotterdam, Nether-lands	11.62
15	Hamburg, Germany	9.30
16	Antwerp, Belgium	8.59
25	Bremen/Bremerhaven, Germany	5.84

Source: Top 50 world container ports. World Shipping Council. Partners in trade. 2015 and The Journal of Commerce annual top 50 World Container Ports, Lloyd's List annual Top 100 Ports, AAPA World Port Rankings and individual port websites.[13]

On the list of the largest seaports in the world there are only two European seaports, namely Rotterdam and Antwerp with similar localization. Each of them belongs to the group of the twenty largest global seaports and enjoys the status of the mega-port (see tab. 1). It is because of the fact that both seaports are well situated relative to the main EU economic centers of production and consumption, and themselves are strongly supported by the Dutch and Belgian economies, which significantly contribute to the world trade. To these countries belongs Germany, the third largest world exporter and importer (tab. 3).

Among the top thirteen trading countries in the world there are seven European countries which together account for almost 25,5 percent of the global export and 18,9 percent of the total world import (in value terms). [17] Though, large countries like Germany, France and the United Kingdom dominate the European trade, a number of smaller countries like Belgium and the Netherlands also show relatively large trade values and intensively use their seaports in overseas trade relations (their share in global export amounted altogether to 6,1 % in 2013; see tab. 3).

Table 3. The world leading exporters and importers (export and import in billion US dollars)

Rank	Exporters	Value	Rank	Importers	Value
1	China	2049	1	United States	2336
2	United States	1546	2	China	1818
3	Germany	1407	3	Germany	1167
4	Japan	799	4	Japan	886
5	Netherlands	656	5	United Kingdom	690
6	France	569	6	France	674
7	Korea, Republic of	548	7	Netherlands	591
8	Russian Federation	529	8	Hong Kong, China	553
9	Italy	501		retained imports	140
10	Hong Kong, China	493			
	domestic exports	22	9	Korea, Republic of	520
	re-exports	471	10	India	490
11	United Kingdom	474	11	Italy	487
12	Canada	455	12	Canada	475
13	Belgium	447	13	Belgium	437

Source: World trade developments. WTO International Trade Statistics. WTO 2014 [17]

Furthermore, the three North Sea countries, namely Belgium, Netherland and Germany have always been strongly oriented to smooth development of waterborne transport and its interconnection on the European scale. As it turned out, it counts in the long term. Because of its favorable position in that area and its good hinterland transport connections, the Netherlands, Belgium and Germany have become important choice locations for many large global European distributors and logistics operators interested in the development of logistic supply chains.

It is worth mentioning that Germany, the Netherlands and Belgium belong to the group of top 10 countries in the world which have the highest score as far as logistics performance index (LPI) is concerned. LPI as a yearly report prepared by the World Bank evaluates the logistical achievements in 155 countries all over the world. It takes into account such factors (criteria) influencing the logistics performance as: 1. efficiency of clearance process, 2. quality of trade and transport related infrastructure, 3. ease of arranging competitively priced shipments, 4. competence and quality of logistics services, 5. ability to track and trace consignments, 6. timeliness of shipments. The LPI is an interactive benchmarking tool created to help countries identify the challenges and opportunities they face in their performance on trade logistics and what they can do to improve their performance. [14]

The Logistics Performance Index is based on a worldwide survey of operators on the ground (global freight forwarders and express carriers), providing feedback on the logistics "friendliness" of the countries in which they operate and those with which they trade. They combine in-depth knowledge of the countries in which they operate with informed qualitative assessments of other countries with which they trade, and experience of global logistics environment. [14.]

The logistical performance index 2014, based on global research, favors the European North Sea countries. Germany is the leading country (rank 1) in the world in terms of efficiency and effectiveness of customs and other procedures at its borders, the quality of transport and the ICT infrastructure permeating logistics. Other European North Sea countries like the Netherlands and Belgium are classified at the top of the international 2014 LPI ranking too, taking 2nd and 3th position respectively. [14, 4] Owing to their transport potential and real achievements in trade and logistics, seaports of those countries attain the leading position among European seaports (tab. 3)

There are six Dutch, German and Belgian seaports among the top twenty European ports with cargo handling of more than 40 million tones. The six biggest European seaports with the total throughput of more than 90 million tones on a yearly basis, apart from Novorossiysk and Algeciras, include Rotterdam, Antwerp, Hamburg and Amsterdam. [2, 18]

They are the real global mega-ports in terms of transport and logistics requirements which constitute the highest forth generation seaport class in Europe (so-called log-ports). The absolute leading position among the European mega-ports belongs to Rotterdam. The great advantage of the port of Rotterdam as well as Antwerp and Hamburg (so-called A-R-A ports) lies in the possibility to handle the deepest maritime vessels used in the world fleet. Therefore, the majority of maritime vessels opt for ports on the North Sea for distributing goods throughout Europe. Rotterdam, Antwerp, Hamburg and Amsterdam are capable of penetrating beyond their national borders deep into their European hinterland, while the other smaller seaports primarily fulfill the national or even only regional functions. [4] Subsequently, they have reached the leading position in the container seaborne transport passing European seaports (see tab. 2).

The port of Rotterdam has obtained the absolute dominant position on the European container market handling yearly ca. 11,7 million TEUs (2013). It is ranked first in Europe but scarcely 11th among the world biggest container seaports. [13, 18] The next three biggest container ports and fierce competitors of Rotterdam include predominantly Hamburg, which takes second position, and then Antwerp and Bremerhaven (see tab. 2). All these mega-ports and simultaneously global container hubs, with a number of container terminals each, not only serve one of the richest and most prosperous regions in Europe constituting their hinterland but also themselves are an integral part of vigorous economic centers of trade as well as manufacturing and consumption areas. These are surrounded by great seaport cities or even constitute an integral part of them. They export and import plenty of goods for the city dwellers as well as for manufacturing and processing centers located close to the port-cities. As a result, transport loco plays relatively important role for the dynamic development of the North Sea container hubs.

3 INLAND WATER TRANSPORT AND ITS ROLE IN THE EUROPEAN TRANSPORT SYSTEM

More than fifty countries around the world have navigable waterways networks of more than 1,000 kilometers. On most of these waterways, the inland shipping sector is underdeveloped. It experiences many barriers and obstacles not only of technical and technological nature. They sometimes seriously limit smooth access of inland shipping to the transport system and reduce automatically its potential role in the domestic and international trade. China takes the lead, with more than 110,000 navigable kilometers. It is remarkable that in China the incoming and outgoing flow of goods across water plays a comparable role to that of northwest Europe. Outside Europe there are more than 30 countries in the world which have opportunities to utilize inland shipping on a much larger scale for the transshipments and carriage of cargo for long distances. [1, 3. 8]

Inland waterway transport plays an important role for the transport of goods in Europe, too. The waterways network in the EU-28 amounts to 41,527 km of canals, rivers and lakes. [9, 10] More than 37 000 km of waterways connect hundreds of cities and industrial regions. Some 20 out of 27 Member States have inland waterways, 12 of which have an interconnected waterway networks. The potential for increasing the modal share of inland waterway transport is, however, significant. Compared to other modes of transport which are often confronted with congestion and capacity problems, inland waterway transport is characterized by its reliability, its low environmental impact and its major capacity for increased exploitation..[6, 15, 19] The European Commission promotes and strengthens the competitive position of the inland waterway transport in the transport system, and tries to facilitate its integration into the intermodal logistic chain in the most efficient way. [1, 3, 6]

Inland waterway transport is a competitive alternative to road and rail transport. In particular, it offers an environment friendly alternative in terms of both energy consumption and noise and gas emissions. Its energy consumption per km/ton of transported goods is approximately 17% of that of road transport and 50% of rail transport. Its noise and gaseous emissions are modest. [1, 8, 9] In addition, inland waterway transport ensures a high degree of safety, in particular when it comes to the transportation of dangerous goods. Finally it contributes to the decongestion of the overloaded road network in densely populated regions. According to recent studies, the total external costs of inland navigation (in terms of accidents, congestion, noise emissions, air pollution and other environmental impacts) are seven times lower than those of the road transport. [1, 6, 8, 15]

The European inland shipping sector appears to be able to deliver export products to the rest of the world in a growing number of cases and consign to the European seaports imported raw materials and semi-finished products to the manufacturing centers. Around 20,000 km of the waterways network in the EU-28 are concentrated primarily in the zone with the busiest waterways, i.e. the Netherlands, France, Germany, Belgium and Austria. Consequently, the share of inland water transport in each of these countries modal split is very impressive compared to other EU countries or the EU as a whole (4,0 % in tones and 6,3 % in tkm). [3] For instance, performance of freight transport in inland shipping sector expressed in tone-kilometers (billion t-km) in 2012 amounted to: Germany 58.5, Netherlands 47.5, Belgium 10.4 and France 8.9. [3, 8]

Via the Rhine and its adjacent rivers and canals the industrial areas of north and south Germany, north Switzerland and northeast France are in fact fully accessible to large vessels. Via the Maas and adjacent navigable waterways in Belgium, Luxembourg and north France they are opened up to larger vessels Directly from the Rhine ships can reach the Danube via the Main-Danube channel. This means that also the larger industrial areas in Austria, Czech Republic, Hungary, Croatia, Serbia, Romania and Bulgaria can be reached with larger vessels across water. Via the Elbe and the Oder the industrial areas in Austria, Germany, Poland and Czech Republic are practically accessible, too.

Canals link the Rhine with the Maas, Rhône-Saône, Marne, and Danube (via the Main) valleys. The Rhine is connected to the Mediterranean Sea by the Rhine-Rhone canal and is joined to the Black Sea by the Rhine-Danube canals. This makes it possible for barges and passenger boats to travel from the North Sea to the Black Sea. As a result, the Rhine is the busiest waterway in the world (more than 330 000 000 tons cargo are being transported recently per year) and cargo is transported all over Europe using these two canals.[1, 9, 10]

The Rhine constitutes the backbone of inland navigation in Europe. It is used by more than two-thirds of all goods carried by inland waterway. New markets are booming; these include the transport of containers, weight-intensive goods, chemicals and passengers. However, the principal cargoes carried on the river are: coal, coke, grain, timber, and iron ore. Nowhere else in the world there are freight flows concentrated as massively as on the Rhine. This made it possible for the Netherlands to become the gateway to Europe. The Rhine reserve capacity (700%) and that of the other waterways (100%) ensures that a significant increase in transport volumes over waterways network can be handled without difficulty for many years to come. [1, 8] It is significant driving force especially for Rotterdam as the chief outlet to the North Sea and simultaneously one of the world's largest seaports.

Among many requirements, the Rhine needs to fulfill both transport and logistics functions of its well developed river ports. Amongst them a primary position is taken by Duisburg. Duisburg, the outlet for the Ruhr industrial region, is the world's largest river port. It is officially regarded as a "seaport" because seagoing river vessels go to ports in Europe, Africa and the Middle East. Numerous docks are mostly located at the mouth of Ruhr river where it joins the Rhine. Each year more than 40 million tonnes of various goods are handled with more than 20,000 ships calling at the port. The public harbor facilities stretch across an area of 7.4 square kilometers (2.9 sq mi). There are 21 docks covering an area of 1.8 km2 (0.7 sq mi) and 40 kilometers (25 miles) of wharf. The area of the Logport Logistic Center Duisburg stretches across an area of 2.65 km2 (1.02 sq mi). [9, 16] With more than 2.8 million TEU it is also the largest inland container port based on 2013 figures. A number of companies run their own private docks in the port of Duisburg, bringing the total cargo volume passing through the port to 70 million tons a year. [16]

The Port of Duisburg lies at the junction of the Ruhr and Rhine rivers, about 190 kilometers from the North Sea in western Germany, just 37 kilometers east of the country's border with the Netherlands, about 160 kilometers southeast of the port of Amsterdam and 16 kilometers north of Germany's port of Dusseldorf. As the world's biggest inland port with connections to the North Sea ports through the Rhine-Herne Canal and the Dortmund-Ems Canal it links perfectly these seaports through other inland ports with their very competitive hinterland. The port has convenient access to Europe's rail, road, air, and water transport networks.

In 2013, over 520 thousand people lived in the area of the port of Duisburg. Located in the heart of the European market, the port of Duisburg serves the area of more than 30 million consumers. However, the current population of the Rhine basin is approximately 50 million. The major cities are all situated on the Rhine or on its larger tributaries, and the development of these cities is strongly dependent on water. Similarly, the activities undertaken within these cities influence the waters of the Rhine and its tributaries. In this sense, the Rhine basin could be regarded as some kind of "mega-city". [16, 18]

The problems that have affected the development of the Rhine basin are similar to those currently affecting water resources managers in large cities. Specific issues include water supply, flooding, water quality, energy production, transport and institutional arrangements. The demand on water for a range of purposes has increased significantly with time. Population growth, industry, agriculture, hydropower generation and other water resources users can be either cooperative or competing users.

4 INLAND WATER TRANSPORT AS A DRIVING FORCE OF THE EUROPEAN NORTH SEA MEGA-PORTS

The contemporary development of the EU transport system is characterized by the still growing road transport subsystem. Road transport takes the biggest share of transport performance in all European countries. In Germany, the Netherlands, Belgium and France however, as it was mentioned earlier, inland shipping accounts for a considerable share of transport performance. Nevertheless, its role and functions in a modern transport system and especially its sustainable development is to some extent still neglected. Indeed, the European Commission tries to support such environmentally friendly mode of transport by promoting its development (NAIADES I and II), but its previous efforts are rather miserable. [1, 9, 10]

Recently, several models have come up with growth projections in the European freight transport. Where they tend to differ is in how this growth will translate across different modalities (see tab. 4). Earlier expectations were that the European inland shipping will grow by at least 38% versus 2005 transport performance until 2030. [6, 15] This dynamic is, however, being slowed down since 2010 (tab. 4). The expected growth rate in the period 2010-2030 is projected to be only 121 percent. [6] Hence, the outlook for the inland water transport compared to other modes of transport, especially road transport, in the perspective of the next sixteen years is rather moderate. It could influence to some extent its position among the other modes of transport on the North Sea ports hinterlands but it should not hamper it in any way

Table 4. Scenarios of modal split in European inland transport performance by 2030

Modes of transport	2010	2015	2020	2025	2030
Trucks*	2048,3	2278,9	2485,6	2666,7	2803
Rail*	427,2	469,5	504,6	535,2	558,9
Inland navigation*	294,2	312,9	331,3	344,3	355,3
Trucks	73,95%	74,44%	74,83%	75,20%	75,41%
Rail	15,42%	15,34%	15,19%	15,09%	15,04%
Inland navigation	10,62%	10,22%	9,97%	9,71%	9,56%

* in billion tonne-kilometers
Source: H. van Essen and others, EU transport GHG: Routes to 2050 ? Modal split and decoupling options; Paper 5.Delft 2009, p. 26 [15]

It is remarkable that nowadays at almost every European seaport, especially within general cargo and container segments, road transport plays a primary role in servicing their hinterland. In German seaports, the focus is also on rail, in addition to road transport. In Rotterdam, however, as well as in Antwerp and Amsterdam the emphasis lies on inland shipping. Its outstanding position in ingoing and outgoing container traffic is evident in the port of Rotterdam (see tab. 5).

Table 5. Modal split in container throughput via port of Rotterdam 2013 – 2010 (%)

Mode of transport	2013	2012	2010
Barge	34,8	35,3	33,0
Rail	10,7	10,7	10,6
Road	54,6	54,0	56,4

Source: Port Statistics. Port of Rotterdam Authority. Rotterdam 2015 www.portofrotterdam.com [15.01.2015]

In the port of Rotterdam inland navigation with ca. 35 percent takes a second place among the inland modes of transport serving the carriage of containers to/from its hinterland. The share of rail transport is relatively stable and accounts for 10,7 percent. Admittedly, the road transport slipped slightly from 56,4 % in 2010 to 54,6% in 2013 but it is still the dominant mode of hinterland transport in Rotterdam. In accordance to the EU sustainable transport policy, the Dutch government and the Port of Rotterdam pursue a policy aimed at decreasing the share of road haulage accomplished through modal shift. Since 1993, the truck share has, according to the "old" method, declined from 66% to some 54%.

It has been already emphasized that the great advantage of the port of Rotterdam as well as Antwerp and Hamburg (so called A-R-A ports) lies in the possibility not only to handle the deepest maritime vessels but also to carry out the hinterland transport owing to its exceptional geographic location at the mouth of the large European rivers Rhine, Meuse and Elbe. As a result, Rotterdam but also Antwerp, Amsterdam and Hamburg, can therefore have unlimited access possibilities for transport by water, such as across the river Rhine. If, for example, the Rhine had been absorbed by the sea within the area up to Hamburg, then Hamburg would have been the largest port in Europe. It would simply be impossible to transport the annual volume of goods from the Rhine (300 million tones) as it is today via railways or the road, via one of the other ports within the Hamburg - Le Havre range. [18, 9]

5 NORTH SEA MEGA-PORTS AND PORT CITIES – IRREVERSIBLE INTERCONNECTION

The North Sea mega-ports are characteristic of not only their close inland waterways connections to hinterland and their role in creating competitive advantages for them but also specific spatial, economic and social relations to the coastal regions and port cities. In fact, all these seaports have been closely connected to the surrounding agglomerations for ages or even incorporated into them. These seaports have constituted an integral part of port city organism since medieval times. It is necessary to mention free ports, free port zones or areas as well as entrepots as the main trade easing institutions typical of North Sea ports merging both organisms.[4]

Seaports as vital transport nodes and significant components of the domestic transport system, integrating it with other transport systems of the neighboring countries, have always played very important economic, social as well as political role in many ambitious and independent cities in this part of Europe. Owing to their position in the transport system and usually well developed widespread areas of activity connected with their transport, industrial and commercial functions, strongly influencing the well-being of the port cities, seaside regions and the whole economy, they have been regarded as a public good of special national interest and strategic importance. As a result, seaports were and are still treated and perceived within the overwhelmingly practiced economic doctrine as a public domain. Consequently, being a part of the important country border territory, usually owned by the state or any other public entity (city, autonomous province, etc.), seaports are subject to strict public control and intervention.

The dominant public form of intervention into the port sector is the establishment of the port administration and management system or model. It is set up by law, usually a special parliamentary harbor act which, based on the type of port area and infrastructure ownership, determines the overall legal and administrative relations between the public bodies (city) and the port. Such harbor act constitutes usually port administration and management entity, providing it with precisely defined operational and strategic goals as well as economic and financial tasks and responsibilities.

Taking into account the above mentioned factors, based on ownership, and influencing the most important attitude of the central and regional public entities towards the main seaports as well as relations between the established management entity and the port businesses, four worldwide existing types of seaport administration and management can be distinguished four models: 1. State model, 2. Autonomous, 3. Municipal, and 4. Private one which are straggling with contemporary challenges in Europe and European transport and port sector.

The municipal model was typical of North Sea ports. It is characterized by direct engagement of the local, mainly city or province authorities or self-government in the direct process of port administration and management. That model, which strongly prefers the unconstrained expansion of private sector in the operational sphere, and promotes there the development of fair competition, is very efficient and popular especially in northern Europe. Even large ports like Rotterdam, Antwerp, Hamburg and Amsterdam were managed in such manner till the mid of the last decade of 21st century.

It has many advantages thanks to good cooperation between the port and its surrounding neighborhood – port cities, which enables eliminating many conflicts, e.g. spatial ones, and barriers usually existing between these areas, organizations and systems. 14 However, that model has some constraints too, which might to some extent hamper the port development and its adjustment to the dynamically changing environment. The local municipal authorities can not afford to fund sufficiently the development of port infrastructure, and the port development plans and strategies are sometimes not fully applied to the national plans. What is more, sometimes the local and national interests differ in kind, limiting the necessary area of cooperation to the detriment of the seaport development. [4]

The existing obstacles and barriers to the development of the North Sea mega-ports which were managed and administrated as typical municipal ports have caused significant changes in their management model. In the middle of this decade most of them were partially autonomous in economic and legal sense. Quasi-autonomous port authorities have been established there and new functional task division between municipality and the state has been set up. That new model and concept of port management was implemented in Rotterdam. In accordance thereto the port of Rotterdam Authority is the manager, operator and developer of Rotterdam port and industrial area.

The Port of Rotterdam Authority is a public corporation (N.V.) with two shareholders: the municipality of Rotterdam and the State. As it is revealed by the statutory objectives, this entity operates in two domains: shipping and the port area.

Its statutory objectives include:
1 the promotion of the effective, safe and efficient handling of shipping in the port of Rotterdam and the approach area off the coast,
2 the development, construction, management and operation of the port area. [18]

The port authority also works closely with the State, the municipality of Rotterdam, local authorities and interest groups. As a public entity it will enforce a sustainable strategy towards the hinterland, promoting without doubt inland water transport and rail. The same attitude towards the inland waterways and inland navigation is being provided by the other quasi-autonomous ports, i.e. Hamburg and Antwerp since 2007.

6 SUMMARY

As far as the North Sea ports are concerned, the beneficial geographic situation at the coast and the combination with transport via the Rhine provide Rotterdam, Amsterdam and Antwerp with a great natural advantage compared with other seaports in this European region. Within other matters (road and rail connections) many services and opportunities are equal in the European seaports. In the nearest future this means little or none physical obstacles to let transport via European rivers grow even faster. As for the Rhine, according to research, less than a quarter of the available capacity will be utilized. Therefore, many decades of sustainable growth of transport via northwest Europe is possible, without the need to invest in extra road infrastructure which is capital-intensive. [5, 18]

It means that the inland water transport will play a significant role in the development of North Sea ports giving them a kind of competitive advantage against other European seaports were this mode of transport is underdeveloped. Among beneficiaries of this tendency there are port cities which will consume a significant part of added value generated by inland shipping connecting seaports to their hinterland. The economic, financial, social and environmental relations between these organisms will be strengthened by providing them with unique opportunity to create long term real spatial order, and in the end by eliminating still existing barriers to harmonized cooperation and development. It is especially necessary in the period of development of the core network of transport infrastructure in the EU where the biggest seaports and city nodes will play very important role in the smooth and efficient transport of passengers and goods within the TEN-T constituting the backbone of the European transport and logistics systems.

BIBLIOGRAPHY

[1] Commission Staff working document. Annex to the Communication from the Commission on the promotion of inland waterway transport „NAIADES". Brussels 17.012006 SEC (2006) 34/3 com (2006) 6 Final

[2] European Ports. Top 20 European Ports, 2013-2011. Port of Rotterdam Statistics 2015

[3] EU transport in figures. Statistical Pocketbook 2014. EC Brussels 2014

[4] Grzelakowski A. S., Inland Water Transport as a Factor Influencing Mega-Ports and Seaport Cities Development (from the European North Sea Perspective). Logistics and Transport. No. 2(11)2010

[5] Grzelakowski, A. S. Transport infrastructure in the face of challenges concerning security and reliability of transport and logistics macrosystems. Logistyka 2014, nr 4

[6] http://ec.europa.eu/transport/modes/inland/index. [02.02.2015]

[7] IAPH Statistics on world ports. Tokyo 2013

[8] Inland shipping. An outstanding choice. The Power of Inland Navigation. May 2009, www.inlandshipping.com

[9] Inventory of Main Standards and Parameters of the E Waterway Network, Economic Commission for Europe. Blue Book. Brussels 2013

[10] NAIADES – European IWT News. www.naiades.info/page/nl. (29.01.2015)

[11] Review of maritime transport, 2014. Report by the UNCTAD secretariat. UN New York and Geneva 2014

[12] The Journal of Commerce annual top 50 World Container Ports, Lloyd's List annual Top 100 Ports, AAPA World Port Rankings and individual port websites.

[13] Top 50 world container ports. World Shipping Council. Partners in trade. 2015

[14] Trade and Logistics Facilitation. International Logistics Performance Index 2014. 2014 Ranking. www.worldbank.org (30.01.2015)

[15] van Essen H. (and others), EU transport GHG: Routes to 205 ? Modal split and decoupling options; Paper 5. Delft 2009

[16] World Port Source – Port of Duisburg. www.worldportsource.com. (31.01.2015)

[17] World trade developments. WTO International Trade Statistics. WTO 2014

[18] www.portofrotterdam.com. (15.01.2015)

[19] www.transportjournal.com. (29.01.2015)

Maritime Pollution and Environment Protection

Determination of Marine Pollution Caused by Ship Operations Using the DEMATEL Method

Ü. Özdemir, H. Yılmaz & E. Başar

Maritime Transportation and Management Engineering Department, Karadeniz Technical University, Trabzon, Turkey

ABSTRACT: Ships have an important role in among the factors causing marine pollution. Marine pollution by ships damages sea life, which effects human health indirectly, in addition it restricts usage of sea for different purposes. Increasing comprehensive and compelling liabilities related with environmental components and subjects day by day are expected results for environmental science and engineering applications according to 60% of our responsibilities of European Union Integration development. Determination of marine pollution caused by ship operations issue is a multiple criteria decision-making (MCDM) problem, and requires MCDM methods to solve it. Therefore, the role of ship factor in maritime pollution and the possible reasons of this argument can be quantitatively evaluated based on expert knowledge and MCDM methodology. To investigate what makes to reduce the first "caused by ship operations " in marine pollution, the decision-making trial and evaluation laboratory (DEMATEL) method approach was applied in this study.

1 INTRODUCTION

Apart from marine accidents, it can be clearly seen that the wastes sourced from routine activities of ships and transported cargo called as hazardous and/or special waste is considered that the marine pollution will be no recyclable and seriously has high levels of hazard to both human health and economy according to the amount, time and repetition of pollution.

Shipping is principal to our well-being, with around 90% of European Union international trade going by sea and more than 3.7 billion tonnes of freight a year being loaded and unloaded in EU ports. If not correctly controlled, the effect on the environment could be destructive, as ships often carry large volumes of hazardous cargo and generate a significant amount of pollutants throughout their life cycles (EMSA, 2008).

Many of pollutants are released by vessels either operationally or accidentally. The most important environmental damages caused by discharging household wastes and bilge water, dumping ballast water and wash water from tankers, emission of exhaust fumes, leaching of anti-foul paints, pollution with toxic materials, removal, introduction of organisms and acoustic and visual disturbances (Robert, 2001).

Shipping cause problem troubles to the environment both on inland waterways and on the ocean. These problems come from six major origins; routine discharges of oily bilge and ballast water from marine shipping; dumping of non-biodegradable solid waste into the ocean; accidental spills of oil, toxics or other cargo or fuel at ports and while underway; air emissions from the vessels' power supplies; port and inland channel construction and management; and ecological damage in consequence of the introduction of outlandish species transported on vessels (OECD, 1997).

The release of oil and other harmful substances (including noxious liquid substances, sewage, and garbage and air pollution) into the marine environment is regulated in great detail in the International Convention for the Prevention of Pollution from Ships (MARPOL). This convention was adopted by the International Maritime Organization (IMO) in 1973. It has been amended a number of times and is being continuously complemented and strengthened to meet the ever-increasing demands of the world community (EMSA, 2008). In the past few decades, international, regional and national regulation over shipping matters such as navigational safety, ship-source pollution and maritime security have grown to such a scope that the global shipping industry today faces a litany of costly regulatory rules. Accordingly, the ship owner's conventional right of free navigation is presently qualified by considerable requisites such as the protection of the marine environment and the promotion of maritime safety. In particular, the emphasis on marine pollution control by relevant coastal and port states has come to fundamentally erode the traditional right of free navigation accruing to maritime states and their shipping interests (Tan, 2006).

Shipping also causes more invisible types of pollution. Recent concerns include the harmful environmental effects of substances in anti-fouling paint used on ships' hulls, and of species which are transported from one sea area to another in ballast water tanks. Also, the bilge water includes a high amount of dirtiness as well as its toxic, corrosive, inflammable / explosive characteristics. Discharging of the bilge water out of the vessel through the pumps and without waiting for a long time is required, but it is pumped directly into the marine environment. Directly discharge process of bilge water is very harmful for marine environment. Both concerns have led the IMO to adopt specific conventions on the topics in 2001 and 2004 respectively. In addition, pollution is caused when ships are constructed and maintained, and when they are dismantled at the end of their life cycles. The latter is of particular concern at the present time, given that much of the world's ship breaking is done in countries where neither workers nor the environment have adequate protection against harmful practices and substances involved in ship recycling. A convention on the safe and environmentally sound recycling of ships is currently being drafted by the IMO and it is set for adoption in May 2009 (Emsa, 2008; Anderson, 2009).

According to the IMO (International Maritime Organization) MARPOL Code, marine pollution caused by ships is affected by factors such as operational pollution and accidental pollution. Operational pollution means the phenomenon that ship-cause marine pollution is not confined to accidents. In fact, the majority of pollutants are released while the ship is on voyage rather than accidentally. In this respect, activities include the chronic discharge of sewage, tank residues, bunker oils and garbage, as well as the exchange of ballast water, emissions from vessels' engines and pollution due to anti-fouling paints on ships' hulls, dumping of garbage and solid waste, resulting of oil and waste water after deck washing operations, pouring of cargo into the sea, giving directly to the sea of raw sewages. In this sense, determination of marine pollution caused by ships operations is a kind of multiple criteria decision-making (MCDM) problem. So, proposed method is developed for selection with decision-making trial and evaluation

laboratory (DEMATEL) as a ship routine operations process on marine pollution.

This research utilizes DEMATEL technique to explain the relationships between the various criteria. DEMATEL is a comprehensive method for building and analyzing a structural model involving causal relationships between complex perspectives. This study aims to utilize the a kind of multiple criteria decision-making (MCDM) method, named decision-making trial and evaluation laboratory (DEMATEL) technique approach to recognize the influential criteria of marine pollution caused by ships routine operations.

DEMATEL has been successfully applied in many situations, such as marketing strategies, e-learning evaluation, control systems and safety problems (Chiu et al., 2006; Hori and Shimizu, 1999).The methodology can confirm interdependence among variables/criteria and restrict the relations that reflect characteristics within an essential systemic and developmental trend. The end product of the DEMATEL process is a visual representation by which the respondent organizes his or her action in the world (Tzeng et al., 2007).

Therefore, there are many techniques available for risk analysis and traditional decision making methods as "ETA (Event Tree Analysis), AHP (Analytical Hierarchy Processes), ELECTRE (Elemination and Choice Translating Reality English), TOPSIS (Technique for Order Preference by Similarity to Ideal Solution) Grey Relational Analysis (GRA), and so on". These models are generally based on a supposition of independence among criteria affecting the process. On the other hand individual criterion is not always exactly independent. Moreover, it should be stressed that using an additive model which ignores the interrelations among criteria is not always successful in explaining the real world problems because of the changing interdependence levels among various criteria.

2 METHODOLOGY

Marine pollution caused by ships usually occur due to combination of coincidental incidents or processes, as a general rule by negligence of one or more independent components that are required to action accurately for the successful finalizations of the system requirements. The process of determination marine pollution caused by ships is required to handle several complicated factors in a better conceivable and logical manner. So, determination of marine pollution caused by ships issue is a kind of multiple criteria decision-making (MCDM) problem. To solve this problem, we used a MCDM method, called DEMATEL. DEMATEL developed by the Science and Human Affairs

Program of the Battelle Memorial Institute of Geneva between 1972 and 1976 was utilized to study and resolve the complex and intertwined problem group (Tzeng and Chiang, 2007). In recent years, the DEMATEL method has become very popular in Japan, because it is especially pragmatic for visualizing the structure of complicated causal relationships with digraphs (Chiu et al., 2006). The digraph portrays a contextual relation among the elements of the system, in which a numeral represents the strength of influence. In this study, decision-making trial and evaluation laboratory (DEMATEL) method is applied because this method generates causal diagrams to describe the basic concept of contextual relationships and the strengths of influence among the criteria (Wu and Lee, 2011).

In this study firstly, marine pollution caused by routine ship operations criteria was determined. The criteria involved in the marine pollution selection have been chosen according to the IMO (International Maritime Organization) MARPOL Convention. In next step, the DEMATEL analysis was applied in order to determine the criteria as follow section 2.1. To establish the network relationships among criteria in influence each other for the marine pollution selection involves a decision making team which includes 2 academics personal, 4 experienced captains. The expert captains have at least 18 years experiences and worked in master positions in well-known worldwide -based shipping companies as well. Two academic personnel who had professional backgrounds have several academic articles and technical books; and were awarded by TUBITAK (The Scientific and Technological Research Council of Turkey) for their scientific researches and projects about maritime sector.

Then, decision making team E1, E2, ..., E6 is constituted to determine the network relationships. And give the performance scores for each expert in terms of all criteria in the evaluation hierarchical structure respectively. A questionnaire was used to find out influential relations from each expert for ranking each criterion on the appropriate marine pollution caused by ships routine operations with a four-point scale ranging from 0 to 4, representing from 'No influence (0),' to 'Very high influence (4),' respectively. For each pairwise comparison, the decision making team have to determine the Intensity of the relative importance between two criteria. The computation of using DEMATEL technique is based upon these six experts' opinions. So there are 6 dimensions, the six 6 X 6 matrices. We utilized the DEMATEL to construct the influence map in accordance with the real situation in which criteria should be interdependent and determine the importance of criteria.

2.1 DEMATEL Method

In generally, the DEMATEL method is used to illustrate the relations between criteria and to reach the main factor/criteria to symbolize the impact of factor (Tzeng et. al, 2007; Wu and Lee, 2011). The DEMATEL method is established on digraphs which can discrete involved factors into cause and effect groups (Yang et. al, 2008). This method has also been individually used in many activities such as safety problems (Liou et. al, 2007) , transportation (Chen and Yu, 2008), supply chain management (Chiu et. al, 2007) and automotive industry (Wu and Lee, 2011).

The DEMATEL method is briefly described as follow (Wu and Lee, 2011; Liou et. al, 2007; Chen and Yu, 2008).

Step 1: Calculating the direct-relation matrix. Suppose we have L experts in this study and n criteria to consider. Firstly, the relationship between criteria requires that the comparison scale be designed as with five levels, where scores ranging from 0 to 4 represent ''no influence'' to ''very high influence'', respectively. Experts are answered the direct influence degree between criterion ''u'' and criterion ''v'', as indicated by Zuv. The initial direct-relation matrix $Z=[Z_{uv}]|Y|$ is determined owing to pairwise comparisons in terms of influences and directions between criteria. Then, as the result of these evaluations, the initial data can be obtained as the direct-relation matrix that is a matrix Z, in which a Z_{uv} is denoted as the degree to which the criterion u affects the criterion v. The scores by each expert will give us a ''n x n'' non-negative answer matrix $X^k = [X^k_{uv}]_{nxn}$, with $1 \leq k \leq L$. Thus, X^1, X^2, . . . , X^L are the answer matrices for each of the L experts, and each element of is an integer denoted by X^k_{uv}. The diagonal elements of each answer matrix X^k are all set to zero. We can then compute the ''nxn'' average matrix Z for all expert opinions by averaging the L experts' scores as follows:

$$[Zuv]_{nxn} = x = \frac{1}{L} \sum_{k=1}^{k} [X^k_{uv}]_{nxn} \qquad (1)$$

Step 2: Normalizing the direct-relation matrix. The normalized direct-relation matrix M can be obtained by formulas:

$$M = Z \times L \qquad (2)$$

$$L = \frac{1}{\max 1 \leq u \leq n \sum_{v=1}^{n} Zuv}, \; u, v = 1, 2, ,,,,,n \qquad (3)$$

Step 3: Calculating the total-relation matrix. After the normalized direct-relation matrix M is obtained, the total relation matrix K can be acquired by formula (4), in which the H is represented as the identity matrix.

$$K = M \times (H - M)^{-1} \qquad (4)$$

213

Table 1. Initial direct matrix Z.

Criteria vs	Bunker oils and bilge waters	Ballast waters	Garbage and Solid Wastes	Sewages waters	Anti-fouling paints	Deck, hold, washing operations (Daily opr.)
Bunker oils and bilge waters	0.04	0.321	0.302	0.412	0.201	0.329
Ballast waters	0.298	0	0.174	0.185	0.294	0.173
Garbage and Solid Wastes	0.185	0.271	0	0.256	0.112	0.304
Sewages waters	0.271	0.184	0.316	0.02	0.307	0.272
Anti-fouling paints	0.148	0.204	0.194	0.301	0	0.207
Deck, hold, vs. washing operations (Daily opr.)	0.174	0.160	0.217	0.122	0.318	0

Table 2. Total influential relation matrix K.

Criteria	Bunker oils and bilge waters	Ballast waters	Garbage and Solid Wastes	Sewages waters	Anti-fouling paints	Deck, hold, vs washing operations (Daily opr.)	R
Bunker oils and bilge waters	0,612	1,028	0,942	1.109	0,611	1,009	5,311
Ballast waters	0,793	0,687	0,374	0,527	0,594	0,673	2,971
Garbage and Solid Wastes	0,682	0,641	0,527	0,752	0.298	0,904	3,804
Sewages waters	0,385	0,607	1,116	0,497	0,617	0,598	3,816
Anti-fouling paints	0,424	0,404	0,574	0,321	0,745	0,633	3,101
Deck, hold, vs. washing operations (Daily opr.)	0,418	0,468	0,546	0,303	0,731	0,587	3,053
D	3,313	3,835	4,079	3,509	3,596	4,404	

Step 4: The sum of rows and columns are separately denoted as D and R within the total-relation matrix K through equations (5) to (7).

$$K = [k_{uv}], u, v = 1, 2, 3, ..., n \quad (5)$$

$$D = (D_u) = \left(\sum_{v=1}^{n} k_{uv} \right) \quad (6)$$

$$R = (R_v) = \left(\sum_{u=1}^{n} k_{uv} \right) \quad (7)$$

The DEMATEL method analysis was used to obtain the initial direct-relation matrix with using pairwise comparison with the total relation matrix with D+R, D-R values and build a critical relative graph of criteria in the cluster effect. "Du" denotes the row sum of ith row of matrix K. Then, Du denotes the sum of influence dispatching from factor v to the other factors both directly and indirectly. Rv shows the column sum of vth column of matrix K. Rv shows the sum of influence that factor u is receiving from the other factors. The sum of row sum and column sum (D + R) shows the index of representing the strength of influence both dispatching and receiving. Furthermore, if (D- R) is positive, then the factor "u" is rather dispatching the influence to the other factors, and if (D - R) is negative, then the factor "u" is rather receiving the influence from the other factors.

Step 5: Determining a threshold value to obtain the digraph. Since matrix K provides information on how one factor affects another, it is necessary for a decision maker to set up a threshold value to reduce some negligible effects. For these reason, only the effects greater than the threshold value is chosen and shown in digraph. In this study, the threshold value is set up by computing the average of the elements in Matrix K. The digraph can be acquired by mapping the dataset of (D+ R, D-R).

3 EMPIRICAL STUDY

An empirical example for the most important criteria selection for the marine pollution caused by ship routine operations is illustrated to demonstrate the proposed method to be more rational and suitable in this section.

Six criteria involved in the marine pollution caused by ship operations selection are used in this empirical study. The criteria have been chosen according to the IMO (International Maritime Organization) MARPOL Convention. A decision making team were invited to answer the questionnaire. The computation of using DEMATEL method is based six decision making team's opinions. Using the 6X6 pairwise comparisons, the averages of their opinions were calculated .Then, the average initial direct matrix Z is obtained based on formula (1) as Table 1.

Normalized initial direct-relation matrix M is calculated through formulas (2) and (3). Sequentially, the total relation matrix K is also derived utilizing refer to (4) shown in Table 2.

Total sum of effects given and received by each criterion is seen in Table 3 using formulas (6) and (7).Table 3 provides the direct and indirect effects of six dimensions. Finally, the threshold value (0.7218) used in Step 5 is to compute the average of the elements in Matrix T. The digraph of these six dimensions is demonstrated and the network relationship map of DEMATEL method was obtained and shown in Fig. 1.

Table 3 shows that a "bunker oils and bilge waters" criterion is the most important dimension with the largest (D + R) value of 8,624 whereas "Anti-fouling paints" criterion is the least important dimension with the smallest value of 6,697. The importance of dimensions can be determined by the (D + R) values. To further investigate the cause-effect relationship of dimensions, ballast waters, garbage and solid wastes and deck, hold, vs. washing operations are net causes based on positive (D - R) values.

Table 3. Sum of influences given and received on each criterion

Criteria	D + R	D - R
Bunker oils and bilge waters	8,624	-1,998
Ballast waters	6,806	0,864
Garbage and Solid Wastes	7,883	0,275
Sewages waters	7,325	-0,307
Anti-fouling paints	6,697	0,496
Deck, hold, vs. washing operations (Daily opr.)	7,457	1,351

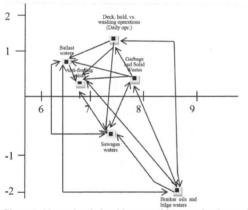

Figure 1. Network relationship map of impacts for the marine pollution caused by ship operations criteria

Bunker oils and bilge waters, and sewages waters are net receivers due to negative (D - R) values. Furthermore, deck, hold, vs. washing operations (Daily opr.), ballast waters and anti-fouling paints are the three most essential dimensions to improve the marine pollution caused ships by further considering the causal relationships. To further investigate the cause-effect relationship of dimensions, "deck, hold, vs. washing operations, (Daily opr.)", ballast waters, anti-fouling paints and garbage and solid wastes are net causes based on positive (D- R) values. Bunker oils and bilge waters, and sewages waters are net receivers due to negative (D-R) values. From Fig. 1, bunker oils and bilge

waters, garbage and solid wastes, and "deck, hold, vs. washing operations (Daily opr.)" are the three most critical dimensions. Specifically, "deck, hold, vs. washing operations (Daily opr.)" directly affects bunker oils and bilge waters and sewages waters. Six dimensions are influenced or mutually influenced by any pair of dimensions except "deck, hold, vs. washing operations (Daily opr.)". In summary, "deck, hold, vs. washing operations (Daily opr.)" is the most important dimension followed by ballast waters anti-fouling paints then garbage and solid wastes criterion. Therefore, "deck, hold, vs. washing operations (Daily opr.)", ballast waters and anti-fouling paints are the three most essential dimensions to improve the minimizing marine pollution caused by ship operations by further considering the causal relationships.

4 CONCLUSION

In this paper, we would like to new framework for determine and evaluated the marine pollution related to ship operations with above mentioned these six criteria. This study presents DEMATEL method for analyzing the importance of criteria correlations and also to describe the contextual relationships among these criteria. Based to the results, six criteria have some interrelations with each other. According to the impact-direction map, we can make accurate decisions. Also, the results showed that bunker oils and bilge waters is a key value factor and powerful influential criteria. However, by further considering causal relationships, improving both bunker oils and sewages waters factor dimensions cannot be effectively strengthen minimizing marine pollution caused by ship operations because these two dimensions are influenced by the other dimensions. In contrast, "deck, hold, vs. washing operations (Daily opr.)", ballast waters, anti-fouling paints and garbage and solid wastes factor are all causal dimensions and have positive impacts on bunker oils and bilge waters, and sewages waters factor such that any improvement on these dimensions would also improve bunker oils and bilge waters, and sewages waters. The proposed framework brings several contributions to marine pollution evaluation and selection. At first, a new model for selecting marine pollution with emphasis on ship operations issues has been developed. At second, the DEMATEL method was applied in selecting marine pollution cause with respect to ship routine operations. This feature is also unique with regard to previous studies. In addition, results of these modeling can be used with other decision making method. Application of other decision support approaches would help to extend the analysis such as analytical network process (ANP).

REFERENCES

Anderson, D. 1998. The Roles of Flag States, Port States, Coastal States and International Organizations in the Enforcement of International Rules and Standards Governing the Safety of Navigation and the Prevention of Pollution from Ships under the UN Convention on the Law of the Sea and other International Agreements. *Singapore Journal of International and Comparative Law*. 2, 557-578.

Chen, C.H. & Yu, W.Y. 2008. Using a Strategies Approach to Analysis the Location Selection for High-Tech Firms in Taiwan. *Management Research News*, 31(4), 228-244.

Chiu, Y.J., Chen, H.C., Tzeng, G.H. & Shyu, J. Z. 2006. Marketing strategy based on customer behavior for the LCD-TV. *International Journal of Management and Decision Making*, 7(2), 143–165.

EMSA, 2008. European Maritime Safety Agency Work Programme 2008. s. 53.

Hori, S., Shimizu, Y., 1999. Designing methods of human interface for supervisory control systems. *Control Engineering Practice 7*, 1413–1419.

Liou, J.J.H., Tzeng, G.H. & Chang, H.C. 2007. Airline safety measurement using a hybrid model. *Air Transport Management*, 13(4), 243–249.

OECD, 1997. Organisation for Economic Co-Operation and Development, The Environmental Effects Of Freight. Paris: Head of Publications Service.

Robert, B. 2001.Protecting the Oceans from Land-Based Activities. Oxford: GESAMP Report and Studies No. 71.

Tan, A.K.J. 2006. Vessel-Source Marine Pollution The Law and Politics of International Regulation. Cambridge: Cambridge University Press.

Tzeng, G.H., Chiang, C.H. & Li, C.W. 2007. Evaluating intertwined effects in e-learning programs:a novel hybrid MCDM model based on factor analysis and DEMATEL. *Expert Systems with Applications*, 32 (4), 1028–1044.

Yang, Y.P.O., Shieh, H. M.,Leu, J.D. & Tzeng G.H. 2008. A Novel Hybrid MCDM Model Combined with DEMATEL and ANP with Applications. *International Journal of Operation Research*, 5(3), 160-168.

Wu, W.W. & Lee, Y.T. 2011. Developing global managers' competencies using the fuzzy DEMATEL method. *Expert Systems with Applications*, 32 (2), 499–507.

Joint-Task Force Management in Cross-Border Emergency Response. Managerial Roles and Structuring Mechanisms in High Complexity-High Volatility Environments

O.J. Borch & N. Andreassen
University of Nordland, Bodø, Norway

ABSTRACT: In this paper we focus on managerial roles and structuring mechanisms within the crisis preparedness system. We elaborate on the challenges of crises management in complex and volatile environments. The coordination and control mechanisms are of importance to safeguard operations including joint operations including several preparedness institutions, especially in cross-border cooperation. We include examples from the maritime Arctic. This paper contributes to the crisis management literature emphasizing the relations between context, managerial roles and the organizational structuring mechanisms needed to facilitate the interplay between several emergency response actors.

1 INTRODUCTION

Crisis management often takes place in challenging contexts. A crisis situation is also often characterized by the need for a broad range of efforts, and at the same time resource scarcity. Thus, we often are in need of capacities from different institutions and even cross-border support. As an example, incidents at sea are challenging compared with land-located incidents due to remoteness, lack of resources and nature. Help from a broad range of actors and countries may be necessary. It is a crucial task to be able to integrate multiple actors into a functioning emergency response system (Sydnes & Sydnes, 2011). Composite crises may include search and rescue (SAR), oil spill recovery, fire fighting, salvage, and actions violent behavior such as terror. To cope with such emergencies, there is a need for a broad range of capabilities from multiple actors and across many jurisdictions (Comfort & Kapucu, 2006).

The crisis management tasks become more difficult in environments characterized as highly complex and volatile (Hossain & Uddin, 2012; Bigley & Roberts, 2001). Dealing with maritime crises in this context increases the need for interaction between actors from several preparedness institutions.

A complicating variable related to the emergency operations is the presence of different formal and informal institutions (Van de Ven & Walker, 1984), as well as cultural differences and a lack of trust between different parts of the preparedness system (Kapucu, 2005; Axelrod & Cohen, 1999; Borch & Arthur, 1995). Increased environmental volatility may also call for dynamic capabilities in the command structure for improvisation and fast reorganization of the available resources (Borch & Batalden, 2014; Turoff et al., 2012). In a traditional hierarchical command structure that we find within the preparedness institutions the need for flexibility, improvisation and fast reorganization may be hampered by "silo thinking" and rigid formal structures (Bigley & Roberts, 2001).

So far, we have few studies emphasizing the contextual influence on emergency management and preparedness system coordination (e.g. Larsson & Hyllengren, 2013; Buck et al., 2006). In particular, we lack studies emphasizing the relations between managerial roles, capabilities needed and the role of structuring mechanisms in joint, cross-institutional operations (Hossain & Uddin, 2012; Turoff et al., 2004).

To increase both effectiveness and efficiency within the preparedness system, we are in need of managerial concepts and command structures for optimal exploitation of joint resources. In this paper, we elaborate on the managerial challenges of coordination and control in composite emergency operations. We take into consideration the contextual challenges facing the command system in high volatility, complex environments, and how this context may influence on the managerial roles and structuring of the preparedness system.

This paper is organized as follows: First, we present relevant theory about managerial roles and mechanisms of coordination and control. We emphasize the interorganizational coordination and the importance of contextual variables. Then, the paper proceeds with presenting the context of the High North as a case illuminating a complex, volatile environment. Next, the case of a standard emergency management model Incident Command System (ICS) is discussed. This standardized emergency management system is implemented in several emergency organizations in a number of countries. The ICS is used as an example in the discussion on how the range of roles and structuring mechanisms may differ in context in contrast to a "one type fits all" approach. In the conclusion chapter, we draw attention to the special managerial demands of joint actions that may include cross-border cooperation. We discuss the implications for future research in this area.

2 THEORY

Within organizations, a broad range of managerial roles has to be matched by adequate coordination and control mechanisms to achieve an effective interplay (Bigley & Roberts, 2001). Organizations dealing with crisis often have to cooperate closely with other preparedness institutions. For such cooperation, bridging mechanisms to match two or more organizations with different managerial systems have to be defined. Due to the character of a crisis, this coordination has to run smoothly from the start and at a very high pace (Comfort & Kapucu, 2006). This calls for both roles and structuring mechanisms being adapted to different settings.

2.1 *Managerial roles*

Emergency management is characterized by a strict interplay between the operational levels from the headquarter down to the on-scene incident command structure close to the incident site. This implies focus on roles and capabilities at both strategic, operational and tactical level. Uncertainties and conflicts over the roles in a between the layers may negatively influence on managers' performance. In complex environments, a large number of aspects towards a broader range of stakeholders have to be considered (Mintzberg, 2009).

Managerial role is a set of actions and responsibilities that are assigned for each of them. Mintzberg (1973) claims that managerial roles within an organization can be classified into three main groups: interpersonal, decisional and informational.

Interpersonal roles include the figurehead, leader, and liaison roles. They arise directly from formal authority and involve basic interpersonal relationships. The figurehead role involves both internal motivation and inspiration, and representing the crisis organization externally towards different stakeholders, for example media, interest groups and next of kin; the leader role constitutes leadership duties towards subordinates and the duty to have overall responsibility for the unit. Contact and coordination outside the vertical chain of command are referred to as the liaison role.

Informational roles include the monitor, disseminator, and spokesman roles. By means of interpersonal contacts these roles are central in an organizational unit. Managers constantly scan the environment for information, pass information further to subordinates and some information to people outside the unit. Constant information flow is critical to be able to allocate resources and achieve efficient mitigating actions with lowest possible risk to the personnel.

Decisional roles include the entrepreneurial action, disturbance handling, resource allocation, and negotiator roles. Information is the basic input for managers to decision making. In the entrepreneurial role, managers seek to improve the unit by adapting it to changing conditions in the environment. They also respond to different pressures and handle ad hoc problems. Improvisation finding new create solutions may be important in the very turbulent situation of a major crisis. In the resource allocator role, managers are responsible for decisions and strategy for resource distribution. To achieve the necessary resources and results, negotiations are important. Gaining control over and acquiring costly resources from other institutions and other countries may be challenging. Negotiations may be seen as an integral task.

Fast decisions may be of special importance to meet dynamism in the environment. In stable organizations, formal duties descriptions may contribute to harmonized action. However, in a crisis organization, the defined standard operating procedures that have functioned well in the past may not be appropriate (Rosenthal et al., 2001). Thus, in organizations facing volatile environments, there is a need for innovation and entrepreneurial, dynamic capabilities related to specific persons or integrated into the present roles (Borch & Madsen, 2007). The operational and tactical management may have to improvise and work on reconfiguration, including new action pattern, repositioning of resources and uplinking to other roles and processes. This means that the discussion on the contents of the managerial roles has to be linked up to the coordination and structuring mechanisms.

2.2 The joint coordination and structuring mechanisms

Coordinating and controlling are essential mechanisms in organizations (Mintzberg, 2009). In general, structuring mechanisms represent a set of procedures for assembling and reassembling various organizational elements into a variety of configurations (Bigley & Roberts, 2001, p.1287). In crisis situations, the traditional control structures may be superseded by the coordinating mechanisms (Hales, 2002). The coordination of tasks refers to a systematic relationship between decisions about resources and processes in order to achieve the desired outcomes (Alexander, 1995; Haas, 1992; Auf der Heide, 1989).

In emergency response systems, these coordination tasks involve institutions with various organizational design. The joint emergency operations may include police and armed forces, coastal guard and rescue coordination centers, fire and rescue services, helicopters and ambulance, other public authorities and private actors. In some emergency systems, voluntary organizations are particular useful on-scene because they can provide great numbers of well-trained people who are familiar with the local areas. The cooperating institutions may have implemented different organizational structures, routines, management roles and control mechanisms. Therefore, achieving coordination between them becomes challenging.

In emergency situation, the speed of adapting to a specific organizational structure is, however, crucial. Organizations need to address these challenges and ensure an effective interplay. As organizations vary, their organizational structure should be flexible and capable of linking up to each other through employing structuring mechanisms such as interdepartmental liaisons, joint procedure sets and multi-functional task forces (Alexander, 1995).

The emergency response systems found include some standard coordination mechanisms. According to Bigley and Roberts (2001) structuring mechanisms consist of at least four basic processes; structure elaboration, role switching, authority migrating, and system resetting. The **structure elaboration** process is initial and important because management should be organized on-scene under demanding circumstances. **Role switching** is the process of assignment and reassignment of personnel to different positions in accordance of the functional requirements of the situation. **Authority migration** happens when critical expertise or capacity in a certain emergency area can be de-coupled from the official hierarchy and moved to another authority when needed. **System resetting** is another process to match changes in working conditions through making the organization look through the structure, competence and routines in the light of the new scene of operation.

2.3 The importance of context

A central assumption in organization theory is that organizations exist in their social and environmental context, and are influenced by it (Bigley & Roberts, 2001). From this perspective, contextual factors influence the firm performance and organizational outcomes. As suggested by Hansen and Wernerfelt (1989), the contextual variables include factors like sociological, political, economic and technological conditions; and human resource factors.

Roles and structuring mechanisms are difficult to configure in large disasters, which often involve multiple hazard, with a range of agent-generated demands, multiple responding agencies, and conflicting goals that cannot be anticipated and reconciled. Coordinative mechanisms depend on complexity of disaster response, recovery and mitigation tasks (Buck et al., 2006). The term complexity have been traditionally associated with a description of the working environment of an organization, and in broader systems also with external environment like weather conditions, ecologies, information networks, and number of stakeholders (Dooley, 2004). Complexity characteristic illustrates the range of factors and dependency relations among the involved actors within the business processes of an organization (Borch & Batalden, 2014). Managerial challenges are linked to the coordination and control of a broad range of physical, cultural or institutional elements in the environment.

Volatile or turbulent environments are characterized by lack of understanding of the cause-effect relations making decision-making a challenging task. Volatility is instability and lack of predictability that will aggravate the uncertainty of outcomes (Borch & Batalden, 2014). Extreme events are regarded as volatile because of rapid changes and unpredictable outcomes (Turoff et al., 2012).

As for resources, the emergency situations are often characterized with limited physical resources as well as competent personnel in and around the organization. This calls for extra links to more institutions. This increases complexity and uncertainty about how these external resources can be integrated into the emergency organization.

The environment may vary between stable and volatile characteristics, and simple or complex. In low complexity and stable environments, the number of external links may be low and management may concentrate on the intra-organizational roles, and coordinate through hierarchical structures and functional specialization. High volatility and complex organizations call for a broader set of roles and more sophisticated coordination mechanisms.

3 THE CASE OF THE ARCTIC AND THE INCIDENT COMMAND SYSTEM

As shown above different contexts may influence the managerial roles and the included coordination mechanism. We illuminate these challenges through a discussion on the roles and structuring mechanisms of the standardized incident command system (ICS). This system that originated from the US fire brigades is now implemented in several national preparedness organizations worldwide.

3.1 The Operational context of the High North

The operational environment of the High North is characterized as both complex and volatile (Borch & Batalden, 2014). In the High North the volatility parameter refers to the difficulties the actor face on predicting nature, and the functionality of resources available, among others due to different cultures, political interests and training.

Another crucial characteristic of the High North which can influence operational environment is the scarcity of resources. Generally, this term refers to the lack of critical resources for survival and growth. In emergency systems, they are capabilities that are needed for response to mitigate the crisis situation. The resource challenges are present both related to equipment, personnel and organizations (Comfort & Kapucu, 2001).

The increased activity in the High North increases the vulnerability related to human safety, environment, physical installations and vessels. The High North is defined as the geographical regions north of the Polar Circle where maritime operations are challenged by long physical distances to civilization, limited harbor infrastructure, low temperatures with ice and icing, polar lows, and vulnerable nature. This calls for extra competence and capabilities for all activity in this region. Earlier studies have increased our understanding on the effects of increased complexity in offshore commercial operations (Gudmestad et al., 1999; Thunem, 2010). Increased knowledge on adequate operational concepts in the Arctic is in demand. In its whitebook "An Innovative and Sustainable Norway" the Norwegian Government states that the main objective now is "to secure an active Norwegian presence in the North and to exploit the resources and transport opportunities in the region". Thus, the maritime industry in the North represents an area of commitment. This ambition calls for a new knowledge about how to increase safety and facilitate effective and efficient exploitation of commercial opportunities in the region.

In addition, the political sensitivity is present as this is a region with shared responsibility between US, Canada, Denmark, Finland, Sweden, Iceland, Norway and Northern Russia. These Arctic states are members of an intergovernmental forum, the Arctic Council. All cooperation, coordination and interaction among the Arctic states are handled through the Arctic Council with participation of other non-governmental associations and non-arctic states observers. The Arctic states are committed to several bilateral and multilateral agreements in relation to certain emergency preparedness activities. The question of emergency preparedness in the High North has a primary focus on search and rescue operations, and on preparedness for pollution caused by extensive maritime activities from shipping, fisheries, offshore petroleum installations and maritime tourism.

Search and rescue operations in the High North are since May 2011 governed by the Agreement on Cooperation on Aeronautical and Maritime Search and Rescue in the Arctic. This is the first and the only one legally binding treaty signed under the auspices of the Arctic Council so far. The responsibility is expanded between all the eight Arctic states. Oil spill response responsibility in the High North is coordinated by the Agreement on Cooperation on Marine Oil Pollution Preparedness and Response in the Arctic since 2013. The United Nations Convention on the Law of the Sea (UNCLOS) is the international treaty created at the third United Nations Conference on the Law of the Sea. These international agreements demonstrate the commitment for joint coordination of emergency response in the Arctic. It is important for all Arctic countries to involve relevant preparedness institutions in the north in the process and to inform them about all conditions of the agreements aiming at coherent coordination.

Unfortunately, climate change and its unpredictable consequences make emergency response in the region complicated. We still lack knowledge on how these composite contextual elements may influence the operational interaction between the institutions within the preparedness system. Resources for providing effective emergency response in the Arctic are limited (Arctic Council, 2009). Oil spill recovery techniques suffer from reduced functionality under severe Arctic environment. There is limited police authority present to deal with violent action. Having limited resources countries depend on each other's assistance. Cooperation of all the personnel resources and units available is a key to success.

A major part of the mobilized forces will be the ships, helicopters and equipment hired by the commercial actors such as the shipping industry and oil companies. Advanced and tailor-made technology is needed to deal with environmental challenges. Understanding the technology and competences for utilizing complex preparedness tools are of critical importance in the High North (Borch & Solesvik, 2013). The units present may have a multi-functional design including safety and security functions.

To sum up, the Arctic context represent a high complexity and volatile environment, where resources available for the emergency response units are scarce. This may imply a broad set of adapted managerial roles and a need for structuring mechanisms facilitating operation in a high complexity-high volatility context.

3.2 Incident command system

Deciding upon the managerial roles and finding the adequate organization of emergency response units is a challenging task. Institutions within the preparedness system have implemented a broad range of standard functions and roles as well as structural configurations to deal with critical incidents. In several countries like Norway, we find that these systems vary across institutions. In Norway, the police with the overall coordinating role has a different organizational structure and positions than the other emergency institutions like the fire brigade and paramedic systems.

The Incident Command System (ICS) is an emergency management tool which includes specific roles and facilitates coordination and control of personnel and equipment at incidents of any types. Originally, the ICS was created for fire departments fighting wildland fires in California in the 70ies. Since then, the ICS approach has turned out to be suitable for a wide range of emergencies, such as fires, hazardous materials spills recovery, and multi-casuality accidents of nearly any size (Bigley & Roberts, 2001; Lindell et al., 2005; Buck et al., 2006). The ICS includes a standard management hierarchy and managing procedures.

In general, the ICS is constructed around five major roles: command, planning, operations, logistics and finance/administration (Lindell et al., 2005). These blocks are supposed to be applied to situations of all sizes. The basic ICS includes a set of rules and practices to guide the actions, standardized job descriptions with a training program for each positions, common terms for equipment and supplies, a structured chain of command from the specialist on the ground to the incident commander, authority commensurate with responsibility and task assignments, span of control limited to the number of people that one person can effectively control, and sectoring of work to ensure efficiency, effectiveness and safety. (Buck et al., 2006).

4 DISCUSSION

Even though some emergency management system like the ICS has been widely implemented, Buck et

al. (2006) highlight the importance of context as a largely un-examined precondition to an effective management system. Bigley & Roberts (2001) call for testing the emergency management models in diverse contexts. When a broad range and different types of institutions are involved, we need to analyze the effects carefully. Not the least, the interplay between very different types of organizations like the military forces, professional emergency institutions, private companies like the ship owners and volunteer organizations may function differently in extreme environments like the Arctic.

4.1 The Arctic context and managerial roles

Operating in the Arctic means that the management has to take into consideration a number of bilateral and multilateral treaties that address different spaces and types of activities. An example is the emerging Polar Code legislation launched by the UN International Maritime Organization introducing strict demands as to vessel configuration as well as competence of the crew. This calls for extra decisional and informational roles.

Within emergency management, there is a need for the additional roles like disturbance handler and negotiator. For some areas, there are disputes over territorial rights, and the region is political sensitive with a lot of military activity taking place. This calls for *additional decisional roles dealing with security and political issues*. There is a need for strong top down communication between top government and operational level to check out acceptable solutions in sensitive areas. In some operational fields like SAR and oil spill recovery there are quite clear agreements on responsibility and host nation support. However, these agreements might not be valid or provide fast enough action when it comes to implementing them through resource acquisition. One example is the sanctions introduced in 2014 against Russia during the Ukrainian crisis.

A high degree of isolation and turbulence especially in winter months in the High North creates difficulties in gaining enough professional capacities. This calls for attracting more volunteers and others with limited training. It means that the recruitment and the educational actions may be of extra significance. The Norwegian Directorate for Civil Protection claims that there is even stronger need for training when the actors have to interact with foreign resources (DSB, 2013). Therefore, the *informational duties integrated into logistics roles* are of special importance in the emergency management system.

There will also be different and conflicting interests in this region. In major international emergencies, every action taken is highly visible. This means an extra focus on the commander's responsibility for the overall operation and *the status of formal authority.*

People coming from different cultures and language groups may have problems understanding each other as well as trusting each other. This calls for additional *cross-cultural liaison roles,* and *the role of rewriting of procedures.* As an example, during the evacuation of the Maxim Gorkiy cruise ship after collision with ice at Svalbard, the Norwegian coast guard as on-scene commander was not able to communicate with Russian military and civilian helicopters and airplanes due to language problems and lack of joint procedures (Hovden, 2012). Such an operational context influences the initial condition of the ICS system and impedes the implementation of the pre-established standardized procedures.

4.2 The High North context and the structuring mechanisms

The complexity of the High North is caused by a large number of stakeholders and the range of natural factors that have to be considered in operations. There are climate challenge with extreme weather, and the need for inclusion a broad range of government institutions. Another complicating factor relates to the fact that responsibilities and interests in rescue operations and surveillance are divided between many separate national institutions. In Norway, a grounding or a fire-fighting operation at sea may involve 10-12 government institutions at local, regional and national level. This calls for extraordinary competence and capabilities for co-ordination and overall governance of the maritime activity. Structuring mechanisms such as *joint authority tools, horizontal staff interaction, frequent multi-level communication, delegated authority and adapted operational procedures* and *standard high tech communication tools* may contribute to solve the coordination challenges.

Volatility is characterized by the lack of understanding of the cause-effect relations during emergency operations. Consequently, the established standard operating procedures may only partly fit the situation (Christensen & Johannessen, 2005; Stacey, 2001). There will never be one task that fits a certain emergency. Turbulence is especially crucial when analyzing the tactical level challenges. The dynamic capabilities related to *acquiring for new resources, teaming old and new resources* and *finding new solutions* in high ambiguity settings are crucial.

To sum up, there is a constant need for on-scene structure elaboration, because of high complexity and volatility, and for readiness for role switching because of the resource scarcity. Authority migrating can be of need because the High North is a multi-actor political arena, so the system should be ready for resetting.

4.3 Managerial roles and structural mechanisms in a high complex and volatile environment

The incident commanders of emergency organizations have to fill several roles to deal with the crisis situation. Within crisis management there are duties related to (Turoff et al., 2004):

- mapping human and equipment resources
- intelligence
- reports and updates the situation,
- operational coordination,
- maintaining or acquiring new resources
- advice and information
- redesign of roles and responsibilities
- setting priority and new strategy

The crisis management system and especially the incident commander roles have to be scrutinized to secure system effectiveness in different contexts.

In a crisis situation it is never certain who will take on which role or a combination of roles. In some emergencies, persons who are qualified for one type of actions may be assigned for other roles. Bigley & Roberts (2001) state that the system in use must be able to expand and contract, change strategic orientation, modify or switch tactics as an incident unfolds. In volatile environments, there may be incidents that are totally new to the commanders. Buck et al (2006) point out that the main concern of today's debates in literature focuses on inter-organizational coordination mechanisms. The critics claim that standardized systems fail to recognize all transformations of the structure and functions of the established organizations during response phase. In practice, this means that the system includes separate units using different command approaches, and even unexperienced actors like volunteer organizations. This coordination problem coupled with the complexities of the organizational environment creates the need for establishing new adapted coordination and control mechanisms.

Studies of the ICS concept have shown that they are most successful in stable, low complexity environments with a common government structure, with pre-established protocols for interorganizational development in times of crisis and when there is a significant interagency pre-training of command staff (Buck et al, 2006). The adjusted coordination approaches for multi-task response is required. Buck et al. (2006) emphasize the following coordinative mechanisms including shared knowledge and technical aspects, a shared vision of response while working problems together, a high level of trust in one another; a trained response community with knowledge of the common system, understanding how to improvise within the common purpose, and collective recognition of capabilities and limitations. If these criteria are not met, there may be significant challenges for an effective emergency response.

Therefore, additional roles and mechanisms that are emerging out from the operational context should be integrated in the new emergency management model facilitating joint coordination (Figure 1).

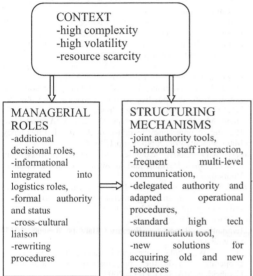

Figure 1. Managerial roles and structuring mechanisms that may contribute to joint emergency system

The operational context of the High North characterized by high complexity, volatility and resource scarcity calls for new managerial roles and structuring mechanisms. The Figure 1 summarizes the additional elements, which should be integrated into the emergency management system when a wide range of actors are involved.

5 CONCLUSIONS

In this paper, we have illuminated the relation between context and the crisis management systems. We have shown how context characteristics and the variety of managerial roles may influence on the need for tailor-made structuring mechanisms. There is the need for preparedness systems with a broader range of managerial tools to face the challenges of complexity and volatility. The dynamic capabilities are needed at all levels of management in order to meet the unpredicted challenges, even if new roles and functions may have to be developed and coordinated. Thus, we are in need of the sufficient knowledge of the operational context and the knowledge on how to reconfigure the emergency system models within the preparedness system.

The findings provide some ideas for further research. There is a need for more in-depth case studies of how the crisis management systems function in different contexts. In particular, there is a need to look into the different roles and adjacent capabilities of the command system. In addition, the importance and limitations of different structuring mechanisms should be studied. Cross-country studies may bring more understanding of institutional issues and how one may bridge institutional differences in joint operations. The development and implementation of joint training programs and best practice competence schemes should also be studied in further detail.

REFERENCES

Alasoini, T. 2011. Linking theory and practice: Learning networks at the service of workplace innovation, Vol. 75. Helsinki: TYKES.

Alexander, E. R. 1995. How organizations Act Together: Interorganizational coordination in Theory and Practice, Gordon and Breach Publishers.

Auf der Heide, E., 1989. Disaster response: principles of preparation and coordination. CV Mosby St.

Axelrod, R. & Cohen, M. D. 2000. Harnessing Complexity: Organizational Implications of a Scientific Frontier. New York, Basic Books

Bigley, G. A. & Roberts, K.H. 2001. The incident command system: high reliability organizing for complex and volatile environments. Academy of Management Journal, vol.44, no.6, 1281-1299.

Borch, O.J & Madsen, E.L. 2007. Dynamic capabilities facilitating innovative strategies in SMEs. International Journal of Technoentrepreneurship, 1(1): 109-125.

Borch, O.J & Solesvik, M. 2013. Collaborative Design of Advanced Vessel Technology for Offshore Operations in Arctic Waters. Computer Science 2013; Volum 8098 (1). ISSN 0302-9743.s 157 - 160.

Borch, O.J. & Arthur, M.B. 1995. Strategic network among small firms: Implications for strategy research methodology. Journal of Management Studies, 32, 4:419-441.

Borch, O.J. & Batalden, B. 2014. Offshore service vessel logistics and entrepreneurial business process management in turbulent environments. Maritime Policy & Management: The flagship journal of international shipping and port research (forthcoming)

Borch, O.J. 1994. The Process of Relational Contracting. Developing Trust-Based Strategic Alliances among Small Business Enterprises. I Paul Shrivastava, Jane Dutton and Anne Huff (eds.): Advances in Strategic Management, JAI Press Inc., Greenwich, Connecticut, 1994.

Buck, D. A., Trainor, J. E. & Aguirre, B. E. 2006 A Critical Evaluation of the Incident Command System and NIMS, Journal of Homeland Security and Emergency Management, Vol.3, Issue 3, Article 1., Available at: http://www.bepress.com/jhsem/vol3/iss3/1.

Comfort, L.K. & Kapucu, N. 2006. Inter-organizational coordination in extreme events: The World Trade Center attacks, September 11, 2001, Nat Hazards, Vol.39, pp.309-327.

Dooley, K. J. 2004. Complexity Science Models of Organizational Change and Innovation., in Poole, Marshall

Scott and Van de Ven, Andrew H. (eds.) Organizational Change and Innovation, Oxford University Press.

Ekvall, G & Arvonen, J. 1994. Leadership profiles, situation and effectiveness. Creativity and Innovation Management, 3, 139-161.

Fukuyama, F. 1995. Trust: The Social Virtues and the Creation of Prosperity. . London: Hamish Hamilton.

Gudmestad, O.T., Zolothukhin, A.B., Ermakov, A.I., Jacobsen, R.A., Michtchenko, I.T., Vovk, V.S., Løset S. and Shkinek, K.N. 1999. Basic of offshore petroleum engineering and development of marine facilities with emphasis on the Arctic offshore, ISBN 5-7246-0100.-1.

Haas P.M. Introduction: epistemic communities and international policy coordination. International Organization 1992; 46 (1): 1–35.

Hales, C. (2002). Bureaucracy-lite and Continuities in Managerial Work. British Journal of Management, 13, 51–66.

Hansen, Gary S. and Wernerfelt, Birger. 1989. Determinants of Firm Performance: the Relative Importance of Economic and Organizational Factors, Strategic Management Journal, Vol.10, No.5, Sep-oct 1989, pp.399-411.

Hatak, I., & Roessl, D. 2010. Trust within Interfirm Cooperation: A Conceptualization. Our Economy, 56(5-6): 3-10.

Hossain, L. & Uddin, S. 2012. Design patterns: coordination in complex and dynamic environments, Disaster Prevention and Management, Vol.21, No.3, pp.336-350.

Hovden, S. T. 2012. Redningsdåden-om Maksim Gorkiy-havariet utenfor Svalbard i 1989. Commentum Forlag, Sandnes.

Kapucu, N. (2005) Interorganizational Coordination in Dynamic Context: Networks in Emergency Response Management, Connections, 26 (2), pp.33-48.

Levin, D. Z., & Cross, R. 2004. The strength of weak ties you can trust: The mediating role of trust in effective knowledge transfer. Management Science, 50(11): 1477-1490.

Litwak, E. & Hylton, L. F., 1962. Interorganizational analysis: a hypothesis on co-ordinating agencies, Administrative Science Quarterly, 6 (4), pp. 395–420.

Louis. Lie, A., 2010. Coordination process and outcomes in the public service: the challenge of inter-organizational food safety coordination in Norway, Public Administration. DOI: 10.1111/j.14-67-9299.2010.01845.x.

Madsen, E.L., Alsos, G.A., Borch, O.J., Ljunggren, E. & Brastad, B. (2007) Developing entrepreneurial orientation. The role of dynamic capabilities and intangible resources. In Gillin, L. M.(ed.) 2007, Regional Frontiers of Entrepreneurship Research 2007, Swinburne University, Melbourne, VIC.

McAllister, D. J. 1995. Affect- and Cognition-based trust as foundation for interpersonal cooperation in organizations. Academy of Management Journal, 38(1): 24-59.

Mintzberg, H. 1973. The Nature of Managerial Work. New York: Harper Row.

Mintzberg, H. 1979. The Structuring of Organizations: A Synthesis of the Research. En- glewood Cliffs, NJ: Prentice-Hall.

Mintzberg, H. 1990. The Manager's Job. Folklore and Fact., Harvard business review, march–april 1990, 42pp.

Minzberg, H. 2009. Managing. Williston, VT, USA: Berrett-Koehler Publishers.

Mooradian, T., Renzl, B., & Matzler, K. 2006. Who Trusts? Personality, Trust and Knowledge sharing. Management Learning, 37(4): 523-540.

Quarantelli, E. L., 1986. Research findings on organizational behavior in disasters and their applicability in developing countries. Preliminary paper # 107. Newark, DE: Disaster Research Center, University of Delaware.

Ripperger, T. 1998. Ökonomik des Vertrauens: Analyse eines Organisationsprinzips. Tübingen: Mohr Siebeck.

Seidman, H. & Gilmour, R., 1986. Politics, position, and power: from the positive to the regulatory state, 4th ed. New York: Oxford University Press.

Sommer, M., Braut, G.S. & Njå, O. (2013) 'A model for learning in emergency response work', Int. J. Emergency Management, Vol. 9, No. 2, pp.151–169.

Stacey, R.D. 2001. Complex responsive processes in organizations: learning and knowledge creation. London, Routledge.

Sydnes, M & Sydnes, A.K. 2011. Oil spill emergency response in Norway: coordinating interorganizational complexity., Polar Geography, Vol.34, No.4, pp.299-329.

The Norwegian Directorate for Civil Protection (DSB). 2013. Evaluation Report – Exercise Barents Rescue 2013. www.dsb.no

Thomas, C.W., 2003 Bureaucratic landscapes: interagency cooperation and the preservation of biodiversity. Cambridge, MA: MIT Press; 2003.

Thunem, Atoosa. 2010. Understanding and Describing Complexity in Safety and Event Analysis of Socio-Technical Systems: The Voyage and Findings. Reliability Engineering & System Safety journal.

Turoff, M., M. Chumer, B. Van de Walle, & X. Yao. 2004. "The Design of a Dynamic Emergency Response Management Information System (DERMIS)", The Journal of Information Technology Theory and Application (JITTA), 5:4, 2004, 1-35.

Turoff, Murray, White, Connie & Plotnick, Linda. 2011. Dynamic Emergency Response management for large Sacle Decision making in Extreme hazardous Events, in Burstein, F, Brezillon, p. and Zaslavsky, A. (eds.) Supporting Real Time Decision-Making, Vol. 13, Springer .

Van de Ven, A. & Walker, G. 1984. The Dynamics of Interorganizational Coordination, Administrative Science Quarterly, vol.29, No.4, pp.598-621.

Wilensky, H.L.,1964. The professionalization of everyone? American Journal of Sociology 1964; 70 (2): 137–158.

Environmental Risk Assessment for the Aegean Sea

I. Koromila & Z. Nivolianitou
Institute of Informatics and Telecommunications National Centre for Scientific Research "Demokritos", Athens, Greece

S. Perantonis, T. Giannakopoulos, E. Charou & S. Gyftakis
System Reliability and Industrial Safety Laboratory, Institute of Nuclear and Radiological Sciences & Technology, Energy and Safety, National Centre for Scientific Research "Demokritos", Athens, Greece

K. Spyrou
Ship Design Laboratory, Department of Naval Architecture and Marine Engineering Department, National Technical University of Athens, Athens, Greece

ABSTRACT: This paper presents a probabilistic model predicting the risk of a possible ship accident occurrence in the Aegean Sea. The proposed model combines the results of the accident probability (P) with the severity of its consequences (S). The P factor is calculated using the Bayesian Network methodology. Static and dynamic information is taken into consideration. The S factor is estimated through a mathematical formula. Under the "consequences", the cost resulting from an oil spill is considered. In order to demonstrate the applicability of this method a sample use cases has been conducted. The whole case study has been run within the framework of the AMINESS project.

1 INTRODUCTION

Oil spill marine accidents occurring near to shore, as the Prestige wreck was, and inside areas of significant environmental interest, like the Exxon Valdez accident, have dramatically affected both the environment and the life in the area. The Aegean Sea is a special marine area in all aspects, owing to its exceptional geographical morphology in conjunction with significant environmental and ecological sensitivity and the huge volume of traffic that leads to increased probability of marine accidents. An oil spill accident in such an area would negatively affect all social, economic, environmental and cultural sectors not only of Greece but also of the greater area of the Eastern Mediterranean basin.

Several researches on risk of marine accidents have been conducted during the last years. A probability risk analysis on tankers grounding using the fault and event tree analysis was performed by Amrozowicz (1997). Jensen et al. (2009) focuses on the events that lead to a collision or grounding and the damage to the ship after the collision or grounding event has happened. Finally, one of the most recent work is that of Pereira et al. (2011) presenting a fuzzy logic based intelligent decision making system aiming at improving the safety of marine vessels by avoiding collision situations.

Although there is a significant number of studies that analyze the cases of collision and grounding, these studies do not take into account the environmental effects of a marine accident. Therefore, the present study uses the collision/grounding probability model and tries to extend it by incorporating elements of environmental assessment.

The structure of the paper is as follows: in section 2 the proposed model is developed, including the methodology of marine accident probability estimation and the mathematical formula of the consequences. In section 3 the evaluation of the model is conducted together with some discussion of the results. Finally, conclusions are drawn in section 4, followed by acknowledgements and the list of references.

2 THE PROPOSED MODEL

2.1 *Methodology used*

Although a reactive approach has traditionally been used in analyzing marine accidents, a shift towards more proactive approaches has been initiated in recent years. The International Maritime Organization (IMO) has already introduced such a proactive approach to safety, called "Formal Safety Assessment" (FSA), using insights from the nuclear and aerospace industries (IAEA, 1992/1995; Papazoglou et al 1992; Nivolianitou and Papazoglou, 1998). According to the IMO, FSA is defined as "a rational and systematic process for assessing the risk related to maritime safety and the protection of the marine environment and for evaluating the costs and benefits of IMO's options for reducing these risks" (IMO, 2007). A typical FSA study compromises of the following five steps; (a) identification of hazards, (b) development of risk analysis, (c) setting of risk control options, (d) making of cost benefit assessment, and (e) production of recommendations for decision-making.

In the context of the present work the risk analysis step has been implemented for the specific identified hazards (i.e. the collision, contact and grounding accidents). According to the FSA guidelines, risk is defined as the probability (P) of an event multiplied by the severity of its consequences (S).

$$R = P \times S \qquad (1)$$

Therefore, the proposed model combines the results of the accident probability (P) with the severity of its consequences (S). The factor P is calculated using the Bayesian network methodology by, firstly, using a simplified model to predict the probability of an accident given the main characteristics of the vessel; namely the ship type, size, age and flag (the static model). A more advanced model taking into account the dynamic factor of the probability is elaborated, along with considerations about the weather, the geographical area and the traffic conditions. As far as consequences (S) are concerned and taking into account the environmental scope of the AMINESS project (Giannakopoulos et al., 2014), considerable focus is to be placed on the estimation of the impact (costs) resulting from an oil spill.

In the final stage of the model, the two factors (P and S) are integrated and produce the final environmental risk of a possible ship accident in the Aegean Sea. A process flow chart is illustrated in Figure 1.

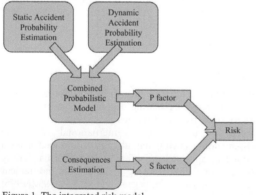

Figure 1. The integrated risk model

In the following sub-sections the Bayesian models are discussed.

2.2 Static Model

The purpose of the static model is to estimate the probability of the collision and the grounding event given the static information of a particular vessel. For this reason, a simple Bayesian network has been implemented. The structure of the network is presented in Figure 2.

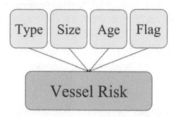

Figure 2. The static model

Each of the input variables has a set of values. Namely, the vessel type could take the values "passenger", "tanker", "general cargo" and "other". Namely, "passenger" is any vessel carrying more than 12 passengers, "tanker" is any vessel carrying liquid cargo, "general cargo" is any vessel carrying bulk cargoes, containers or vehicles and "other" is any other vessel, such as research vessel, yacht, tug, and fishing. The variable of the size could be 'small', 'medium' and 'large' depending on weight of the cargo (deadweight-DWT) for tankers and general cargoes and the length for a passengers. All other vessels are considered as 'small'. The age of the vessel is separated into 'new', 'middle' and 'old'. A vessel built no more than five years is considered 'new', while a vessel built more than twenty-six years is 'old'. Finally, the vessel flag could take the values "low risk", "medium risk", "high risk" and "very high risk", according to PARIS MOU flag list 2012.

Such vessel characteristics remain constant during the voyage and are easily extracted from the ship Automatic Identification System (AIS). In order to train the static Bayesian network, 200 marine accidents from the Greek Ministry of Mercantile Marine accident database have been utilized.

2.3 Dynamic Model

For the dynamic model, a more complicated Bayesian network is developed, taking into consideration weather, geomorphological and traffic conditions. Figure 3 illustrates an overview of the proposed dynamic model.

Briefly explained, the input variables are the "navigation area" and the "weather". The "navigation area" variable includes the following values: "restricted waters", "traffic" and "environmental sensitivity". The "weather" variable is composed of the "sea state", the "wind speed", the "swell" and the "visibility". It also takes into account weather conditions such as fog or rain, and time indications such as daytime or nighttime.

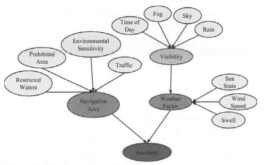

Figure 3. The dynamic model

These dynamic variables change during the voyage. The National Meteorological Office of Greece provides data on both the current and the future weather conditions. As far as the geomorphological characteristics and the environmental sensitive areas in the Aegean Sea are concerned, the required data are collected from the appropriately constructed AMINESS database. The vessel density per navigation area is calculated using the AIS vessels sign. In order to train the dynamic Bayesian network, the AMINESS database has been used that stores several types of marine data, namely; AIS trajectories, geomorphological information, areas of environmental interest and weather data.

2.4 Consequences Model

When considering the accident scenarios of collision, contact and grounding, consequences can result in impacts on humans (crew, passengers, and third parties), on ships, on the environment, in industry, and on the waterways regarding their present and future use. In the context of the present work only the environmental impact has been taken into consideration.

The main categories of costs associated with the environment and especially with oil spills are cleanup costs, environmental damage, and socioeconomic effects (Kontovas et al, 2010). The first cost group covers the clean-up costs of an eventual spill, research costs and other specific costs, such as the loss of cargo and/or vessels and the repairs needed after accident There is a general agreement (Etkin, 1999; Grey, 1999; White and Molloy, 2003; Kontovas et al, 2010) that the main factors influencing the clean-up cost of oil spills are related to (a) the type of oil, (b) the location, (c) the weather and sea conditions, (d) the amount of spilled oil together with the rate of spillage, and (e) the clean-up response. Referring to the second cost

group, oil spills may impact the environment causing physical smothering of organisms, chemical toxicity or ecological changes (ITOPF, Technical Information paper 13). Finally, the socioeconomic effects consist of property damage and income losses (Liu and Wirtz, 2006).

The total cost of an oil spill can be derived by using at least four different methods (Kontovas and Psaraftis, 2008). These are the following:

1 Adding up all relevant cost components (clean-up, socioeconomic and environmental).
2 Using a model that estimates the total cost.
3 Estimating clean-up costs through modeling and then assuming a comparison ratio between environmental and socioeconomic costs.
4 Assuming that the total cost of an oil spill can be approximated by the compensation eventually paid to claimants.

Several pieces of research dealing with the estimation of the total cost of an oil spill have been conducted during the last decades. There are some scientists that propose for complicated mathematical formulas to calculate oil spill costs (Cohen 1986, Etkin 2000 & 2004, Shahriari and Frost 2008) and others who estimate clean-up costs through actual modeling (Vanem et al 2007, Jean-Hansen 2003). The most recent and commonly accepted methodology is to estimate the total cost by using compensation data (International Oil Pollution Compensation Fund-IOPCF annual reports). Grey (1999), Yamada (2009), Psarros et al. (2009) and Kontovas et al. (2010) are some of the related works implementing regression analysis using historical oil spill data from tankers, reported by the IOPCF.

In the present work the advocated model for estimating the total cost of an oil spill is the one proposed by Kontovas et al. (2010), which is considered as the most appropriate for our work, as the proposed by them formula has recently been adopted by the IMO/MEPC as the basis for further discussion on environmental risk evaluation criteria. The mathematical framework is stated in equation 2.

$$S = 51,432 \times V^{0.728} \tag{2}$$

where S=the consequences measured in US dollars; and V=the volume of oil spilled measured in metric tons.

In the present model three cases of the V factor are considered; zero, mean and total outflow. Zero outflow implies no consequences, mean outflow means that approximately half of cargo is lost, while in total outflow all the cargo is lost.

3 MODEL EVALUATION

In order to demonstrate the applicability of the method presented in this study, five areas of interest have been extracted through a local maximum estimation procedure that was applied combining both the static and the dynamic models. These areas are briefly explained below.

Area 1: The Kafireas Street is located between Euboea and Andros islands. It is a narrow passage that links the Piraeus port with the ports of Northern Aegean Sea and the major ports of Black Sea. Intense marine traffic combined with frequent adverse weather conditions make this area risky.

Area 2: The Kithera Street is an area with dangerous navigational conditions, located between Elafonissos and Kithera Islands. Vessels crossing this passageway carry hazardous cargoes from the Central Mediterranean Sea to the Suez Canal or the Black Sea and vice versa.

Area 3: Ikaria Island is located in the Central Aegean Sea. Although there are no geomorphological restrictions in this area, numerous dangerous vessels pass through the island's region.

Area 4: Alonissos Island is located in Northern Aegean Sea. The area is part of the NATURA 2000 ecological network and thus it is of special environmental interest. This environmental sensitivity in conjunction with the highly hazardous transported cargoes makes the Alonissos area extremely risky.

Area 5: Cyclades is a group of small islands located in Central Aegean Sea. Dangerous vessels and adverse weather conditions led to the choice of such an area.

For each of the above areas, a three-month dataset on vessel trajectories (June, July, and August of the year 2009) was used to extract an accumulated average of the respective deadweight. For the scope of the present study, only oil and chemical tankers have been used. Table 1 shows the average deadweight and the respective expected cost in the case of an accident in each area of interest.

Table 1. Average deadweight and the respective expected cost for each area.

Area	Average Vessel's Deadweight [tons]	Average Expected Cost [million USD]
Kafireas Street	9000	39
Kithera Street	5500	27
Ikaria Island	14500	55
Alonissos Island	9000	39
Cyclades	12000	48

A possible marine accident within the area of Alonissos Island would negatively affect the country's economy resulting in a cost of 9 million US dollars. Most importantly, such an accident would have also significant impact on both marine and human life. The expected cost of an accident in Kithera Street or in Kafireas Street is not that high, compared to the other areas. This is because the area is mainly crossed by small tankers that would cause

only small oil spills and would lead to low clean-up cost. Finally, as far as the other two areas are concerned, the average expected cost of an accident is quite high. This result seems to be rational, since these areas belong to the Central Aegean region where large tankers carrying oil products from the Black Sea to the Central Mediterranean Sea or the Suez Canal and vice versa.

4 CONCLUSIONS

This paper develops a methodology to predict the environmental risk of a possible marine accident. Two Bayesian models were constructed, taking into account the vessel's static and dynamic information. The former refer to the vessel's type and size, while the latter include the weather conditions and the characteristics of the navigation area. These models assess the probability of an accident at a specific area and after taking into account the consequences (i.e. cost of an oil spill) estimate the final environmental risk. Five areas of interest were chosen to demonstrate the model's applicability. The average volume of oil transferred through each area and the respective expected cost were estimated. The results indicated significant differences in the environmental costs per region. A step forward aiming to reduce the probability of an accident, would be to propose designated vessel routes through the Aegean Sea.

ACKNOWLEDGEMENTS

This work was carried out in the framework of the project "AMINESS: Analysis of Marine Information for Environmentally Safe Shipping" which was co-financed by the European Fund for Regional Development and from Greek National funds through the operational programs "Competitiveness and Enterpreneurship" and "Regions in Transition" of the National Strategic Reference Framework Action: "COOPERATION 2011 – Partnerships of Production and Research Institutions in Focused Research and Technology Sectors".

REFERENCES

Amrozowicz, M., Brown, A.J., Golay, M. 1997. A Probabilistic Analysis of Tanker Groundings. 7th International Offshore and Polar Engineering Conference, Honolulu, Hawaii.

Cohen, M.A. 1986. The costs and benefits of oil spill prevention and enforcement. Journal of Environmental Economics and Management 13: 167–188.

Etkin, D.S. 1999. Estimating clean-up costs for oil spills. In: Proceedings. International Oil Spill Conference, American Petroleum Institute, Washington, DC.

Etkin, D.S. 2000. Worldwide analysis of marine oil spill cleanup cost factors. Arctic and Marine Oil Spill Program Technical Seminar. Vancouver, BC, Canada.

Etkin, D.S. 2004. Modeling oil spill response and damage costs. 5th Biennial Freshwater Spills Symposium, New Orleans, the United States.

Giannakopoulos, T., Vetsikas, I., Koromila, I., Karkaletsis, V., Perantonis, S. 2014. AMINESS: A Platform for Environmentally Safe Shipping. 7th International Conference on Pervasive Technologies Related to Assistive Environment (PETRA 2014), Rhodes, Greece.

Grey, C. 1999. The cost of oil spills from tankers: an analysis of IOPC fund incidents. The International Oil Spill Conference 1999, Seattle, USA.

International Atomic Energy Agency (1992) IAEA: Procedures for Conducting Probabilistic Assessments of Nuclear Power Plants (Level 1). Vienna, Safety Series No. 50-P-4. ISBN 92-0-102392-8.

International Atomic Energy Agency (1995) IAEA, Procedures for Conducting Probabilistic Assessments of Nuclear Power Plants (Level 2). Accident Progression, Containment Analysis and Estimation of Accidents Source Terms. Vienna, Safety Series No. 50-P-8. ISBN 92-0-102195-X, ISSN 0074-1892.

International Maritime Organization (2007) IMO, Formal Safety Assessment: Consolidated Text of the Guidelines for Formal Safety Assessment (FSA) for Use in the IMO Rule Making Process. MSC/Circ.1023–MEPC/Circ.392. London (MSC 83/INF.2).

Jensen, J., Soares, C. G. and Papanikolaou, A. 2009. Collisions and Groundings. In Papanikolaou, A. (Eds.). Risk-Based ship design: Methods, Tools and Applications: 213-231. Athens: Greece.

Kontovas, C., Psaraftis, H., Ventikos, P. 2010. An empirical analysis of IOPCF oil spill cost data. Marine Pollution Bulletin 60: 1455–1466.

Kontovas, C.A., Psaraftis, H.N. 2008. Marine environment risk assessment: a survey on the disutility cost of oil spills. 2nd International Symposium on Ship Operations, Management and Economics. Athens, Greece.

Liu, X., Wirtz, K.W. 2006. Total oil spill costs and compensations. Maritime Policy and Management 33: 460–469.

Nivolianitou, Z. and Papazoglou, I. A. 1998. An auditing methodology for safety management of the Greek process industry. Reliability Engineering and System Safety 60: 185-197.

Papazoglou I. A., Nivolianitou Z., Aneziris O., Christou M. 1992. Probabilistic safety analysis in chemical installations. Journal of Loss Prevention in the Process Industries 5: 181-191.

Perera, L. P., Carvalho, J. P., Guedes Soares, C. 2011. Fuzzy logic based decision making system for collision avoidance of ocean navigation under critical collision conditions. Journal of Marine Science and Technology 16: 84-99.

Shahriari, M. and Frost, A. 2008. Oil spill cleanup cost estimation – developing a mathematical model for marine environment. Process Safety and Environmental Protection 86: 189–197.

Vanem, E., Endresen, Ø., Skjong, R. 2007. Cost effectiveness criteria for marine oil spill preventive measures. Reliability Engineering and System Safety 93: 1354–1368.

White, I.C., Molloy, F. 2003. Factors that determine the cost of oil spills. International Oil Spill Conference. Vancouver, Canada.

Yamada, Y. 2009. The cost of oil spills from tankers in relation to weight of spilled oil. Marine Technology 46: 219–228.

Psarros, G., Skjong, R., Endersen, O., Vanem, E. 2009. A perspective on the development of Environmental Risk Acceptance Criteria related to oil spills. Annex to International Maritime Organization document MEPC 59/INF.21, submitted by Norway.

Probabilistic Meta-models Evaluating Accidental Oil Spill Size from Tankers

J. Montewka, F. Goerlandt & X. Zheng
Dept. of Applied Mechanics, Aalto University, Espoo, Finland

ABSTRACT: In this paper, we introduce two probabilistic meta-models evaluating accidental oil spill volume for two types of tanker: VLCC and Aframax and two types of accident: collision and grounding. The meta-models utilize two engineering models, one for ship damage estimation due to an accident, another for evaluation of an oil spill size given the damage extent. Both models are coupled and executed for a large number of accidental scenarios, thus two extensive datasets are developed, one for collision and one for grounding accident. Subsequently, the datasets are used to develop two casual, probabilistic meta-models by applying the Bayesian Networks. The obtained results are compared with the real-life cases, and good agreement is found.

1 INTRODUCTION

There are several types of model available for the estimation of damage extent and resulting oil spill size that a tanker is likely to face as a result of collision or grounding accident. First type is based on statistical data about the accidental oil spills that happened in the past, (Eide, Endresen, Brett, Ervik, & Røang, 2007; IMO, 2003; Montewka, Ståhlberg, Seppala, & Kujala, 2010), however the challenge with using these models is that they often treat correlated variables as independent random variables.

Second type of models attempts to capture the mechanism governing an accident and resulted oil spill, see for example (Friis-Hansen & Ditlevsen, 2003; Gucma & Przywarty, 2008; Krata, Jachowski, & Montewka, 2012; Papanikolaou, 2009; Smolarek & Mazurek, 2013). However, this modeling practice requires a significant number of input parameters that usually cannot be estimated without uncertainties, moreover these It is time consuming, which does not allow for quick analyses.

Third type of models is based solely on expert judgments, see (Helle, Lecklin, Jolma, & Kuikka, 2011; Lehikoinen, Luoma, Mäntyniemi, & Kuikka, 2013). Although these models provide good estimates in general, it cannot provide any solid basis for the cause-effect analyses.

One major shortcoming of these modeling practices, except the aforementioned type two, is that the inherent feature of an accident an following consequences, causality, is not reflected properly. Another issue is that hardly ever uncertainty of the model is addressed.

These two issues can be tackled, to some extent, by adopting appropriate modeling techniques, for instance combining the engineering models with probabilistic causal models. Hence, the causality of the analyzed process and the relevant correlation among variables are retained, inputs and output are linked in a probabilistic fashion, and two-way inference is feasible, i.e. causes → effect and effect → causes. Moreover, the uncertainty associated with the input parameters and model itself can be assessed, and the most effective ways of uncertainty reduction can be proposed.

Therefore, in this paper we introduce two probabilistic meta-models evaluating accidental oil spill volume for two types of accidents: collision and grounding, involving two types of tanker: VLCC and Aframax.

The meta-models utilize two engineering models, one for ship damage estimation due to an accident, another for evaluation of an oil spill size given the damage extent. Both models are coupled and executed for a large number of impact scenarios. As a result two extensive datasets are developed, one for ship-ship collision and one for grounding accident. Subsequently the datasets are used to develop two probabilistic meta-models with the use of Bayesian Networks.

The proposed meta-models allow not only for the estimation of the size of an oil spill given a limited set of input variables, most of them are available from the regular messages transmitted within the Automatic Identification System (AIS). They can also assist in evaluating the required changes in input variables in order to reach a predefined level of an output variable.

The models can be further used in the process of environmental risk assessment for the sea areas or in any studies focusing on the assessment of the required oil combating capacity. In principle the models could be used in the process of risk-based ship design, linking the probability of an oil spill with specific tanker design parameters.

The remainder of the paper is organized as follows: Section 2 introduces the overall idea of the meta-models shown here. All the sub models involved are scrutinized in Section 3 and the resulting meta-models are described along with their uncertainties in Section 4. Section 5 presents the results of benchmarking, whereas Section 6 concludes.

2 PROBABILISTIC META-MODELS

The probabilistic meta-models presented here, turn the deterministic models into probabilistic, to support decision-making in the presence of uncertainty, by informing the potential end-users about the most probable size of accidental oil spill at sea for a selected type of tanker. Two types of accident are considered, collision between two ships and ship grounding, with respect to Aframax and VLCC type of tanker, see Table 1 for the main characteristics of these.

Table 1. The main parameters of two tankers considered in this study: Aframax and VLCC.

Tanker type	DWT [tons]	Cargo [m3]	LOA [m]	B [m]	V [kn]	T [m]
Aframax	115000	131000	250	44	15	13.6
VLCC	316000	353000	332.6	60	16	21.0

The size of an oil spill is evaluated with the use of a set of deterministic, engineering models. Since the input parameters of these models are burdened with uncertainty, the appropriate modeling techniques are applied with an attempt to represent the existing uncertainty and show the ways to reduce it in the future. The uncertainty comes from the lack of knowledge about the analyzed domain, and the processes involved. In our case, the deterministic models evaluating the extent of accidental damage of a ship and the resulting size of an oil spill can be accurate, if they are fed with the accurate values of the input parameters. However, this is an ideal case. In reality, a range of values can describe the

parameters (a.k.a aleatory uncertainty) or their values are simply not known (a.k.a. epistemic uncertainty). In the first case, the multitude of runs of deterministic models (i.e. Monte Carlo simulations - MC) may give some indication about the spread of the results. However this method alone does not help in the presence of epistemic uncertainty, especially when several alternative hypotheses (AH) may exist. In this case, the AH testing shall be carried out for the relevant AHs. In practice this means development of a set of models, reflecting all anticipated AHs. Then, by running MC for all the models, we obtain a large dataset of inputs and corresponding outputs, which by the adoption of Bayesian Networks is made into a probabilistic model. Such a model allows probabilistic reasoning about the analyzed domain, reflecting the existing background knowledge and assessing the associated uncertainty through the value of information analysis. Performing sensitivity analysis, one can measure the strength of influence of the parameters included in the model on the outcome.

2.1 Bayesian Networks

Bayesian networks are selected as a risk-modelling tool as these have a number of favorable characteristics. BNs can contextualize of the occurrence of specific consequences through situational factors, which represent observable aspects of the studied system. They furthermore allow integration of different types of evidence through various types of probabilities and provide a means for performing sensitivity analysis. It is also rather straightforward to incorporate alternative hypotheses in the model. Bayesian networks are relatively widely used tools for risk modeling (Aven, 2008; Fenton & Neil, 2012)

BNs represent a class of probabilistic graphical models, defined as a pair $\Delta=\{G(V,A),P\}$. (Koller & Friedman, 2009), where $G(V,A)$ is the graphical component and P the probabilistic component of the model. $G(V,A)$ is in the form of a directed acyclic graph (DAG), where the nodes represent the variables $V=\{V_1,...,V_n\}$ and the arcs (A) represent the conditional (in)dependence relationships between these. P consists of a set of conditional probability tables (CPTs) $P(V_i|Pa(V_i))$ for each variable V_i, $i=1,...,n$ in the network. $Pa(V_i)$ signifies the set of parents of V_1 in G: $Pa(V_1)=\{Y\in V|(Y,V_i)\}$. A BN encodes a factorization of the joint probability distribution (JDP) over all variables in V:

$$P(V) = \prod_{i=1}^{n} P(V_i \mid Pa(V_i)) \qquad (1)$$

By the adoption of Bayesian Networks as modeling tools, we develop probabilistic meta-models that can perform forward (predictive) reasoning, from new information about causes (explanatory variables) to updated beliefs about the effects (response variables), following the directions of the network arcs. Alternatively, backward (diagnostic) reasoning is possible, from the effects of interest to the most probable causes, where the information in the model is propagated against the direction of the arcs.

2.2 Concept of Meta-models

A concept of two meta-models presented here is depicted in Figure 1. Therein the flow of background knowledge (BK) is depicted, along with its sources. Traffic data and characteristics of tankers form inputs (impact scenario) to the oil spill model in case of collision. For a grounding model, additional parameters describing a rock and its location with respect to a tanker are needed. This BK is organized and propagated through several engineering models, adopting some central assumptions and entertaining various alternative hypotheses when applicable, accounting for epistemic uncertainty. Also the input variables are described as distributions to account for the aleatory uncertainty. This logic is then turned into a computer code, and MC simulations are executed to cover the modeling space to the largest extent. As a result, a dataset is obtained containing a set of basic input parameters, which are linked with the output - size of an oil spill given an accident - through a set of intermediate variables. These input parameters, describing an impact scenario, are as follows:

- for collision: speeds and masses of two colliding ships, collision angle, impact location, loading conditions of a struck ship, bow half entrance angle;
- for grounding: tanker's speed, depth of a rock with respect to ship's draft, tip radius, apex angle and eccentricity of a rock that a tanker runs over.

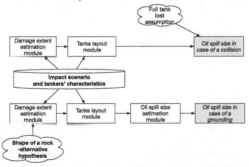

Figure 1. A concept of meta-models estimating accidental oil spill size.

Based on the dataset, with the use of Bayesian Networks, two probabilistic meta-models are developed. They link the impact scenario with the

oil outflow, without a necessity to include all the intermediate variables. The dataset is valid for a specific type of tankers, i.e. VLCC and Aframax.

3 UNDERLYING ENGINEERING MODELS

The first step of the modeling process adopted here is to find relations between impact scenario and the extent of damage that a tanker is likely to suffer as a result of ship-ship collision or grounding. An engineering model mimicking this situation is briefly described in Section 3.1 with relevant references given for the scrutiny.

Subsequently, based on the size of damage and the layout of a tanker, the number of damaged tanks is estimated along with the resulting quantity of cargo oil that is spilled. These are described in sections 3.2 and 3.3.

3.1 Damage extent estimation models

To determine the extent (longitudinal and transversal) of damage due to collision between two ships or a ship running aground we adopted two models proposed in (van de Wiel, 2008). Therein the polynomial regression model uses a set of predictor variables to link the impact scenario variables to the longitudinal and transversal damage extents. These predictor variables are representative of the impact scenario, which can be described through the vessel masses, the vessel speeds, the impact angle, impact location along a struck ship hull, bow half-entrance angle. In case of a collision between two ships, the expected damage length and damage width are expressed as:

$$y_L = exp\left(h_L\left(x \mid \hat{\beta}^l\right)\right) \qquad (2)$$

$$y_T = exp\left(h_T\left(x \mid \hat{\beta}^l\right)\right) \qquad (3)$$

where:

$$h_L\left(x \mid \hat{\beta}^l\right) = \sum_{i=1}^{5} \hat{\beta}_0^l + \sum_{j=1}^{5} \hat{\beta}_{i,j}^l x_j^i \qquad (4)$$

$$h_T\left(x \mid \hat{\beta}^l\right) = \sum_{i=1}^{5} \hat{\beta}_0^l + \sum_{j=1}^{5} \hat{\beta}_{i,j}^l x_j^i \qquad (5)$$

The maximum and minimum longitudinal damage positions, y_{L1} and y_{L2} depend on the damage length y_L, the relative impact location l and a weight factor θ:

$$y_{L1} = (1-\theta)y_L + (1-l)L \qquad (6)$$

$$y_{L2} = -\theta y_L + (1-l)L \qquad (7)$$

In case of a tanker running aground, expected damage length and damage width are expressed as:

$$y_L = exp\left(h_{L}\left(\boldsymbol{x} \mid \hat{\beta}^{l}\right)\right) \tag{8}$$

$$y_T = exp\left(h_{T}\left(\boldsymbol{x} \mid \hat{\beta}^{l}\right)\right) \tag{9}$$

where:

$$h_{L}\left(x \mid \hat{\beta}^{l}\right) = \sum_{i=1}^{6}\hat{\beta}_0^l + \sum_{j=1}^{6}\hat{\beta}_{i,j}^l x_j^i \tag{10}$$

$$h_{T}\left(x \mid \hat{\beta}^{l}\right) = \sum_{i=1}^{6}\hat{\beta}_0^t + \sum_{j=1}^{6}\hat{\beta}_{i,j}^t x_j^i \tag{11}$$

Damage positions and are obtained as follows:

$$y_{L1} = (1-\theta)y_L + (1-l)L \tag{12}$$

$$y_{L2} = -\theta y_L + (1-l)L \tag{13}$$

For the in-depth description of the method, the coefficients ($\hat{\beta}$) and their values, the reader is referred to (Goerlandt & Montewka, 2014; van de Wiel, 2008). Once the expected extent of damage is obtained, the number of tankers affected by the damage needs to be estimated to get some understanding about the amount of oil that will be lost. The adopted procedure is described in the following sections.

3.2 Tanks layout

The tank arrangement of a tanker is determined based on the procedure introduced in (Smailys & Česnauskis, 2006). The main parameters related to tank dimensions, volumes and locations of transverse and longitudinal bulkhead are shown in Figure 2. The ship length between perpendiculars, L_{PP}, is divided into three parts: L_A, L_T, L_F. Their meaning is as follows: the horizontal distance from the aft perpendicular to the bulkhead of the aft most side of a tank (L_A); the horizontal distance from the fore perpendicular to the bulkhead of a fore most tank (L_F); the length of a tank (L_T). The latter once multiplied by a number of tanks (n) yields the length of a cargo section of a tanker, therefore $L_{PP}= L_A + nL_T + L_F$. The ship width is taken as $B=mB_T+2w$, where m is the number of tanks in the transverse direction, B_T is tank compartment width and w stands for double hull width. The ship depth comprises the following components: $D=D_T+h$, where D_T is the tank depth and h double bottom height.

The volume of a given tank (V_i) can be expressed as: $V_i=C_iB_TL_TD_T$, where C_i is a volumetric coefficient, taking the actual shape of a tank into consideration compared with a rectangular block,

see Table 2. Since the main parameters of ships are known, along with the tank arrangement and the necessary coefficient – see Table 2 -, the parameters of interest (B_T, L_T, D_T) can be easily calculated. The tanks layout of Aframax is 2x6, whereas for the VLCC it is 3x5. This means, that Aframax has two lines of tanks on each side ($m=2$), with six tanks in each line ($n=6$). The VLCC has three lines of tanks ($m=3$), two on both sides and one in the center, with five tanks in a line ($n=5$).

The double hull width w and double bottom height h are estimated according to relevant rules for classification of ships (Det Norske Veritas, 2013). The positions of transverse bulkheads (TBH) and longitudinal bulkheads (LBH) can be determined as:

$$TBH_l = L_A + lL_T , \quad l = 0,1,2,...,n \tag{14}$$

$$LBH_k = w + kB_T , \quad k = 0,1,2,...,m \tag{15}$$

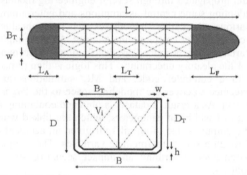

Figure 2 Definition of tanker and tank parameters, (Goerlandt & Montewka, 2014)

Table 2. Coefficients concerning tank arrangement, based on (Smailys & Česnauskis, 2006).

Tanker type	Section	Length [-]	C_i Side tank	C_i Center tank	Tank layout
VLCC	L_F	0.05L	0.77	0.85	3x5
	L_T	0.76L	1.00	1.00	
	L_A	0.19L	0.91	0.94	
Aframax	LF	0.05L	0.765	N/A	2x6
	LT	0.855L	1.00	N/A	
	LA	0.195L	0.92	N/A	

3.3 Oil spill size estimation model

3.3.1 Collision

In the procedure adopted here estimating the amount of oil spilled in case of a collision, it is assumed that all oil in a tank is lost, once the tank is breached. More accurate models for evaluating the oil spill exist (Sergejeva, Laarnearu, & Tabri, 2013; Tavakoli, Amdahl, & Leira, 2010), but their application requires information regarding the damage height and opening size and shape. As the damage extent model by van de Wiel and van Dorp (van de Wiel & van Dorp, 2011) only provides crude information regarding the limits of the breached area, no detailed information for these factors is available. Thus, the conservative assumption of a complete loss of oil from damaged tanks is made, as in (van Dorp & Merrick, 2011).

Figure 3 presents a collision scenario, where two tanks are breached. This results in an oil spill of a size equal of the volume of these two tanks (filled in black).

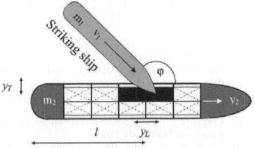

Figure 3. Definition of oil outflow in a collision scenario, (Goerlandt & Montewka, 2014).

3.3.2 Grounding

In case of a grounding accident, only a certain portion of oil escapes from the damaged tank. The amount of oil that is lost is estimated based on the principle of hydrostatic pressure equilibrium between the oil in damaged tank and the surrounding seawater. To find the height of oil column that satisfies the pressure equilibrium, the following equation is applied:

$$g\rho_s T = g\rho_o h_o \qquad (16)$$

where g is the gravitational acceleration 9.81 m/s², ρ_s is the density of sea water 1.025 ton/m³, T is a draught of tanker [m], ρ_o is the density of oil 0.847 ton/m³, h_o is the post-accidental height of oil column in a tank [m], which eventually yields:

$$h_o = \rho_s T / \rho_o \qquad (17)$$

The oil volume that is lost from a damage tank (V_{oi}) can be estimated as follows:

$$V_{oi} = C_i B_T L_T \left(D_T - h_o\right) \qquad (18)$$

3.3.3 Linking impact scenarios with oil spill size

Once the procedures for the quantification of accidental damage extent in a given impact scenario and the resulting oil spill size estimation have been settled, we perform Monte Carlo simulations, where a number of impact scenarios are executed and extensive dataset is obtained.

Since the aim of this study is to develop a meta-model that links an impact scenarios with the oil spill size, the main intention is to link the input and

output parameters recorded in the dataset. All the other intermediate parameters, despite they are recorded, will not be used for the development of meta-models.

An exemplary collision scenario is depicted in Figure 3, where the striking ship, denoted with a subscript 1, collides with the struck tank, denoted with a subscript 2, with a striking angle φ. The collision results in a damage length y_L along the longitudinal direction of the struck tank and a damage depth y_T in the transverse direction of the struck tanker. Comparing y_L with the positions of transverse bulkheads TBH, we can know how many tank compartments are broken in longitudinal direction; comparing y_T with the positions of longitudinal bulkheads LBH, we can know whether only tank compartments on either side, both sides are broken or on both sides and in the center ones for the VLCC are broken.

The positions of damage extent are defined as y_{L1} (starting point) and y_{L2} (ending point). The relative impact location l is measured from the aft of the struck tanker, and is expressed as a share of tanker length, where 0 means that a tanker is struck in her stern, 0.5 stands for a striking location in the amidships and 1 striking location in the bow.

The input variables used to model the process described above are listed in Table 3. Then we performed over 1 million MC simulations, which provide us with the results covering the modeling space. The number of MC repetitions depends on the number of different combinations of input variables. To develop the meta-models both the explaining and explanatory variables need to be discretized, and we already anticipated the number of states that the variables will take. There are 6 input variables taking not more than 5 states each, therefore there are approximately 5^6 possible combinations that the input variables can take. By performing over 1.5 million simulations, we repeated each combination 100 times. This seems to be enough to obtain stable distribution of output variable given all the inputs.

Table 3. Impact scenario parameters and their ranges adopted for modeling a collision accident.

Variable	Definition	Unit	Range
m_1	Striking ship mass	Tonnes	[0,200k]
v_1	Striking ship impact speed	Knots	[0,24]
v_2	Struck ship impact speed	Knots	[0,18]
ϕ	Impact angle	Degrees	[0,180]
l	Relative impact location		[0,1]
η	Striking ship bow half entrance angle	Degrees	[14,23]

If a tanker grounds when it comes to an obstruction and tank compartments are damaged, the same parameters as in the collision cases are used to describe the damage position and damage length. Subsequently the damage extent is compared with the positions of TBHs and LBHs.

The shape of a ground that a tanker is hitting is represented as a cone shaped rocky pinnacle with a rounded tip. It is assumed to be strong enough not to be damaged in the course of grounding accident, and all the kinetic energy is taken by a ship structure during grounding. This is a major assumption here, since the energy dissipation due to heave and roll motions of a tanker after running an obstacle are not accounted for. The model does not account for the post-accidental strength of a tanker. Therefore it does not estimate the spills resulting from a hull breakdown. The rock is described by four parameters listed in Table 4. Ands an exemplary grounding scenario and rock parameters are depicted in Figure 4.

Table 4. Impact scenario parameters and their ranges adopted for modeling a grounding accident.

Variable	Definition	Unit	Range
V	Struck ship velocity	Knots	[0,18]
O_d	Obstruction depth from mean low water	Meters	[0,20]
O_a	Obstruction apex angle	Degrees	[15,75]
O_r	Obstruction tip radius	Meters	[0.5,10]
C	Rock eccentricity	-	[0,1]

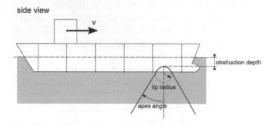

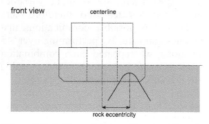

Figure 4. Definition of a grounding scenario and parameters of a rock.

4 PROBABILISTIC META-MODELS

Four probabilistic meta-models are presented here, for collision and grounding accidents for two types of tankers (Aframax and VLCC). The models link in a probabilistic fashion the parameters describing impact scenario, as specified in Table 2 and 3, with the output in terms of the volume of accidental oil spill.

Based on the engineering models, which are described earlier, an extensive training dataset is obtained, which is used to parameterize the meta-models.

4.1 Procedure

The following procedure is adopted to determine the probabilities of oil spill sizes, conditional to the impact situational variables:

1 For all combinations of input variables and the tanker loading condition TLC:
   For j=1→100:
   a) Sample a value in the considered discrete state of input variables and TLC
   b) Determine damage length y_L and damage depth y_T,
   c) Determine limits of the longitudinally breached area y_{L1} and y_{L2},
   d) Compare y_{L1} and y_{L2} with the locations of the transverse bulkheads TBH,
   e) Compare y_T with the locations of the longitudinal bulkheads LBH,
   f) The volume of spilled oil is calculated as

   $$V_{oil} = \sum_{i=1}^{N} V_i,$$ with N the number of tanks

   enclosed in the area encompassed by the comparisons in step d. and e, and V_i the cargo tank volume.

2 Count the relative occurrence frequency of V_{oil} for each of the discrete classes of the BN-variable Oil outflow, see Figure 5 and 6.

4.2 Models

Probabilistic meta-models are depicted in Figure 5-8. The input parameters are described as distribution, likewise the output. Models allow for two-ways reasoning, form the causes to the effect and in the revers direction, from the effect to the most probable causes.

By doing the forward reasoning, one can estimate the size of an oil spill given a set of variables describing an impact scenario. However, by performing the reverse reasoning, the following question can be addressed: given the maximum oil spill size allowed, what are the input parameters ?

Therefore the potential areas of application of the models presented, are bound by these two questions:
− given all the input parameters, what is the probable oil spill size ?
− given the oil spill size, what are the probable values of the input parameters ?

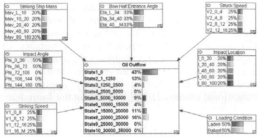

Figure 5. A probabilistic meta-model for Aframax tanker colliding with another ship.

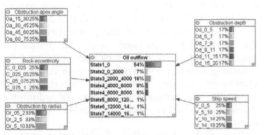

Figure 6. A probabilistic meta-model for Aframax tanker running aground

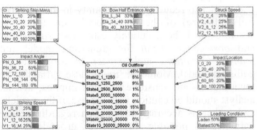

Figure 7. A probabilistic meta-model for VLCC tanker colliding with another ship.

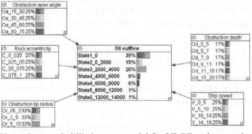

Figure 8. A probabilistic meta-model for VLCC tanker running aground

4.3 Sensitivity analysis

The purpose of a sensitivity analysis is to investigate the effect of changes in the assigned probabilities of the network variables on the probabilities of a specific outcome variable. In a one-way sensitivity analysis, every conditional and prior probability in the network is varied in turn, keeping the others unchanged. A sensitivity-value approach presented by Coupé and van der Gaag (Coupé & van der Gaag, 2002) is applied.

A sensitivity function is defined which describes an outcome Y as a function of the parameter $z = p(Y = Y_i \mid \pi)$, where Yi is one state of the outcome (a certain interval of oil spill size), and is a combination of states for the parent nodes of Y. For a network with no observations on any of the network variables, a linear sensitivity function is found:

$$f(z) = u_1 z + u_2 \tag{19}$$

where the constants u_1 and u_2 are identified based on the model. The first derivative of the sensitivity function at the base value describes the effect of minor changes in the original parameter value on the value of the output, leading to a numerical sensitivity value:

$$f'(z) = u_1 \tag{20}$$

The sensitivity of a input parameter on the model outcome is considered by $max|u_1|$. The results of sensitivity analysis for the meta-models are presented in Table 5 and 6.

Table 5. Sensitivity analysis for probabilistic meta-models estimating accidental oil spill in case of ship-ship collision.

Variable	VLCC	Aframax	Qualitative ranking
Loading conditions	0.724	0.708	High
Impact location	0.373	0.469	High
Striking speed	0.036	0.038	High
Impact angle	0.022	0.022	High
Bow half entrance angle	0.014	0.057	Moderate
Struck speed	0.006	0.009	Low
Striking ship mass	0.002	0.002	Low

Table 6. Sensitivity analysis for probabilistic meta-models estimating accidental oil spill in case of grounding accident.

Variable	VLCC	Aframax	Qualitative ranking
Ship speed	0.82	0.618	High
Obstruction depth	0.282	0.556	High
Rock eccentricity	0.184	0.098	Moderate
Obstruction tip radius	0.092	0.1	Low
Obstruction apex angle	0.007	0.011	Low

For the meta-model addressing oil spill resulting from a collision, there are two dominating variables, towards which the model is very sensitive. These are *loading conditions and impact location*. It is not surprising that the former variable is the top one, however, an analyst is usually quite confident about it. If the model is to be used as a tool for planning oil spill response capacity for a given sea area, it is rather well known what is the share of tankers sailing under ballast and complementary share of those which are loaded. Despite its high influence on the model, this variable can be estimated with low uncertainty.

However, when it comes to variable *impact location*, it is more problematic, since we simply do not know which part of the tanker will be hit by another ship. Whether it is going to be the cargo space, engine room or forecastle, since the damage to forecastle and cargo space results in an oil spill, the damage to engine room does not. Since, there is no reliable material to be based on, therefore, any estimate of this variable will be associated with relatively high uncertainty.

The other variables affecting the outcome of the model, however to smaller extent, are: *striking speed* and *impact angle*. In fact, none of these variables can be precisely estimated. There were several studies on impact scenarios and several models were proposed, mostly based on statistics, trying to link the actual speed and courses with the impact speed and angle. For an extensive analysis of these models see (Goerlandt, Ståhlberg, & Kujala, 2012; Ståhlberg, Goerlandt, Ehlers, & Kujala, 2013). However, the main conclusion from the aforementioned studies, is that the uncertainty related to the estimates of these parameters is high.

Bow *half entrance angle* is somewhat affecting the model, however this parameter could be estimated, quite accurately based on the ship types navigating in the area of interest. Therefore, the uncertainty associated with it is small.

The other two parameters of the models, namely *struck speed* and *striking ship mass* are of low importance for the model presented, therefore do not need any accurate estimates.

Grounding meta-models, are highly sensitive to two parameters: *ship's speed* and *obstruction depth*, see Table 5. Since, these two can be reasonably well estimated, the accuracy of the model does not suffer. However, the next parameter on the list, called rock eccentricity is highly uncertain. The remaining two parameters do not have any profound effect on the outcome, therefore any precise estimates are not required.

4.4 *The value of information analysis*

The value-of-information analysis identifies the most informative variables, with respect to the output variable. For this purpose the concept of Shannon entropy - H(X) - is utilized, as follows, (Kjræulff & Madsen, 2012):

$$H(X) = -\sum_{i=1}^{n} p(x_i) \ln p(x_i) \qquad (21)$$

where X is a random variable with n states $\{x_1, ..., x_n\}$, and $p(x_i)$ is the probability of outcome x_i. However, in a model, where the response variable (X) is conditionally dependent on a number of explanatory (parental) variables (Y), a measure of the uncertainty of X given an observation of Y needs to be estimated. This is done by applying the conditional entropy $H(X|Y)$, following the formulae:

$$H(X|Y) = \sum_{i,j} p(x_i, y_i) \ln \frac{p(y_i)}{p(x_i, y_i)} \qquad (22)$$

where $p(x_i, y_j)$ is the probability that $X=x_i$ and $Y=y_j$. Conditional entropy is calculated for each pair of variables ($X|Y$) that exists in a model. It determines the variables among which the probability mass of the uncertainty associated with the output is scattered. This analysis can be seen as a tool for screening a model for variables, which could most effectively reduce the model uncertainty by analyzing the potential usefulness of additional information, before the information source is consulted.

The results of the value-of-information analysis with respect to the output of the models developed in this paper are gathered in tables 7 and 8.

In order to reduce the uncertainty in the outcome (oil spill size) of the meta-models, the following variables shall be accurately estimated: *ship speed* and *obstruction depth* for grounding model and *loading conditions* and *impact location* for the collision model. Moreover, to all these variables the meta-models are sensitive to. This implies, that an analyst should be caution with the use of the models, and gather as accurate information regarding these variables as possible. In case of grounding, this should not be difficult. Since the information about ship speed is available from the AIS, whereas the obstruction depth can be taken from the sea chart for an analyzed sea area. One solution to decrease the uncertainty associated with *rock eccentricity* is to conduct a survey of available grounding reports to find the patterns of grounding damages. If these can be found, the distribution of *rock eccentricity* could be updated.

For collision meta-model, the issues of sensitivity and uncertainty seem more complex, since one of the relevant variable, impact location, can't be improved regarding its uncertainty. One solution to this problem, is to take a conservative assumption of impact location always falling within the cargo space. Another relevant variable loading conditions, as elaborated earlier, can be estimated quite accurately; therefore its effect on the total uncertainty of the outcome can be minimized.

Table 7. Results of the value-of-information analysis – grounding meta-model.

Variable	Aframax H(X)	Aframax H(X|Y)	VLCC H(X)	VLCC H(X|Y)
Oil spill	1.41		1.51	
Ship speed		0.32		0.47
Rock eccentricity		0.07		0.09
Obstruction depth		0.25		0.03

Table 8. Results of the value-of-information analysis – collision meta-model.

Variable	Aframax H(X)	Aframax H(X|Y)	VLCC H(X)	VLCC H(X|Y)
Oil spill	1.65		1.51	
Loading cond.		0.48		0.46
Impact location		0.47		0.49

5 BENCHMARKING

In this section, we present the results of a benchmark, which can be seen as an evaluation of the level of agreement of the meta-models with the real life cases. For this purpose, we selected three accidents involving Aframax and VLCC tankers, where the relevant input parameters and outputs are known. Two of them are collisions: Aframax m/t Nassia with m/s Shipbroker (cargo ship), and VLCC m/t Maersk Navigator with m/t Sanko Honour, one is a grounding accident of VLCC m/t Showa Maru in Malacca Strait. The accident data are obtained from the extensive tankers casualty database; see (Devanney, 2015). The basic information about the analyzed accidents is given in Tables 9 and 10.

Based on this information, we define the impact scenarios and set up our a'priori probabilities for the input variables in the meta-models. Subsequently by running the models, we obtain the posterior probabilities for the size of an oil spill given the impact scenario. Since the meta-models are of probabilistic nature, we obtain the most probable distribution of oil spill size given an impact scenario, rather than a single value. Finally, these distributions are compared with the actual size of the oil spills as reported by (Devanney, 2015).

Table 9. Available data on two collision accidents used for benchmarking the meta-models, (Devanney, 2015).

Type of struck tanker	Aframax 132517 DWT	VLCC 233000 DWT
DWT of striking ship	25400	96545
Striking speed	Unknown	Unknown
Impact angle	Unknown	Abt 90 deg
Struck speed	Unknown	Unknown
Impact location	Forward part of cargo section	Midship
Loading conditions	Loaded	Loaded
Bow half-entrance angle of striking ship	Flat bow	Flat bow
Oil spill size [tons]	23500	29400

Table 10. Available data on grounding accident used for benchmarking the meta-models, (Devanney, 2015).

Type of tanker	VLCC 237695 DWT
Ship speed [kn]	11.5 kn
Obstruction depth [m]	3-4 meters less than ship's draft
Obstruction tip radius [m]	2-5
Rock eccentricity	In the centerline - 1
Obstruction apex angle	Unknown, but anticipated to be sharp (less than 60 deg)
Oil spill size [tons]	4500

The a'priori distributions of input parameters and resulting distributions of oil spill size are depicted in Figures 9-11.

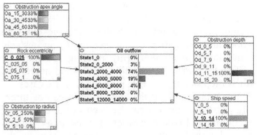

Figure 9. An impact scenario for VLCC tanker running aground, as defined in Table 10, and the modeled oil spill size. The actual oil spill size was 4500 tons.

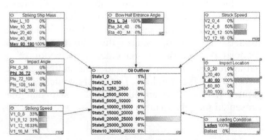

Figure 10. An impact scenario for VLCC tanker colliding, as defined in Table 9, and the modeled oil spill size. The actual oil spill size was 29400 tons.

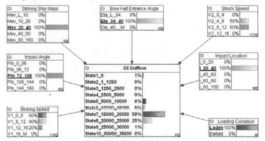

Figure 11. An impact scenario for Aframax tanker colliding, as defined in Table 9, and the modeled oil spill size. The actual oil spill size was 23500 tons.

239

6 CONCLUSIONS

This paper introduces two probabilistic meta-models evaluating accidental oil spill volume for two types of tanker: VLCC and Aframax and two types of accident: collision and grounding.

The meta-models are organized on the basis of two engineering models, one for ship damage estimation due to an accident, another for evaluation of an oil spill size given the damage extent. To organize the available background knowledge into a dataset, which is further used to develop meta-models accounting for the existing uncertainties we adopted MC simulations and causal probabilistic modeling techniques, aka Bayesian Networks.

The potential areas for the improvement have been defined, which are associated with the better definition of the impact scenario for the collision model and estimation of the location of an obstacle with relation to a tanker, while running aground.

The benchmarking results show good agreement of the meta-models results with the real life cases.

The application of the meta-models can be three-fold: first, the models can be used in the process of environmental risk assessment for sea areas. Second it can be applicable in any studies focusing on the assessment of the required oil combating capacity. Third, it can be used in the procedures of risk-based ship design, where information about oil spill size is required e.g. for evaluating the cost-effectiveness.

Nevertheless, the caution is advised when the models are used, since the modeling space associated with the post-accidental behavior of a damaged ship due to wave and wind actions is not covered by the models. The post-accidental hull failure will lead to an oil spill of a size that the presented models do not anticipate.

ACKNOWLEGMENT

This work presented here is carried out within the EU-funded FAROS project (Grant no. 314817). The financial support is acknowledged.

The presented Bayesian Network models have been developed using GeNie modeling environment developed at the Decision Systems Laboratory, University of Pittsburgh, available from http://genie.sis.pitt.edu.

REFERENCES

Aven, T. (2008). Risk analysis: assessing uncertainties beyond expected values and probabilities. Chichester, UK: Wiley.

Coupé, V. M. H., & van der Gaag, L. C. (2002). Properties of sensitivity analysis of Bayesian Belief networks. Annals of Mathematical and Artificial Intelligence, 36, 323–356.

Det Norske Veritas. (2013). Rules for classification of ships. Høvik.

Devanney, J. (2015). Center for tankship excellence. Retrieved from www.c4tx.org.

Eide, M. S., Endresen, O., Brett, P. O., Ervik, J. L., & Røang, K. (2007). Intelligent ship traffic monitoring for oil spill prevention: risk based decision support building on AIS. Marine Pollution Bulletin, 54(2), 145–8.

Fenton, N., & Neil, M. (2012). Risk assessment and decision analysis with Bayesian networks. CRC Press.

Friis-Hansen, P., & Ditlevsen, O. (2003). Nature preservation acceptance model applied to tanker oil spill simulations. Structural Safety, 25(1), 1–34.

Goerlandt, F., & Montewka, J. (2014). A probabilistic model for accidental cargo oil outflow from product tankers in a ship-ship collision. Marine Pollution Bulletin, 79(1-2), 130–144.

Goerlandt, F., Ståhlberg, K., & Kujala, P. (2012). Influence of impact scenario models on collision risk analysis. Ocean Engineering, 47(-), 74–87.

Gucma, L., & Przywarty, M. (2008). The Model of Oil Spills Due to Ships Collisions in Southern Baltic Area. TransNav, International Journal on Marine Navigation and Safety Of Sea Transportation, 2(4), 415–419.

Helle, I., Lecklin, T., Jolma, A., & Kuikka, S. (2011). Modeling the effectiveness of oil combating from an ecological perspective--a Bayesian network for the Gulf of Finland; the Baltic Sea. Journal of Hazardous Materials, 185(1), 182–92.

IMO. (2003). Revised interim guidelines for the approval of alternative methods of design and construction of oil tankers. Regulation 13F(5) of Annex I of MARPOL 73/78 Resolution MEPC.110(49). International Maritime Organization.

Kjærulff, U. B., & Madsen, A. L. (2012). Bayesian Networks and Influence Diagrams: A Guide to Construction and Analysis: A Guide to Construction and Analysis (Google eBook) (Vol. 2012, p. 399). Springer.

Koller, D., & Friedman, N. (2009). Probabilistic graphical models: principles and techniques (1st ed.). The MIT Press.

Krata, P., Jachowski, J., & Montewka, J. (2012). Modeling of Accidental Bunker Oil Spills as a Result of Ship's Bunker Tanks Rupture – a Case Study. TransNav, International Journal on Marine Navigation and Safety Od Sea Transportation, 6(4), 495–500.

Lehikoinen, A., Luoma, E., Mäntyniemi, S., & Kuikka, S. (2013). Optimizing the recovery efficiency of Finnish oil combating vessels in the Gulf of Finland using Bayesian Networks. Environmental Science & Technology, 47(4), 1792–9.

Montewka, J., Ståhlberg, K., Seppala, T., & Kujala, P. (2010). Elements of risk analysis for collision of oil tankers. In Risk, Reliability and Safety (pp. 1005–1013). London: Taylor & Francis Group.

Papanikolaou, A. (Ed.). (2009). Risk-Based Ship Design: Methods, Tools and Applications (p. 376). Springer.

Sergejeva, M., Laarnearu, J., & Tabri, K. (2013). Hydraulic modelling of submerged oil spill including tanker hydrostatic overpressure. In C. Guedes Soares & J. Romanoff (Eds.), Analysis and Design of Marine Structures (pp. 209–217). LONDON, UK: CRC Press, Taylor&Francis Group,.

Smailys, V., & Česnauskis, M. (2006). Estimation of expected cargo oil outflow from tanker involved in casualty. Transport, 21(4), 293–300.

Smolarek, L., & Mazurek, J. (2013). Oil Spill Models: A State of the Art of the Grid Map as a Function of Wind, Current and Oil Parameters. TransNav, International Journal on Marine Navigation and Safety Od Sea Transportation, 7(1), 19–23.

Ståhlberg, K., Goerlandt, F., Ehlers, S., & Kujala, P. (2013). Impact scenario models for probabilistic risk-based design for ship--ship collision. Marine Structures, 33, 238–264.

Tavakoli, M. T., Amdahl, J., & Leira, B. (2010). Analytical and numerical modelling of oil spill from a side damaged tank. In Proceedings of the Fifth International Conference on Collision and Grounding, ICCGS. Espoo, Finland: Aalto University.

Van de Wiel, G. (2008). A Probabilistic Model for Oil Spill Volume in Tanker Collisions and Groundings. Delft University of Technology.

Van de Wiel, G., & van Dorp, J. R. (2011). An oil outflow model for tanker collisions and groundings. Annals of Operations Research, 187(1), 279–304.

Van Dorp, J. R., & Merrick, J. R. W. (2011). On a risk management analysis of oil spill risk using maritime transportation system simulation. Annals of Operations Research, 187(1), 249–277.

Negative Impact of Cruise Tourism Development on Local Community and the Environment

J. Kizielewicz
Gdynia Maritime University, Gdynia, Poland

T. Luković
University of Dubrovnik, Dubrovnik, Croatia

ABSTRACT: Cruise ship tourism is one of the fastest growing segments of the tourism industry. Cruise Line International Association estimated that 6.4 million European residents booked cruises in 2013. Some coastal cities look at the famous marine tourist destinations, which are home ports or ports of call for huge cruise vessels, with a great envy seeing in them a chance for social and economic development and creation of new jobs. However, next to the positive aspects of development of cruise ship tourism, marine tourist destinations faced with a number of problems related to mass tourism. Congestion and littering of the cities increase in the price of services and products, noise, degradation of natural and cultural environmental, discomfort for holiday tourists and lots of others are the most important adverse effects of development of cruise ship tourism. These and other problems cause that local authorities and tourist entities in famous marine destinations, wonder to work out the effective tourism policy, in order to develop tourism in accordance with the principles of sustainable development. The main purpose of the article is an attempt to identify the negative aspects of development of cruise ship tourism.

1 INTRODUCTION

In many tourist regions in the world, tourism is a major sector of the economy and it has a significant impact on GDP of these regions and countries and creation of new jobs. Local authorities, tourist organizations and hotel and travel entrepreneurs solicit about clients through serious engagement in promotional activities to encourage hundreds of thousands of people to visit their cities and regions. The uncontrolled influx of thousands of tourists can lead to environmental and cultural degradation of tourist centers and increase in dissatisfaction among the local communities. Unfortunately, the activity of above mentioned entities, for the sustainable tourism development is not so impressive. The reason for this is, on one hand, the lack of effective methods and tools and on the other hand, the lack of determination and conflict of interest. Development of mass tourism generate significant benefits both to local budgets, but also to hotel and travel enterprises, port authorities and many other entities, who do not want to give up those revenues.

Cruise Line International Association CLIA reports that "cruise industry has the most significant growth in the whole tourist market. Since 1980, average annual average growth in passenger numbers in the activities of cruising was 7.6%". (CLIA 2014) In addition, cruise travels have become tourist products which are widely available, nowadays, what is the result of a fierce competitive struggle about clients undertaken by the cruise ship-owners. The prices of an average cruise travel became available to broad masses of tourists, what also have a significant impact on the spread of this form of tourism and development of mass tourism at sea.

The main objective of the research in this article is an attempt to identify the negative aspects of the development of cruise ship tourism. Moreover, there was formulated a hypothesis that uncontrolled development of cruise ship tourism can cause a number of negative effects for a local community of marine tourist destination. In order to verify the research hypothesis, two marine tourist destinations are presented, i.e.: the city of Dubrovnik in Croatia and the city of Gdansk in Poland.

In the article, there were applied two research methods i.e. an exploration method and "desk research" method. There were also showed a few data from the reports issued by international cruise ship organizations and sea port authorities.

2 THE PHENOMENON OF MASS TOURISM

„Mass tourism can be defined as a form of tourism in which large numbers of tourists (the masses) travel to similar places at similar times". (Lück ed. 2007).

Mass tourism is treated as the opposite to individual tourism which is a more sophisticated form of tourism, identified as a form of qualified tourism.

The emergence of mass tourism is caused by several factors. First of all, the increase of the societies' wages what made collecting money for realization of the higher level of needs that is, for example relaxation and entertainment. Secondly, the development of technology, particularly in passenger transport and the emergence of the airlines in the 1960s caused quick and cheap travel to different continents for millions of people. To satisfy tourists' needs tour-operators began to prepare tourist packages and decrease the prices of tourist offers. Thirdly, there is observed the fashion for recreation and intensive promotional campaigns. In the case of cruise ship tourism, an important role is attributed to the film entitled "The Love Boat" which "was a hit ABC series that aired from 1977-1986 set aboard the cruise ship Pacific Princess. The show focused on the comedic tales of the crew and a revolving set of passengers" (Fox News 2015). Thanks to this production, cruise travels have become dreams of lots of people around the world. „Pleasure travel by ship, once considered the preserve of the wealthy elite, became comparable in price to mass-market resort holidays (Dickinson & Vladimir 1997).

Cruising is a mixture of sea transport, travel, tourism and hospitality industries (Pallis & Lekakou 2004) and is defined as a holiday activity of passengers who pay for itineraries (and potentially other services onboard) of at least one night stays onboard with a capacity of at least 100 passengers. It is common that, for statistical purposes, cruising is defined as the round trip which represents a tourist travel for several days according to a specified itinerary (elaborated plan of a journey) for cruises. (Luković et. al. 2015)

According to the Florida-Caribbean Cruise Association FCCA, globally 21.7 million passengers decided to purchase a cruise in 2014, "with 11.9 million sourced from North America and 9.8 comprised of international passengers. This represents an approximate 2% increase over 2013's figures and is commensurate with the added capacity". (FCCA 2014)

In order to meet the growing demand for cruise travels „cruise ship-owners, order larger and larger cruise vessels for more than six thousand passengers and two thousand crew" (Kizielewicz 2014), such as: Oasis of the Seas for 6630 passengers and 2160

crew (Royal Caribbean 2015). Moreover, "the current cruise ship order book from 2014-2016 includes 34 new ships (22 ocean-going vessels and 12 riverboats/ coastal vessels) from FCCA and/or CLIA Member Lines, with over 60,000 berths and a capital investment value over $14 billion" (FCCA2014).

The more passenger seats on cruise ships are, the lower price of cruise tickets are, because the fixed costs, associated with the maintenance and operation of cruise ships, are broken down on a larger number of passengers.

Therefore, on the market the leading cruise ship-owners i.e. Carnival Cruise Line, Royal Caribbean Line and Norwegian Cruise Lines, every few years order modern cruise vessels. Wide range of catering services, cultural and entertainment facilities, sports, leisure and beauty treatments are offered for passengers on average cruise ships. "Travel writers have compared them to theme parks (Boorstin 2003). Wal-Mart stores (Hilton 2002) and McDonald's restaurants. (Grossman 1991)". (Weaver 2005)

In view of the above considerations, it can be clearly concluded that cruise ship tourism is considered to be the mass tourism. A few hundred to a few thousand passengers are usually on each cruise vessel and mostly these are people who benefit from tourist packages bought at the travel agencies or directly at cruise line offices. Hundreds of coaches and tourist guides are involved to serve passengers visiting the ports of call during cruise travels and this requires efficient organization and precise logistics from tour operators on land.

3 NEGATIVE ECONOMIC CONSEQUENSES OF CRUISE TOURISM DEVELOPMENT ON TOURIST DESTINATIONS

Sightseeing of the marine cruise destinations is a main motive of most cruise travels. Cruise vessels call at ports from where cruise passengers are taken up by coaches or shuttle buses to the most attractive corners of a region. Sometimes, passengers decide to organize their trips themselves and they rent bicycles or sailing ships or even go for walk. It is worth to know that there are some regions where even a dozen or more cruise ships are handled at the same time.

"Passengers and crew during 2013 were 114.87 million visits to the mainland to generate 52.31 billion of direct spending in the destination and source markets in the world. This also includes costs for goods and services to the shipping companies for the purpose of cruises". (CLIA 2014)

In addition to the significant economic benefits generated by development of cruise ship tourism, the huge mass of tourists disembarking cruise ships can be a reason of significant logistic problems for the tourist sector, catering services and local transport. In addition, too many tourists visiting tourist desstinations interference seriously in the functioning of the city and its community. It may seem that a tourist sector, catering services and shopping centres should be satisfied with such marketing issues because they have guaranted a huge demand for products and services which they offer. In the long term, it can have a negative impact. Nowadays, in the era of information technology, access to the Internet and via the media, exchange of exprience through Facebook, Twitter or Tripadvisor, cruise passengers may create a bad echo in the world and discourage potential future tourists.

Sustainable development of tourism in coastal regions is not an easy task. These are the areas which are characterised by strong seasonality and all entities, who are engaged in tourist services, are interested in generating the highest revenues in order to survive the remaining part of the year. On one hand, they are interested in increasing the number of incoming tourists. Local authorities promote tourist attractions of their regions at the international tourist fairs. Tour-operator fight for attention of cruise lines and make deals for supporting their ships at ports of call. But, the others are very enthusiastic for the reason that there is a great demand for their services and they have guaranted a full occupancy in restaurants, bars, museums, entertainment centers, etc.

Among the most important negative economic aspects, related to the development of mass cruise tourism, there could be mentioned:
- Strong seasonality in generating revenues by tourist companies, catering units, shopping and cultural-entertainment centres;
- A significant reduction in employment in tourist and catering services after the end of the season;
- Low wages of workers employed seasonally in the tourist and hotel industry;
- An increase of prices of basic consumer goods and services during the tourist season, what is a great problem for local residents;
- Specialization of trade sector on selling of law quality and cheap souvenirs, which are imported from abroad, mainly China and Thailand, rather than the original products of the native arts and crafts;
- Commercialization of culture;
- A huge influence of tourist industry on the economy (local or national) - so-called the tourist monoculture;
- Realization of direct and indirect tourist investments by local authorities instead of municipal investments for local communities.

The development of tourist economy, on one hand, carries a powerful revenues to budgets of local tourist destinations and creates thousands of jobs,

but it should not be forgotten about the whole range of negative economic impacts of uncontrolled development of tourist market, which in the long term, may bring disastrous consequences.

4 NEGATIVE SOCIAL IMPACTS OF DEVELOPMENT OF CRUISE TOURISM ON TOURIST DESTINATIONS

The attitude of local communities to the development of tourism in their locality and/or region usually follows due to a specific process. In the initial phase, when residents do not recognize nuisances related with the influx of thousands of tourists and they are encouraged to cooperate by the tourist sector, they see mainly direct or indirect advantages for themselves. Direct, when they are involved in supporting tourists, and indirec – when they take benefits thanks to the fact that a region is enriched by cash spent by tourists. This stage can be called as the level of euphoria. Meanwhile, in the second stage, when tourists start visiting a tourists destination, local residents tolerate their presence, mainly due to the financial benefits achieved and this level is defined as the level of apathy. Then, there is a moment, when influence of tourists' presence on local communities starts to be noticed and that usually causes the residents' frustration. This stage is the level of irritation. At last, reluctance to visitors and hostile attitude appear in residents' behaviour. This is the level of antagonizm. This last stage, in which an attitude of the local community is so negative and closed that tourists decide to choose other tourist destinations, it is just called the final level. (Cooper et. al. 1998)

The reasons for reactions, mentioned above, there are negative effects of tourists' presence. It is known that most tourists behaves irrationally when they travel, otherwise than in a place of thier own residence. This supports the fact that they are in a different environment, in another country and among unknown people and even under the influence of alcohol. Then, tourists are more likely to enjoy casual sex, abuse drugs and alcohol, make horrible noise and they are rude to residents.

It is worth to point out that overcrowded tourist centers attract thieves, rapists and murderers. Therefore, just in the high season police ststistics show lots of burglaries, theft, robberies, and even rape and crimes. In the face of such negative phenomena, it is hard to be surprised, that local communities often have negative attitudes to development of tourism industry in their place of residence.

Moreover, the influx of cosmopolitan communities to the tourist destinations can cause numerous adverse effects, that may influence upon the residents' behaviour. Weaker individuals, especially children and young people at school, are exposed to the phenomenon of prostitution, drug abuse or robberies.

„Tourist influence the behaviour of the host population by their example. This is an area where tourism development is at a distinct disadvantage when compared with the use of alternative industries as means to economic development". (Cooper et. al. 1998)

5 THE NEGATIVE INFLUENCE OF CRUISE TOURISM ON NATURAL AND CULTURAL ENVIRONMENT

"The environment, whether it is natural or artificial, is the most fundamental ingredient of the tourist product. However, as soon as tourism activity takes place, the environment is inavitably changed and modified either to facilitate tourism or during the tourism process". (Cooper et. al. 1998)

It is obvious that it is not possible to develop tourist industry without any impact on the environment. This impact can be: direct or indirect, but it is always present. The direct effects are felt as soon as tourist traffic in high season becomes huge, through the increase of pollution of air, water, increase of the level of noise and amount of garbage, and even troubles with delivery of energy through increasing of consumption.

Tourist visiting seacoast beaches are responsible for littering and devastating. Tourists using popular forms of scuba diving on coral reefs affect the environment of flora and fauna. Tourists fishing for fish and other marine creatures reduce their population, and for example, visiting unique cultural relics can lead to their devastation by graffitti, littering or destruction.

„Environmental preservation and improvements programmes are now an integral part of many strategies and such considarations are treated with much greater respect" (Cooper et. al. 1998) by the local, regional and state authorities. In order to protect unique natural areas, which have not been degraded, yet or those which are at risk, local authorities seek protection by the creation of national parks, nature reserves, landscape or establishing the areas of quiet and prohibition of fishing. Moreover, activities aiming to preserve the historical monuments through the restrictive regulations concerning protection of cultural heritage have been undertaken, as well. That is all for making restrictions in their construction and operation. It sometimes happens, that in order to protect these monuments and improve the comfort of visiting, the limits in the number of visitors are placed in museums and historic sites in high season. The development of mass tourism always involves a significant loss for environment in the places visited.

Risks relating to ecology in marine environments are becoming extremely high due to: population growth on the coast, unsustainable and large scale fisheries, exponential growth of merchant fleet, pollution from diverse sources and content, climate change, elimination of high-value biodiversity habitats, etc. Cruising as a growing tourism industry produces additional pressures on the marine environment and sustainable development which emphasizes this problem. (Rogers & McLain & Zulo 1998)

"Activities of passenger ships, cruise ships, such as anchoring at the port or sailing, produce numerous harmful emissions that have a wide range of environmental impacts. It is the emission of harmful substances that produce cruisers in the form of harmful materials, vapours, liquids, particles and energy, such as: (Rogers & McLain & Zulo 1998)
– waste (communal, hazardous, floating, Persistent Organic Pollutants)
– gases (Sox, NOx, Volatile Organic Compounds, particles)
– nutrients
– bacteria, viruses and pathogen organisms
– biocides
– hydrocarbons (oil and derivates)
– invasive and alohtone species
– noise
– light".

Table 1. Cruise Ship Report Card Grade Chart 2014

Cruise Line	Sewage Treatment	Air pollution reduction	Water quality compliance	Transparency	2014 Final Grade
Disney	A	B-	A	F	C+
Princess	B-	B	A-	F	C
Holland America	B+	C-	A	F	C
Norwegian	A	D-	A	F	C
Celebrity	A	D	N/A	F	D+
Regent Seven Seas	C+	F	A	F	D
Royal Caribbean	A	F	N/A	F	D
Carnival	F	D	A	F	D
Cunard	A	F	N/A	F	D
Seabourn Cruises	A	F	N/A	F	D
Oceania Cruises	C	F	C+	F	D-
Silversea	F	F	A	F	D-
MSC Cruises	D	F	N/A	F	F
P&O Cruises	D-	F	N/A	F	F
Costa	F	F	N/A	F	F
Crystal	F	F	N/A	F	F

*e.g. A grade means the highest.
Source: *Cruise Ship Report Card Grade Chart 2014*, Friends of the Earth, www.foe.org/cruise-report-card (2014.12.30)

"Friends of the Earth's Cruise Ship Report Card compares the environmental footprint of 16 major cruise lines and 167 cruise ships according to four environmental criteria: Sewage Treatment, Air Pollution Reduction, Water Quality Compliance and Transparency. (Friends of the Earth 2014) (Table 1.).

Pollution caused by cruise ships is very burdensome for tourist destinations. They can only be prevented by the restrictive international rules and active politics for sustainable development carried out both by ports authorities, as well as local governments in the tourist destinations. "Cruise ships generate an astonishing amount of pollution: up to 25,000 gallons of sewage from toilets and 143,000 gallons of sewage from sinks, galleys and showers each day. Coastal environment and marine life are at risk from the threats of bacteria, pathogens and heavy metals generated in these waste streams". (OCEANA 2015).

International Convention for the Prevention of Pollution from Ships (MARPOL) is one of the most important international marine environmental conventions. It was designed to prevent and minimize pollution of the seas, including dumping, oil and air pollution. It was adopted by the International Maritime Organization (IMO) in 1973 and updated in 1978 and therefore is often referred as MARPOL 73/78. (MARPOL 1978)

Uncontrolled and intensive tourism development leads to environmental degradation, destruction of cultural relics, and the deterioration of the image of the tourist destination, as friendly to tourists, which in turn, can lead to decrease the importance of a destination on the tourist market.

6 A CASE STUDY: THE CITY OF DUBROVNIK

Dubrovnik is one of the most important marine tourist destinations in southern Europe. For years, it is at the forefront of ports, that support the greates numbers of cruise passengers anually. For example, in 2013, Dubrownik was visited by 1 136 663 cruise passengers. (CLIA EUROPE 2014) (Table 2.)

The city of Dubrovnik is visited by ships owned by the largest cruise liners such as: Royal Caribbean Cruise Lines, Holland America Line, Celebrity Cruises, and also Norwegian Cruise Line, P&O Cruises and the others. In 2013, 711 cruise ships were handled here. This make Dubrovnik port the third in the Mediterranean and the tenth in the world. (Dubrovnik Cruise Port 2015)

Dubrovnik is situated on the picturesque coastline of the Adriatic Sea. Dubrownik, as a historic city walled with 15 towers, bastions and the fort, that was built in the period from the 9th to the 17th century, is a great tourist attraction

Year	Calls	Passengers
2000	168	126.841
2001	279	205.095
2002	343	264.902
2003	480	395.342
2004	504	457.334
2005	553	510.641
2006	574	603.047
2007	606	667.769
2008	700	850.828
2009	628	845.603
2010	707	916.089
2011	681	985.398
2012	654	950.791
2013	711	1 136 663

Source: Statistics of the Dubrovnik Cruise Port, Dubrovnik Port Authority 2015.

A standard sightseeing tour programme includes visiting the Luža Square with the Sponza Palace, where the collection of manuscripts is presented. There is also an exhibition describing the history of Dubrovnik. Tourists visiting Dubrovnik love walking through the main historical Stradun Street, which is also called Placa Street. Unfortunately, during the tourist season, this street, which is only 300 meters long, more than 15 thousand or even 20 thousand cruise passengers are. (Fig.1) This situation spun out of control and led to the severity of the negative effects of seasonality. Huge crowds of tourists, passing through the town during the day, cause traffic jams, horrible noise and pollution of the city.

Figure 1. Stradun Street in Dubrovnik in summer
Source: T. Luković, 2012.

Table 2. The number of cruise vessels and passengers visiting the port of Dubrovnik in the period from 2000 to 2013

Cruise harbour Dubrovnik

Figure 2. Stradun Street in Dubrovnik in winter
Source: T. Luković, 2012.

As a rule, cruise ships dock in the port of Dubrovnik in early morning hours or before noon for the period from 8 to 12 hours and this time cruise passengers are picked up on a few hours' trips in the city and a region.

In addition, souvenirs for tourists and other goods, which are sold in the shops in the city center of Dubrovnik, are generally in poor quality for inadequate high prices and they are mostly imported from other countries.

There are reasons why local community living in Dubrovnik complain about crowds in the city, communication problems, queues in shops and high prices. In addition, the owners of property in the Old City of Dubrovnik sold most of their flats and houses to private companies dealing with accommodation and catering services. After tourist season, the Old Town become empty and it is really hard to find any facilities for anyone. (Fig.2) Municipal authorities estimates that only about 1.5 thousand inhabitants live in the Old Town, nowadays. (Horwath Consulting Zagreb 2011)

The second group, who is disappointed with this state of things, these are tourists who have chosen Dubrovnik for a few days stays, because they can not find any places in restaurants, pubs, cafeterias. They are also tired with queues in shops and overcrowded streets and awful noise. This, of course, raises their reasonable anger and dissatisfaction. Tourist offer in Dubrovnik is mainly addressed to one-day tourists, what is also poorly rated by other tourists.

The local authorities of the city of Dubrownik have tried to find a suitable solution how to develop tourism industry in accordance with the sustainable rules and how to oversee this type of phenomena in the future. But, so far, all attempts have not been succesful.

In 2011, Horwath Consulting Zagreb prepared the document called "Strategija razvoja turizma Dubrovačko-neretvanske županije." regarding the development of tourism in the city of Dubrovnik. In above mentioned document, interesting findings

from the survey conducted among tourists visiting Dubrovnik were presented. According to the authors of the study, tourists visiting Dubrovnik were very satisfied with staying and visiting of the city. (Horwath Consulting Zagreb 2011). The only question is whether these were exclusively cruise passengers, but also tourists staying more than one day in Dubrovnik.

7 A CASE STUDY: THE CITY OF GDAŃSK

Gdansk, a city with a thousand-year tradition, is one of the most attractive tourist cities in Poland. The City of Gdansk is located in the north of Poland at the Baltic Sea Coast. In the middle ages, the city was an important centre of culture and arts, and thanks to its convenient position on the communication routes, was also a significant trade centre. Nowadays, Gdansk is visited by more than 2 million tourists per year, including a significant group of cruise passengers from ships calling at ports in Gdynia and Gdansk. (Table 3)

First cruise ships appear in Polish sea ports usually in May, and the last leaves the Polish ports at the end of September. Therefore, the tourist season is short and lasts just four or five months of the year. Larger cruise ships dock in Gdynia, which is a an incoming port, and smaller ones – in the port of Gdansk, which is a port of call.

Table 3. The number of cruise vessels and passengers visiting the port of Gdansk and port of Gdynia in the period from 2000 to 2013

| Year | Gdańsk | | Gdynia | |
	Calls	Passengers	Calls	Passengers
2000	14	3642	72	57610
2001	17	3486	74	56460
2002	14	3609	53	26666
2003	7	3367	95	58411
2004	28	7359	82	72977
2005	32	8353	94	88723
2006	29	9703	89	94135
2007	39	12193	87	89088
2008	36	13276	89	123521
2009	40	16753	96	134884
2010	26	8378	85	125005
2011	21	6787	56	78418
2012	29	8294	69	108628
2013	31	10508	57	80528

Source: Own elaboration on the base of: Statistics, Statystyka, Port of Gdansk Authority S.A., Gdansk 2014.;
http://www.portgdansk.pl/zegluga/cruise-to-gdansk [2014.01.14]; Zestawienie statków turystycznych, które zawinęły do Portu Gdynia w latach 2000-2013, Port of Gdynia Authority S.A., Gdynia 2014.
http://www.port.gdynia.pl/pl/wydarzenia/wycieczkowce [2014.01.14]

A few hundred to a few thousand passengers usually leave the cruise ships to visit the coastal cities. There have been years when even three cruise ships with almost 10,000 passengers onboard were handled in the port of Gdynia at the same time. In view of the fact that this phenomenon has a seasonal nature, supporting so large number of passengers is a great challenge for tourist sector. There are some problems with delivering an adequate number of buses with relevant standard, employment of tourist guides with knowledge of foreign languages, and the fact that, cruise ships stay in the ports for the period from 8 to 12 hours, when trips to the most attractive objects in the region must be organised.

Cruise passengers, visiting the Gdansk Region, are particulary interested in sightseeing of the Old Town of Gdansk, the Mediaeval Castle of the Teutonic Knights in Malbork, Hel Pennisula and the most interesting attractions on Kashubian Region. In the season, next to cruise passengers, the greatest tourist attractions of the Gdansk Region, are visited by thousnad of other tourists and one-day visitors from various corners of the World. On one hand, the tourist industry takes care about increasing the number of incoming visitors and tourist organisations promote the tourist valors of the Gdańsk Region on the international tourist fairs in order to encourage potential tourists to come, but on the other hand - the great concentration of demand in a small area in the season causes serious logistical problems.

Figure 3. Długa Street in Gdansk in summer
Source: J. Kizielewicz, 2014.

Figure 4. Długa Street in Gdansk after the tourist season
Source: M. Wieliczko, 2015,
http://www.wieliczko.com.pl/oferta-bank-zdjec/

In recent years, there is observed a trend for organization of various entertainment and cultural events, that, unfortunately, are also organized in high tourist season, which intensifies further the negative effects of mass tourism. For example, in Gdańsk Region during the tourist season in 2014, the Red Bull Air Races that attracted 70 thousand participants and the Open'er Festival with 80 thousand participants were organized.

Of course, the phenomenon of mass tourism in the Gdansk Region, it is not so burdensome, as is in the case of famous tourist destinations in the Mediterranean Sea Region, such as for example the city of Dubrovnik, described above. However, negative effects of concentration of tourist demand at one time and place are noticeable, as well.

8 CONCLUSION

As a result of analysis carried out, the research hypothesis that uncontrolled development of cruise ship tourism may cause a number of negative effects for marine tourist destinations and their communities has been verified positively, because:

- In spite of a large number of positive benefits gained thanks to development of cruise ship tourism, there is a whole range of negative consequences that can touch both a tourist destination and its local community.
- Thousands of cruise passengers visiting the coastal cities in tourist season cause their congestion, littering and an increase of noise intensity, traffic jams and environmental pollution.
- Local traders in coastal cities, in order to generate profits, focus their activities on commercial sales of cheap and poor quality souvenirs and other products, instead of unique hand-made and original local products.
- The great number of tourists in the tourist destinations has got an important influence upon increasing of demand for goods and services. It also causes the increase of prices and long queues in the shopping centers, what arouses huge resentment among the local communities.
- With the increase of a number of tourists in coastal regions, there has been observed the rapid increase in the level of crime and other pathological behaviors with participation of local residents (prostitution, drug addiction, alcoholism)

Not from today, local authorities and tourist industry have struggled with the problems associated with the effects of mass tourism. "The role of the state, regional and local governments are to facilitate the normal development and define the rules for development in order to develop the retained all the characteristics of sustainable development, which special attention should be given" (Kizielewicz & Luković 2013). On one side, a large tourist demand is an objective and, on the other hand, there is a problem how to balance it. Of course, the best method is to extend the tourist season through organization of attractive tourist events out of the season and decreasing prices of hotel and tourist services. In many cases, such solutions were an antidotum for balancing of a tourist demand, however, in Northern European countries, the weather conditions are factors restricting significantly these kind of activities. Strong winds, rain, low temperatures in autumn-winter period, discourage tourists effectively to avoid these areas. Tourism industry must be developed in accordance with the idea of sustainable development, otherwise, in the long term negative effects may be irreversible and lead to reduced attractiveness of tourist destinations.

REFERENCES:

Boorstin, J. 2003. Cruising for a Bruising? Fortune 143–150, June 9.
CLIA Europe, 2014. The cruise industry, Contribution of Cruise Tourism to the Economies of Europe 2014 Edition, 10. Cruise Line International Association, [for:] MedCruise, Cruise Europe and individual port data.
CLIA, 2014. The Global Economic Contribution of Cruise Tourism 2013, Business Research & Economic Advisors, SAD http://www.cruising.org/
Cooper CH., Fletcher J., Gilbert D. & Wanhill S. 1998. Tourism. Principles and Practice, (2nd edition), 147-176, Longman, New York, 31.
Dubrovnik Cruise Port Information, 2015. Dubrovnik Travel Experience, http://www.dubrovnik-travel-experience.com/dubrovnik-cruise-port-information.html
Dubrovnik Port Authority, 2015. Statistics of the Dubrovnik Cruise Port, Dubrovnik.
FCCA, 2014. Cruise Industry Overview – 2014, State of the Cruise Industry, 1, Florida Caribbean Cruise Association, Florida .
Friends of the Earth, 2014. Cruise Ship Report Card Grade Chart 2014, www.foe.org/cruise-report-card
Grossman, L. 1991. Carnival Cruise Discusses Investing in Seabourn Line, The Wall Street Journal (September 18):B7.

Hilton, S. 2002. Carnival Attraction: Overcoming the Fear of Affordable Cruises, Whether it's More about Fun than Finesse. The San Francisco Chronicle (September 15):C1.
Horwath Consulting Zagreb, 2011. Strategija razvoja turizma Dubrovačko-neretvanske županije, 106. Dubrovnik.
http://www.royalcaribbean.com/findacruise/ships/class/ship/home.do?shipClassCode=OA&shipCode=OA&br=R
IMO International Maritime Organization, 1978. Konwnecja MARPOL, International Convention for the Prevention of Pollution from Ships, http://www.imo.org/KnowledgeCentre/ReferencesAndArchives/HistoryofMARPOL/Pages/default.aspx
Kizielewicz, J. & Luković, T., 2013. The Phenomenon of the Marina Development to Support the European Model of Economic Development, (in:) TransNav - The International Journal on Marine Navigation and Safety of Sea Transportation, Volume 7, Number 3, p. 466. September, Gdynia.
Kizielewicz, J. 2014. Cruising w regionie Morza Bałtyckiego - stan i perspektywy rozwoju, [in:] porty morskie i żegluga w systemach transportowych, Dąbrowski J. & Nowosielski, T. ed., InfoGlobMar 2014, 184, Instytut Transportu i Handlu Morskiego, Uniwersytet Gdański, Gdańsk.
Lekakou, M. B., Pallis, A. 2004. Cruising The Mediterranean Sea:Market Structures And Eu Policy Initiatives, Journal of Transport and Shipping, Vol 2.
Lück, M. (ed.), 2007. The Encyclopedia of Tourism and Recreation in Marine Environments, CABI International, 306, Cambridge.
Luković et al., 2015. Nautički turizma Hrvatske, Sveučilište u Dubrovniku.
OCEANA 2015. Cruise Ship Pollution: Overview, Protecting The World Oceans, http://oceana.org/en/eu/home
Port of Gdansk Authority S.A., Gdansk 2014. Statistics, http://www.portgdansk.pl/zegluga/cruise-to-gdansk
Port of Gdynia Authority S.A., 2014. Zestawienie statków turystycznych, które zawinęły do Portu Gdynia w latach 2000-2013, Gdynia.
http://www.port.gdynia.pl/pl/wydarzenia/wycieczkowce
Rogers, C.S., McLain, L., Zulo, E. 1998. Damage to Coral Reefs in Virgin Islands National Park and Biosphere Reserve from Recreational Activities, 405-410 Coral Reefs,. 2.
Then/Now: The cast of 'The Love Boat', http://www.foxnews.com/slideshow/entertainment/2014/05/28/thennow-cast-love-boat/#slide=
Weaver, A. 2005. The McDonaldization thesis and cruise tourism, [in;] Annals of Tourism Research, Vol. 32, No. 2, pp. 347, Elsevier Ltd.
Wieliczko, M. 2015. http://www.wieliczko.com.pl/oferta-bank-zdjec/

Vessel Traffic Service (VTS)

Improving Safety of Navigation by Implementing VTS/VTMIS: Experiences from Montenegro

S. Bauk
University of Montenegro, Maritime Faculty, Kotor, Montenegro

N. Kapidani
Maritime Safety Department of Montenegro, Bar, Montenegro

ABSTRACT: Vessel Traffic Service (VTS) is a proven system for organizing maritime traffic in coastal areas. It enhances safety and efficiency of navigation, and contributes to the protection of marine environment. VTS services are mainly operated on national basis. The related requirements of individual countries and authorities vary widely. The specification activities of administrative and operational bodies with reference to VTS recently led to Vessel Traffic Monitoring and Information Services (VTMIS), which comprises VTS and additional information services that are required for above mentioned purposes. In the first part of the paper are presented some aspects of implementation of VTMIS in Montenegro such as: intuitional and legal frameworks, human resources requirements, and system specification. Also, it is stated the importance of networking of national VTMIS services at regional (Adriatic and Mediterranean) as well as at EU level. In the second part of the paper some challenges in this domain will be touched, like: solid state radar employment, providing VTS/VTMIS compatibility with SafeSeaNet of the latest generation, including Long-Range Identification and Tracking (LRIT) ships system, etc. All these dimensions supporting safe, efficient and effective navigation will be considered with the purpose of ensuring excellence in navigation at the national, regional, and ultimately at the global level.

1 INTRODUCTION

According to IMO, VTS is a service implemented by a competent authority, designed to improve safety and efficiency of vessel traffic and to protect the environment. The service should have the capability to interact with traffic and respond to traffic situations developing in the VTS area [1]. In Montenegro competent authority for VTS is Maritime Safety Department (MSD), which is a part of Ministry of Transport and Maritime Affairs (MTMA).

The main purpose of VTS is to improve the safety and efficiency of navigation, safety of life at sea and the protection of the marine environment and/or the adjacent shore area, worksites and offshore installations from possible adverse effects of maritime traffic [2]. Implementation of a VTS was undertaken as there were concerns about the levels of safety and as a result of reviewing the existing safety measures.

Montenegro's coast length is only 294 km long. The inland sea area of Montenegro is 362 km², the territorial sea area is 2099 km², and the epicontinental shelf area is 3885 km² (Figure 1).

Figure 1. Montenegrin geographical map including its coast (Source: Web)

Almost half of the coast belongs to Boka Bay that is very attractive for cruisers. In recent years number of cruisers arrivals and number of passengers are increasing considerably (Table 1).

Table 1. Data on cruising ships arrivals in Boka Bay, Montenegro (Source: Port of Kotor Authorities)

YEAR	CRUISE ARRIVALS	Aver. LOA	Aver. PAX/VSL	PAX
2009	262	113.25	287	75128
2010	310	136.30	468	145185
2011	317	153.31	598	189426
2012	343	163.31	715	245400
2013	387	168.43	821	317746

Legend: Avg. LOA – average length over all; Avg. PAX/VSL – average number of passengers per vessel; PAX – average number of passengers per year

Boka Bay is also attractive for pleasure crafts and mega yachts and there are plans to build new marinas and to expand existing ones. Having in mind specific configuration of Boka Bay (Figure 2) and the rest of Montenegrin coast, it becomes obvious that the adequate system for monitoring of ships in order to provide safety at sea, pollution prevention, preservation of biological diversity in Boka Bay and wider, should be implemented. This has crucial importance for the sustainable development of Montenegro and the surrounding sea and coastal area.

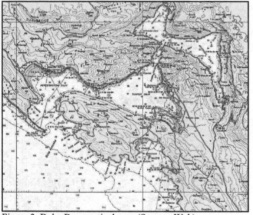

Figure 2. Boka Bay nautical map (Source: Web)

The specific activities of administrative and operational bodies with reference to VTS recently led to Vessel Traffic Monitoring and Information Services (VTMIS), which comprises VTS and additional information services that are required for the above mentioned purposes.

The VTMIS requires a considerable financial investment. It could not be provided through regular budget of maritime authorities in Montenegro. Therefore EU has recognized importance of VTMIS and dedicated 2.8 million of euros through Instrument for Pre-accession Assistance (IPA) funds, for the project of VTMIS establishment and improvement of authorities' response to marine pollution incident(s) - Contract No. IPA 2012/299-349.

However, if it is not effective, the main purposes of the VTMIS will not be achieved. Thus, the local needs and requirements have been carefully studied from the beginning in order to justify the investment. In the preliminary assessment phase, all relevant problems in the maritime area have been defined and carefully analyzed. During preliminary assessment phase many meetings have been held with maritime stakeholders.

2 VTMIS IN MONTENEGRO: BASIC FACTS AND REQUIREMENTS

The institutional frame for implementing VTMIS in Montenegro is organized mainly in three levels of public maritime governance, which can be distinguished as follows:
– the *legislative level* represented by the Parliament;
– the *political level* represented by the Ministry of Transport and Maritime Affairs, Ministry of Interior Affairs, Ministry of Defense, and Ministry of Sustainable Development and Tourism;
– the *implementation level* represented by the Maritime Safety Department, Harbor Master, Port Authorities, Navy-Coast Guard, Maritime Police, and Hydrographical Service.

Maritime safety responsibilities are divided between the MTMA and institutions that are part of it. MSD like an administration department has responsibilities of maritime safety, ship registry, safety of navigation and protection of the marine environment, and the Harbor Master Offices have functions on port traffic, seafarers licensing, port state, ship formalities, and port and ship security.

2.1 Key legislative issues

The national legislative framework, including local laws, rules and recommendations, should be amended in such a way that the relevant requirements of the SOLAS Convention, IMO Resolution A.857(20) and any other applicable international rules, regulations and recommendations are incorporated. The needs for such amendments have to be assessed in the preliminary studies.

Qualified maritime legal experts should be provided by the competent authorities to cooperate with any consultant that they may appoint in describing and analyzing the present framework and in outlining the renewed one.

2.2 Common equipment requirements

In determining the types of services to be provided, certain consideration should be given to the quality of the traffic image, communications capability and other sensor equipment as appropriate. While VHF should be the primary medium of communication, any available device within the maritime mobile service may be used in accordance with the radio regulations.

The equipment performance parameters are strongly dependant upon the services to be provided which influences the mean time between failures and the availability of the service [3]. The information on availability and reliability methods can be found in [4]. In order to ensure adequate availability and reliability, vital parts of the VTS, e.g. all operational VTS communication services, should have back-up systems with a power source independent from the normal power supply. The need for redundant sensors and even an alternative site for the VTS centre should be considered, as well.

3 VTMIS RADAR SYSTEM

In principle VTS radars typically function like ships radars, but in most cases they need to operate simultaneously on short and long range, preferably without the need for operator adjustments [5]. Therefore, their performance requirements are generally different to the requirements for marine navigational radars.

Furthermore, weather related phenomena such as ducting will influence VTS radars more than ships' radars. This can have a significant influence on the performance – either positive or negative.

Good clutter suppression is needed for sea clutter and, in most parts of the world, for rain clutter as well [5]. In addition, the need to observe small targets in rough weather conditions is essential, especially if objectives include detection of targets for SAR and/or security purposes.

Important radar parameters influencing its coverage and characteristics are: frequency band, radar location, antenna height, local weather, transmitted power, antenna features, receiver sensitivity, receiver dynamic features, processing capabilities, system losses, etc.

The parameters are often dependent on each other, and different vendors may choose different methods to solve the same issue. Therefore, it is recommended that overall radar performance requirements are specified, taking the local weather into consideration, rather than specifying radar parameters.

For determination of radar coverage and range performance a combination of site inspections and radar system performance calculations, has been conducted by experts with a sound operational and technical knowledge about the subject [5].

Calculations of performance have been focused on the smallest targets of interest in bad weather

conditions. In the case of Montenegro VTMIS radars will be used also for SAR purposes where detection of small non-metallic targets, such as fiber glass, wooden, rubber or rigid hull inflatable boats, are identified by the MSD as a target of concern. This fact has determined advanced level of capability of VTS X-band radar system and selection of horizontal antenna polarization.

The introduction of AIS further develops VTS into a modern information system, and the presentation of radar information needs to follow the trend, putting new demands on the radar performance. Antenna side lobes and ghost targets (multiple reflections) may lead to false and dangerous results when radar returns and AIS plots are associated. High precision is therefore required to allow for unambiguous correlation of position obtained from two information sources. The VTS authority should define the requirements for the availability of the radar service. Recommended availability for the radar service is 99.9%.

3.1 Solid state radars

Previously employed magnetron radars have been associated with kilowatts or even megawatts of transmitted peak power and required periodic replacement of magnetrons that is not cost effective in terms of maintenance. High voltages circuitry in magnetron radars could cause flash-over that could influence the overall availability of the radar service.

Solid state radars have ability to obtain very high quality radar images with very low power levels with no replacement of tubes and no high voltages. Careful design with complex circuitry with high pulsed currents and efficient temperature management ensure high reliability and low cost, long term maintenance as in many cases only air filter cleaning is required on a regular (yearly) basis.

3.2 Solid state radars for VTMIS

Coherent solid state radar technology has been available for decades, but it was not present at harbor surveillance, VTS and related applications for cost and technical reasons. Technically, the main challenge is that dynamic requirements to radar in littoral waters and build up regions are much higher in comparison to other radar applications. Those challenges have now been met and combined with well-renowned advantages from the Terma Scanter product range [6]. Methods are further refined and implemented on a new technology platform. The result is a software-defined radar series, tailored to individual market segments, virtually unrestricted by dynamic constraints. The digital radar concept with software-defined functionality makes the set-up of the radar easy. Furthermore, interference rejection against disturbance from radars on ships passing nearby the radar has also proven effective, and the dynamic range has proved to be sufficient to eliminate any artefacts from a high number of large buildings and other structures in an operational area. Operational tests have been performed with impressive results.

In a world with asymmetrical threats, detection of small targets is the overall scope of surveillance radar. This detection capability is required in extreme environments and in heavy weather conditions. Hence, there is the need for the adaptation to the surroundings arises. In urban areas with an increasing radio infrastructure, interference from and against other systems is an important parameter, while reliability and required maintenance are important in rural areas with remote sites. At sea, sub clutter visibility and adaptation to a moving platform are important parameters [7].

The SCANTER 5000 series by Terma comprise a new generation of fully coherent, frequency diversity and time diversity, solid state radars with software defined functionality for professional applications such as VTS and coastal surveillance (Figure 3). Extremely high resolution with small range cell size and high pulse compression factors has been achieved utilizing 32-bit floating point calculations throughout the signal processing chain, providing lossless processing virtually unrestricted by dynamics. Frequency diversity and time diversity are standard features for the new series of radars [8].

A complete radar system is illustrated in Figure 3.

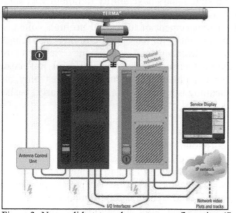

Figure 3. New solid state radar system configuration (Source: [8])

The transceiver acts as the central system component configured with plug-in modules. Peripheral units are added as required by the individual application.

Communication as well as signal and track distributions are preferably provided on single or redundant IP network. Additional serial communication lines are available for easy integration into new or existing systems. The video

outputs are available in analogue, digital and IP network formats.

The new radar series has been designed to comply with IALA V-128 requirements and recommendations. Performance exceeds the advanced examples in IALA V-128 by arbitrary 20% for radar in the most powerful 200 watts configuration. This is of course subject to physical constraints such as atmospheric propagation and the curvature of the earth [8].

Very important feature is that that the transmitter power level can be controlled in sectors as illustrated in Figure 4. In combination with the ability to select sub-bands for transmission, it will also increase robustness against interference between radar stations. The power is adapted to that needed in sectors, reducing undesirable illumination of e.g. populated areas, eventually reducing the cost of spectrum pricing and further enhancing lifetime of the Solid State Power Amplifier (SSPA).

The receiver will automatically adapt to varying power levels, giving undisturbed images to the radar operators [9].

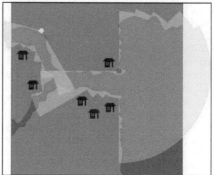

Figure 4. Sector power transmission (Source: [9])

4 TOWARD ADAPTING LRIT, SAFESEANET AND SINGLE WINDOW IN MONTENEGRO

Last year, Maritime Safety Department (MSD) of Montenegro participated for the first time in the experts working groups meetings for CleanSeaNet, LRIT, SafeSeaNet and National Single Window organized by European Maritime Safety Agency (EMSA) and European Commission. Attendance at these meetings provided an opportunity to agree with the representatives of EMSA the next steps regarding the implementation of the CleanSeaNet and EU LRIT system in Montenegro and to agree terms for the trainings which EMSA representatives will hold in Montenegro. Although the membership in some of these systems is exclusively intended for the EU Member States, Norway and Iceland, MSD is allowed to attend these meetings as an observer, which is the result of the recognition of the efforts of

MSD and MTMA in adapting EU standards in the fields of maritime safety and prevention of sea pollution from ships. This is also the result of intensive cooperation and participation of MSD and MTMA in numerous international projects which are financed by the EU. These meetings were also an opportunity to exchange experiences about the system for monitoring maritime traffic with European partners, which is of great significance bearing in mind that MSD is currently implementing VTMIS.

4.1 On VTMIS implementation in Montenegro

Recently, it has started the installation of the VTMIS equipment for monitoring and managing maritime traffic in Montenegrin littoral zone. Italian company "Elman" will complete setting three sensory stations along the coast (at Obosnik, Crni Rt, and Mavrijan). In Figure 5 is shown station at Obosnik.

Figure 5. VTMIS station at Obosnik in Montenegro (Source: Web)

Within the project of VTMIS implementation in Montenegro, the following equipment has been installed: contemporary solid state radar sensor station, AIS base station, VHF simplex transmitters, VHF duplex transmitters, meteorological and hydrographical stations, radio direction finder, etc. This should contribute considerably to the maritime traffic safety along Montenegrin part of the Adriatic Sea, belonging coastal area and surrounding sea and coastal areas.

4.2 Montenegro accessing EU LRIT Data Center

The European Commission gave approval to Montenegro for participating into the CleanSeaNet and EU LRIT Data Centre. So far, this system has been designed exclusively for members of the EMSA (EU countries, Norway and Iceland) and Montenegro is the first "third country", outside the EU, which will be connected to this system, since Montenegro fulfilled all the necessary conditions to

participate in it, due to the assessment of the European Commission and EMSA. These days MTMA and EMSA shall sign a contract on the use of LRIT system by Montenegrin side. MSD is in charge of the implementation of the system in Montenegro, including its use. Thanks to this, Montenegro will join another European system for maritime safety and thus achieve substantial savings in comparison to the previously established national LRIT Data Centre. In LRIT a ship in transit sends a position via its ship borne equipment and Iridium or Inmarsat (C and D$^+$) satellite communication networks. The messages include the ship borne equipment identifier, positional data as latitude and longitude, and the date and time of transmission. Minimum of four position messages are to be sent per day [10].

4.3 Providing access to SafeSeaNet

Montenegrin MTMA and MSD will continue their endeavors toward providing access to SafeSeaNet intended for vessel traffic monitoring in order to enhance:
– Maritime safety;
– Port and maritime security;
– Maritime environment protection;
– Efficiency of maritime traffic and transport.

When accessing SafeSeaNet most users see the web interface. It is map-based graphical interface (GIS), and it is this that makes the system easy to understand and operate, and that makes it possible to users to quickly obtain what they need [11].

4.4 Providing access to National Single Window prototype

Montenegrin responsible bodies, primarily MTMA and MSD will do efforts to provide the access to the common National Single Window (NSW) prototype in order to become familiar with this system. NSW primarily addresses the need for efficient and collaborative electronic transactions between different maritime governmental and business entities. It is an environment which has to support interoperability among highly heterogeneous environments including administrative bodies in ports and in wider administrative context, ships, and cargo issues on local, national, regional, and international levels. It is merging in optimal manner maritime and trade administrative procedures at different operational and national levels.

Therefore, it is important to identify a proper methodology and taxonomy for developing NSW system which will crucially support a seamless integration and functionality of above listed heterogeneous systems into unique environment like NSW [12-17]. In order to be in line with EU developments regarding implementation of NSW and fulfilling requirements from reporting formalities directive [18-21], MSD had taken initiative to join as observer to ANNA project. ANNA project is an EU Member States driven project (www.annamsw.eu), in close co-operation with the European Commission, to support the effective implementation of the EC Directive 2010/65/EU (i.e. Reporting Formalities for Ships arriving in/departing from EU ports) [22].

5 CONSLUSIONS

It might be concluded that implementation of VTMIS on national level is not enough at all. Further steps have to be taken in order to interconnect national VTMIS services in regional and EU levels.

There are some initiatives to improve cooperation in execution of various maritime functionalities in the Adriatic-Ionian sub-region. One of proposal is to design and develop a new system called Adriatic Regional Server (ARS). This server will in the beginning serve for exchanging of AIS data between Adriatic-Ionian countries (Italy, Croatia, Slovenia, Bosnia and Herzegovina, Montenegro and Greece) and later on it will be improved in order to allow exchange of VTMIS data. Based on up-to-date technology, this is intended to provide an overall picture of targets of interests for the whole Adriatic-Ionian sub-region.

The design and development of ARS could be supported by funding provided within the framework of the Commission Implementation Decision (of 2012) concerning the adoption of the Integrated Maritime Policy Work Program: "Test projects on cooperation in execution of various maritime functionalities at sub-regional or sea-basin level".

Montenegro national VTMIS system is specified in the manner to support and to be compatible with existing regional and EU systems like SafeSeaNet and ADRIREP mandatory ship reporting system [23]. The Montenegro VTMIS system shall also comply with IALA Recommendation V-145 [24] on the Inter-VTS Exchange Format (IVEF) that will be the standard for exchanging VTMIS data in the future.

REFERENCES

[1] IMO Guideline for Vessel Traffic Services Resolution A.857(20), 1997.
[2] IALA Vessel Traffic Services Manual. Ed. 5, International Association of Marine Aids to Navigation and Lighthouse Authorities, Saint Germain en Laye, 2012.
[3] IALA Recommendation V-119 – Implementation of Vessel Traffic Services September, 2000 – Revised December, 2009.
[4] IALA Guidline 1035 - Availability and Reliability of Aids to Navigation, 1989 - Revised December, 2004.
[5] IALA Recommendation V-128 June 2007, On Operational and Technical Performance Requirement for VTS Equipment – Ed. 3, 2007.
[6] Pedersen J.C., SCANTER 5000 and 6000 Solid State Radar: Utilisation of the SCANTER 5000 and 6000 series next generation solid state, coherent, frequency diversity and time diversity radar with software defined functionality for security applications, Waterside Security Conference (WSS), 2010, pp. 1-8.
[7] Lokke M., Small target detection with SCANTER 5000 & 6000 radar series, Radar Symposium (IRS), 2011, pp. 403-408.
[8] Pedersen J.C., A Next Generation Solid State, Fully Coherent, Frequency Diversity and Time Diversity Radar with Software Defined Functionality, 17th IALA Conference - Session: e-NAVIGATION AND EMERGING TECHNOLOGIES, 2010, pp 74-85.
[9] Pedersen J.C., First customer installation and site trials with the new SCANTER 5000 Series of radars from Terma A/S, URL:http://www.terma.com/media/155660/pla_white-paper.pdf, last access on 31st December 2014.
[10] EMSA, LRIT Cooperative Data Centre – How it works, URL: http://emsa.europa.eu/lrit-home.html, last access on 31st December 2014.
[11] EMSA, SafeSeaNet, URL: http://emsa.europa.eu/ssn-main.html, last access on 31st December 2014.
[12] EMSA, Operational Projects – National Single Window (NSW) prototype, URL: http://emsa.europa.eu/2014-07-02-10-35-18/nsw.html, last access on 31st December 2014.
[13] Fjortoft K.E., Hagaseth M., et al., Maritime Single Windows: Issues and Prospects, International Journal on Marine Navigation and Safety of Sea Transportation, Vol. 5, No. 3, September 2011, pp. 401-406.
[14] Kapidani N., Bauk S., Implementation of VTS/VTMIS in Montenegro, Proc. of the 7th International Conference Ports and Waterways – POWA, Zagreb, Croatia, 2012, pp. 1-10.
[15] Kapidani N., Bauk S., Strengthening Maritime Safety in Montenegro According to the Response on Oil Spill Pollution, TECHNO-EDUCA 2012, Zenica, B&H, 2012, pp. 96-101.
[16] Government of Montenegro, Ministry of Transport Maritime Affairs and Telecommunication, Transport Development Strategy of Montenegro, 2006.
[17] United Nation Convention on the Law of the Sea, 1982.
[18] UN Economic Commission for Europe, The Single Window Concept, April, 2003, pp.1-4. URL:http://www.unece.org/fileadmin/DAM/trade/ctied7/ece_trade_324e.pdf, last access on 17th February 2015.
[19] Directive 2002/59/EC of the European Parliament and of the Council of 27 June 2002, (repealing Council Directive 93/75/EEC), Establishment a Community Vessel Traffic and Information System, 2002.
[20] Directive 2009/17/EC of the European Parliament and of the Council of 23 April 2009, Amending Directive 2002/59/EC establishing a Community vessel traffic monitoring and information system, 2009.
[21] IALA Recommendation V-103 – December 2009, On Standards for Training and Certification of VTS Personnel, 2009.
[22] Directive 2010/65/EC of the European Parliament and of the Council of 20 October 2010 on reporting formalities for ships arriving in and/or departing from ports of the Member States and repealing Directive 2002/6/EC, 2010.
[23] IMO Resolution MSC.139 (76), Mandatory Reporting System, 2002.
[24] IALA Recommendation V-145 June 2011, On the Inter-VTS Exchange Format (IVEF) Ed. 1, 2011.

Evolutionary Methods in the Management of Vessel Traffic

A. Łebkowski
Gdynia Maritime University, Poland

ABSTRACT: The paper presents evolutionary method as one of the artificial intelligence method used in agent system that can be used to control the movement of ships in a given area. It discusses ways of defining the limits of navigation which occur in the navigation environment. The problem of deriving optimal (safe, shortest, cheapest) route of ship was a topic of many papers. Following the changing standards specified in e-Navigation strategy, proposed agent system bases its operation on agent platform comprising of specialized computer programs – agents, realizing specific goals. Paper describes system structure, methods of solution of collision situations as well as ways of communication and data exchange with environment. Described properties of the system predetermine it for ship movement control systems in congested waters and at open sea.

1 INTRODUCTION

According to the International Convention for the Safety of Life at Sea from 1974 (with subsequent changes), each merchant ship shall possess electro-navigation equipment which should provide adequate communication on sea, which directly affects level of safety of navigation.

The equipment is as follows: radar system supplied with ARPA, AIS system, GPS system, gyrocompass, ECDiS and communication equipment using GMDSS. Listed devices with their advantages and disadvantages can be used by navigators to provide safety of navigation in various levels.

Similarly, in order to increase the safety of navigation of commercial vessels through better organization of the data system, it unifies the standards associated with the exchange of data between vessels and shore stations. The Maritime Safety Committee (MSC) decided in November 2014 to include, in the work programmes of the NAV and Radiocommunications and Search and Rescue (COMSAR) Sub Committees, a high priority item of "Development of an e-navigation strategy", with the NAV Sub Committee acting as coordinator. The presented system is one of the proposals for the modernization of the current vessel traffic management strategies.

One navigator using data from electro-navigation equipment will provide high safety level meaning he will provide large distance between passing objects while another navigator, using identical data could even spell marine disaster. In both cases the significant impact on navigator's decision has their training level, experience, interpretation of imprecise law of the International Regulations for Preventing Collisions at Sea (COLREGS) and hydro meteorological conditions on given waters. Navigator entering congested waters is required to complete certain actions, such as: maintain certain safety conditions; acquaint himself with local pilotage, tides and currents; recognize readings from AIS, radar system and ARPA; recognize fishing vessels traffic; check own vessel's size and manoeuvring capabilities including velocity change capability; check communication with VTS traffic station. Widespread use of navigator's decision support systems, especially in congested areas would certainly have great impact on safety level at sea.

2 STRUCTURE OF AGENT NAVIGATOR'S DECISION SUPPORT SYSTEM [1-7]

Author suggests using the described agent system to supplement navigator's decision in congested waters and at open sea. System consists of IT platform on which the agents run. Agent platforms can be installed on many vessels and in VTS centres controlling traffic on given waters and the satellite AIS stations. The goal of agent system is calculating optimal, or possibly closest to optimal solution of route (safe, shortest, cheapest) which solution would consist of waypoints which given coordinates and velocities on legs connecting these waypoints. To make the work of system possible, there should be developed a way for agents to communicate within a platform and a way to communicate between agent platforms in whole system. The structure of suggested system is shown in Figure 2.

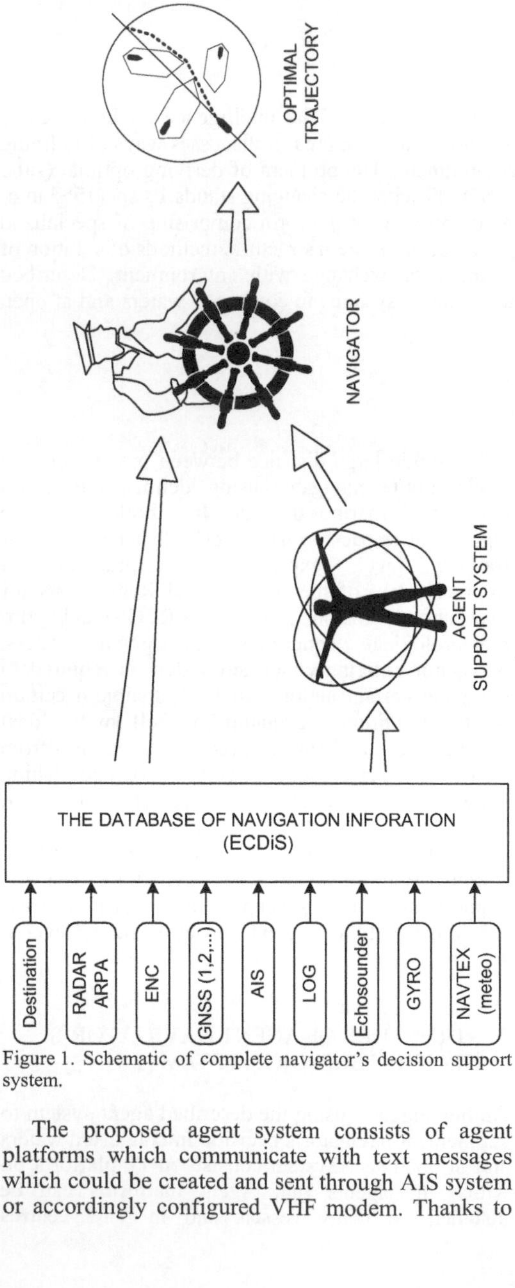

Figure 1. Schematic of complete navigator's decision support system.

The proposed agent system consists of agent platforms which communicate with text messages which could be created and sent through AIS system or accordingly configured VHF modem. Thanks to

agent platforms installed on vessels, shore VTS stations and satellite AIS modules, it is possible to exchange information on navigational situation on given waters, Each platform installed on merchant vessel, shore VTS station or satellite AIS modules consists of the agent platform. Each agent seeks to accomplish best outcome of his job – the biggest benefit for ship owner. These benefits mean minimising the cost of ship operation while maximising safety level of cargo and crew.

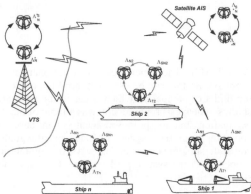

Figure 2. Structure of agent navigator's decision support system.

Figure 3 shows simplified structure of single agent platform consisting of three agents: trajectory agent Λ_T, navigation agent Λ_{SN}, negotiation agent Λ_N. The task of trajectory agent is to define ship route between entry point (actual ship location) and destination. Simultaneously, trajectory agent is responsible for trajectory correction in case of possible collision. The goal of navigation agent is to gather data on actual navigational situation around the ship and its in depth analysis for possible collisions. The job of negotiation agent is to negotiate with other agents installed on their platforms for solutions of collision situations which are satisfactory for all parties, and to reception of information from shore VTS stations or satellite AIS modules.

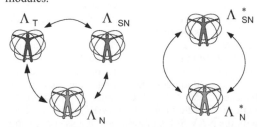

Figure 3. Simplified structure of ship's agent platform and VTS or satellite AIS agent platform.

The structure of agent platform installed on shore VTS station or satellite AIS modules is different than these installed on ships. It consists of navigation agent and negotiation agent which has higher decision-making factor than negotiation agents on ships. Thanks to this property, negotiation agents can forward orders related to traffic organization on given waters to trajectory agents installed on ship's agent platforms [5,6,7].

Implementation of agent system in proposed form allows automatic negotiation between ships operating in same neighbourhood in order to decide on area in which ship route or anti-collision manoeuvre can take place. Thanks to information sharing on intended ship movements which trajectories lead to collision threat, a compromise solution can be negotiated.

Actions of agents on a platform should have a possibility of taking independent decisions without navigator's intervention. Agents must possess ability to exchange information with navigator, who accepts final decision on a particular manoeuvre. Agents are also responsible for communication with other agents working on the same platform and whole agent system so that dynamic changes in the navigation environment could be interpreted accordingly and proper and optimal decisions could be made. Author suggests using an agent system using automatic model of negotiations between "ships" close to formal model of negotiations conducted by navigators. These negotiations depend on many difficult to define factors, such as each navigators' negotiation skills and language barriers. The negotiation process between agent platforms installed on ships is conducted automatically, excluding any communication problems. In case an interest conflict occurs between ships – their trajectories cross, negotiations are made by agents using games theory including positional and dynamic games. Applying games theory proposed by John Nash [8,9] to anti-collision problem on sea, an agreement can be made on deciding a route satisfactory for both parties. In order to exactly calculate ships' routes, a permissible range of solutions is determined for each agent platform – ship. A developer compromise concerns the determination of the area which can be established in the course of negotiations between agent platforms based on ships performing the negotiations. During the negotiations, the relative position of both ships is taken into account, including the COLREGS regulations. Thereby both vessels can achieve measurable gains in the form of clearly defined areas on which they can perform maneuvers. Such process results in benefits like shorter route of passage or safe passing of other vessel for both the ship with the right of way and the ship required to give way. After the negotiation process both vessels possess the knowledge of the domain of feasible solutions, in this case for the optimal route of passage calculated by the evolutionary algorithm.

Let's assume that $P \in [0,1]$ means a probability of successful negotiation between negotiation agents Λ_N^1 and Λ_N^2 based on each ships' agent platforms. We define an objection S1 of agent Λ_N^1 against agreement N1 as another possible agreement N2 according to interpretation: N2 is an alternative agreement proposed by player Λ_N^1 and 1-P is a probability of closing negotiations if player Λ_N^1 will stand by his objection S1. Agent Λ_N^1 will object to agreement N1 only if it will strongly prefer the outcome of agreement N2 over N1. Agent Λ_N^2 can counter-object an agreement N2 if it clearly prefers solution N1 to N2. Assuming the above definitions of objections and counter-objections, the Nash solution to controversial situation between two ships is a set of all agreements N1 characterized by that to each objection of any agents to N1, the second agent can counter-object. Leading further arguments, it can be demonstrated that such solution gives a set of all possible agreements which maximize the product of the von Neumann-Morgenstern utility function of both agents. The solution also possesses certain desirable axioms namely Pareto efficiency, symmetry and independence of irrelevant alternatives. What's most important, it is the only solution having these properties. Another feature of Nash solution is that agent achieves the worse results, the higher its risk aversion [5,6,7,10,11]

3 MODELLING THE NAVIGATIONAL ENVIRONMENT [5,6,7]

To make the action of agent system produce its intended effects possible, each component of agent platform – agent, should receive properly prepared and formatted data. The most important element impacting the calculation of optimal ship's route is proper interpretation of navigational limits occurring in the marine navigational environment. The navigational agent is tasked with providing and interpreting navigational situation around the ship. IT algorithms which were used to construct this agent convert data from AIS, radar system, ARPA, GPS, log, echo sounder and electronic maps to properly interpreted polygons. These polygons represent geographic coordinates of: shore lines, shallows, fairways, regions excluded from navigation, underwater rocks, shipwrecks, dump sites, fisheries, beacons of significant size (lighthouses, stakes, lightships, buoys); other moving objects such as ships, icebergs, areas of adverse weather conditions and other obstacles which could lower the safety level of own ship. Around navigational limitations which position changes dynamically areas are created called the domains, which size depends on COLREGS, given waters (open sea or congested), traffic flow, waterway markings, own ship size, navigator's decided safe distance, relative object velocity, safe manoeuvres time, hydro meteorological conditions and state of the sea, position error as read from instruments, bearing on the passing navigational object and, in case of three-dimensional domains, also bathymetry. An example of the shape of the domain is shown in Figure 4. Depending on the area of navigation structure and shape of the safe areas so-called domain is variable and determined in a fuzzy ways. The domain size is dependend on the scaling factor S, for example the domain size for the encountered dynamic objects such as other ships may change and varies from 0.1 for basic size fairway, through 0.3 in congested water, 0.8 at the approaches to the port and 1 on the open sea.

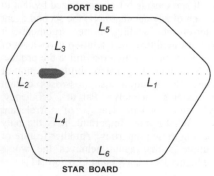

PORT SIDE

STAR BOARD

Figure 4. The shape of the domain for foreign ships, where: L1 – distance ahead of the ship ; L2 – distance astern; L3 – distance from the portside; L4 – distance from the starboard; L5 – the length of the left side of the domain; L6 – the length of the right side of the domain.

Navigation agent agrees permissible area for maneuver uses evolutionary programming during negotiation with other agents on agent platforms. Optimal safe passage choice subprogram includes domain size of passing object which longitudal size L1 is descibed by formula:

$$L_1 = \left(L_j + D_s + \left(V_0 + V_{RELj}\right)T_{sm} + W + HM + U\right)A \quad (1)$$

where:

L_j – length of the vessel expressed in nautical miles [Nm],

D_s – safe distance defined as the shortest distance in which it can pass other vessels expressed in nautical miles [Nm],

V_0 – actual speed of own vessel expressed in knots [kt],

V_{RELj} – relative speed of met j-th vessel expressed in knots [kt],

T_{sm} – safe maneuver time in minutes [min], specifying the time to perform calculations and make decision on maneuver,

W – the ratio of the relative speed and the speed of

the met ship and own ship ($W = V_{RELj}/V_0$),

HM – parameter describing the impact of weather conditions on the domain size,

U – ship position error,

A – scaling factor depends on the type of water area.

HM parameter specifying the influence of meteorological conditions on the domain size is determined in the following relationship:

$$HM = \frac{BS}{12}D_b \quad (2)$$

where: BS – represents the current interference situation described by Beaufort scale.

Determination of other domain sizes is based on COLREGS rules, especially regulations 14,15,16,17,18 and 19 as well as data supplied by navigational situation agent basing on ARPA readings and AIS data. Remaining domain sizes are implemented on following relationships:

$$L_2 = L_3 = 0.25 \cdot L_1 \,;\, L_4 = 0.45 \cdot L_1 \,;\, L_5 = L_6 = 0.7 \cdot L_1 \quad (3)$$

In navigational situations when met ship should give way according to regulation 15, but it did not and did not enter into negotiations with navigational agent, it will be treated by navigational agent as ship with forced right of way. Its domain will be described by following relationship:

$$L_1 = L_j + B_j + D_s \quad (4)$$

where: B_j – length of other ship in meters [m]

4 VERIFICATION OF THE AGENT SYSTEM

In order to verify the operation of agent system, a networked simulator of navigational environment was used. The simulator consists of master station and slave workstations connected to the master by LAN IP network. The master station has software which allows creation of navigation situtations. Workstations run software which simulate presence of ship in nauticial navigation environment. Networked simulator bases on ODE solving procedures taken from "Numerical Recipes in C" by W.H.Press et. al. [12]. Implementation of Runge-Kutta method with constant step size. Developed mathematical model of ship object covers dynamic properties of ship's hull, main engine with one variable pitch propeller, rudder fin and two transverse thrusters – bowthruster and sternthruster. Ship dynamics model takes into account influences of hydrometeorological interference such as wind, swell and currents as well as changes in ship's dynamics due to shallow water.

In the first simulation four ship approach each other from bearings 0°, 90°, 180°, 270° (Figure 5).

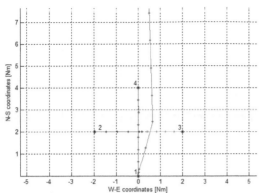

Figure 5. Navigational situation of ships passing – agent system active on ship #1.

Dynamic objects parameters simulated in this situation are presented in table 1.

Table 1. Parameters of the movements in navigational situation

Number of ship	Speed [kt]	Course [deg]
1	3	0
2	7	90
3	5	270
4	10	180

Agent system was active only on ship #1. It can be observed that ship performed correct anti-collision maneuver passing the ship #4 on the port side. Ships #2 and #3 collided.

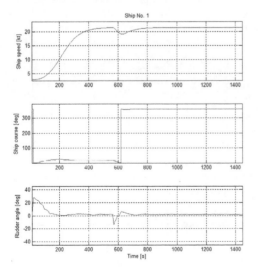

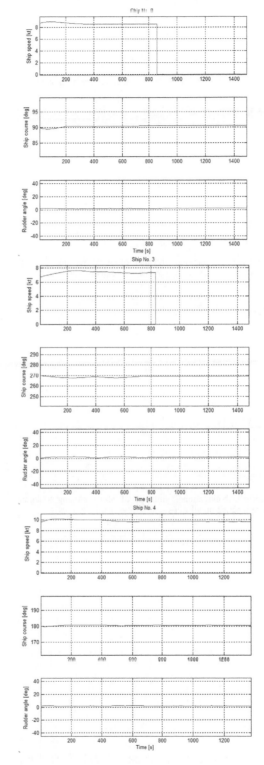

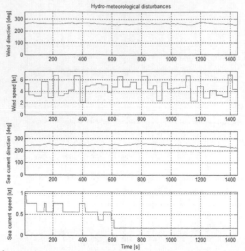

Figure 6. Navigational situation of ships passing – ships motion parameters and hydrometeorological conditions.

In the following simulation the agent system was active on all vessel (Figure 7).

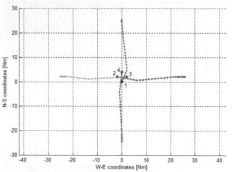

Figure 7. Navigational situation of ships passing – agent system active on all ships.

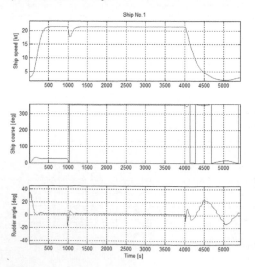

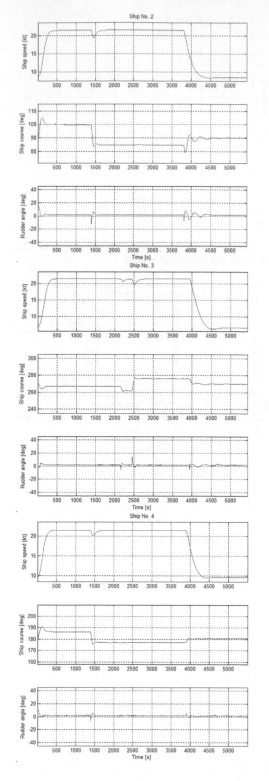

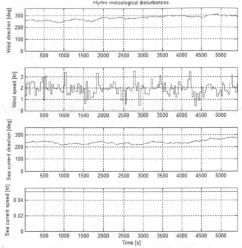

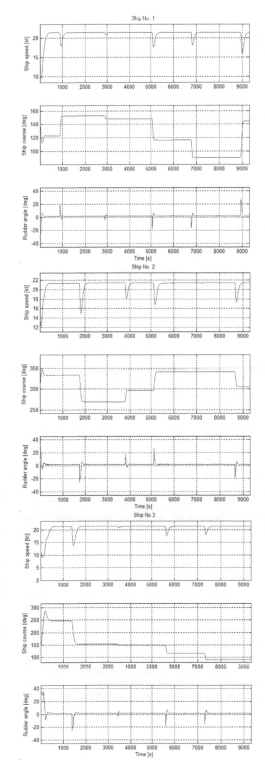

Figure 8. Navigational situation of ships passing – ships motion parameters and hydrometeorological conditions.

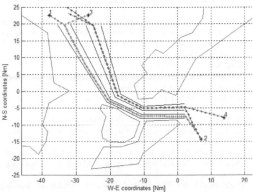

Figure 9. Navigational situation of ships passing – ships motion parameters and hydrometeorological conditions.

Basing on performed simulations it can be concluded that the system correctly performed anti-collision maneuvers for all 4 vessels and then directed all ship along their intended courses. All four ships motion parameters and hydrometeorological conditions of simulated situation are presented in Figure 8.

In the next situation there are four ships. Additional obstacle in calculation of intended passage route are static navigational restrictions – landmass and fairway (Figure 9).

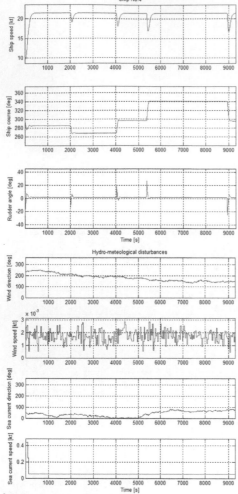

Figure 10. Navigational situation of ships passing – ships motion parameters and hydrometeorological conditions.

5 CONCLUSIONS

- The problem of conducting optimal navigation (safe and inexpensive) in marine navigation environment is a complicated issue of dynamic optimization.
- Carried out simulation studies showed, that the use of agent system results in the positive effects on vessel traffic control systems. The use of evolutionary methods as part of individual agents allows to reduce the risk of collisions and reduces

the time between the appearance of a dangerous situation and decision taken by the navigator.
- Application of proposed agent system would largely improve safety level at sea and limit the operating cost of the ship. It should be emphasized that proposed solution could be used in all situations occurring in marine navigation environment not limited, like most studies, to open sea scenarios without taking into account static limits. Application of agent system using evolutionary methods of determining ship's route as navigator's decision support system will, largely, relieve the workload of navigator.

REFERENCES

[1] Lisowski J., Optimal and game control algorithms of ship in collision situations at sea. Control and Cybernetics, No. 4, Vol. 42, 2013, p. 773-792.
[2] Lisowski J., Optimization-supported decision-making in the marine game environment. Solid State Phenomena, Trans Tech Publications, Switzerland, Vol. 210, 2014, p. 215-222.
[3] Lisowski J., Game strategies of ship in the collision situations. TransNav – The International Journal on Marine Navigation and Safety of Sea Transportation, Vol. 8, No. 1, 2014, pp. 69-77.
[4] Lisowski J., Computational intelligence methods of a safe ship control. 18th International Conference in Knowledge Based and Intelligent Information and Engineering Systems KES2014, Gdynia, Elsevier Procedia Computer Science, No 35, 2014, pp. 634-643.
[5] Łebkowski A., Control of ship movement by the agent system. Polish Journal of Environmental Studies Vol.17, No. 3C, 2008.
[6] Łebkowski A., Negotiations between the agent platforms. Scientific Publications Gdynia Maritime University, Gdynia 2013.
[7] Łebkowski A., The Hybrid Control System of Moving Object in a Dynamic Environment, PhD dissertation, Gdansk University of Technology 2006.
[8] Nash J.F., Equilibrium points in N-person games, Proceedings of the National Academy of Sciences of the United States of America 36, 48-49, 1950.
[9] Nash J.F., Non-Cooperative Games, Annals of Mathematics 54, 286-295, 1951.
[10] von Neumann J., On the theory of Games of Strategy, Volume IV, Annals of Mathematics Studies 40, A.W. Tucker and R.D. Luce, Princeton University Press, Princeton 1959.
[11] von Neumann J. and Morgenstern O., Theory of Games and Economic Behavior. New York: John Wiley and Sons 1944.
[12] Press W.H., Teukolsky S.A., Vetterling W.T., Flannery B.P., Numerical Recipes in C, The Art of Scientific Computing. ISBN 0–521–43108–5, Cambridge University Press 1992.

Vessel Traffic Service (VTS)

Information, Communication and Environment – Marine Navigation and Safety of Sea Transportation – A. Weintrit & T. Neumann (eds.)

Supporting Voice Communication Between Navigator and VTS by Visual Solutions – Exploring the Use of the "Route Suggestion" Functionality within VTS

A. Brodje & R. Weber
Chalmers University of Technology, Sweden

D. Camre & O. Borup
Danish Maritime Authority, Denmark

T. Porathe
Norwegian University of Science and Technology, Norway

ABSTRACT: The present paper describes an experiment where a VTS uses the functionality of "route suggestion" to visually support communication between navigators and VTS Operators. The idea of "route suggestion", a crucial part of current developments within e-navigation, was first introduced with the EfficienSea project and has since been matured within the MONALISA and ACCSEAS projects. Within the ACCSEAS project the functionality has been augmented through the notion of "intended routes", where vessels transmit sections of their planned routes using ECDIS in order for navigators to better understand the intentions of other, nearby, vessels. Also, "route suggestions" has been developed for use by shore stations, such as VTS, to transmit route segments to individual vessels within its area of responsibility. As part of the ACCSEAS project, the latter was tested in a simulator consisting of two full mission bridges and two VTSs, each of the two working in pairs, bridge-VTS. Five different scenarios were used, all set in the entrance of the river Humber, UK. In total, nine Pilots/Masters (Unlimited licence) and two VTS Operators participated in the experiment which ran over four days and was divided into two different simulations. ECDIS and VTS system were represented by the shipboard as well as the shore-based E-navigation Prototype Display system, as developed by the Danish Maritime Authority within the MONALISA and ACCSEAS projects. Data was collected using questionnaires, observation and video protocols. The observation protocol and the video protocol were coded and micro coded using MaxQDA and the questionnaires were used for comparison between the different participants following the analysis of protocol data in MaxQDA. Results indicate that there were no differences in acceptance of the tested "route suggestion" functionality between participants of different age, nor previous experiences as navigators/VTS Operators. Further, the results strongly indicate that the acceptance of the "route suggestion" functionality was due to the VTS Operators bridging the introduction of the new functionality by voice communication. Thus, the "route suggestion" functionality served as a graphical means of supporting voice communication between navigator and VTS Operator.

1 INTRODUCTION

The main purpose of this study was to explore how the introduction of visual communication tools in the Vessel Traffic Service (VTS) domain could support communication between navigator and VTS. Specifically, we wanted to learn more about the effects of introducing the "route suggestion" functionality and how this affects communication between VTS Operators (VTSOs) and navigators on board vessels sailing through a VTS area.

In order to explore the use of the "route suggestion" functionality within VTS, we set up a simulation as part of the ACCSEAS project, which we used for studying the effects on communication between navigators and VTSOs.

1.1 *Background*

Approximately 90% of all world trade is carried by sea on a merchant navy consisting of some 74,000 vessels above 100 gross-tonnage. To this figure should be added some 27,000 vessels not concerned with moving cargo, yet in size equal to merchant vessels such as cruising vessels, and some 9,000 larger military vessels. The estimated value of world trade within shipping was about 9.2 trillion dollars in 2004 (Stopford 2009). The value of any loss of either cargo or ship is considerable, let alone possible loss of life and often severe consequences for the environment. In Europe alone, being one of the major consumer markets of the world, there were about 24,000 calls to ports in 2010 rendering in some 580,000 individual vessel movements within the confined waters of primarily ports and their approaches. Of these movements, the European Maritime Safety Agency (EMSA 2011) reports that 644 vessels were involved in 559 accidents such as sinkings, collisions, groundings and other significant accidents.

A recurring conclusion in accident reports is that of miscommunication between vessels or between vessels and shore stations (EMSA 2008, EMSA 2009, EMSA 2010, EMSA 2011). In an effort to reduce the risk of miscommunication between vessels and between vessels and shore stations, such as VTS, the International Maritime Organization (IMO) first introduced regulation on training of communication skills in the Standards of Training, Certification and Watchkeeping (STCW) in 1978 (IMO 1978). The latter was later complemented by the Standard Maritime Communications Phrases (SMCP) in order to introduce a standard language for communication between vessels and vessel and shore station (IMO 2001).

As reported by Chawla (2015), the navigator is expected to communicate with a great number of different actors involved in the navigational process, commonly having different cultural backgrounds with different mother tongues, making English the preferred language for communication between vessels and between vessel and shore. Yet, as described above, despite the introduction and training in SMCP and the use of a common language, miscommunication is a recurring reason for maritime accidents.

VTS has been established by many coastal states as a risk mitigating system for monitoring and interacting with maritime traffic in confined waters and is normally a 24-7 service (IALA 2008). A VTS centre is manned by VTS-operators (VTSOs) who normally have some kind of maritime background. The latter plays an important role in the daily work of the VTSO, who very much tend to rely on previous experiences as a mariner in both their use of available sensors as well as in their decision making (Brodje et al. 2010).

In their role, VTSOs use an array of sensors, such as radar, AIS (the acronym of Automatic Identification System), VHF radio, CCTV and meteorological sensors. Often, the information these sensors gather is integrated into a VTS-systems, where vessel movements are presented onto electronic charts in the form of radar and AIS targets, sometimes presented in a fused mode. Also, VTS-systems normally have the capacity to present weather information and other information such as ship specifics and port information. From this information, the VTSOs build a situational awareness (SA) (Endsley 1988) of the maritime situation within their area of responsibility, denoted a VTS-area (Brodje et al. 2010). Using this SA, VTSOs interact with vessels for the purpose of distributing information (but also gathering, in building or updating the SA), giving assistance in navigational matters, intervening in hazardous situations or sometimes even organising the traffic, hence sharing the SA with the Officers of the Watch (OOWs) navigating in the VTS-area.

The above process of sharing SA is commonly dual, in that both the OOWs as well as the VTSOs share and update each other, thus building a shared SA. The mean for this sharing is commonly VHF radio, used for voice communication between OOWs or maritime pilots onboard vessels, in turn distributed onboard various vessels around a VTS-area, and a VTSO on-shore. By using VHF all those navigating in a VTS-area are able to listen in to the information shared between any vessel and the VTS, thus building an even greater shared SA.

However, since voice communication often involve updating OOWs and pilots on meeting vessels, expected congestion, and routes and it is not uncommon that VTSOs use local area names as geographic references. This makes it more difficult for the receiver of such information to understand the message of what is being communicated (Brodje et al. 2012). Bridging this gap is one of the aims of e-Navigation.

1.2 e-Navigation

e-Navigation "is the harmonized collection, integration, exchange, presentation and analysis of marine information onboard and ashore by electronic means to enhance berth to berth navigation and related services for safety and security at sea and protection of the marine environment" (IALA 2015). The concept, or rather a vision, was initiated in the IMO with the aim of integrating current and future navigational tools and systems in order to "enhanced navigational safety [...] while simultaneously reducing the burden on the navigator" (IMO 2014).

e-Navigation is based upon interconnecting different navigation systems and sensors, allowing these to share information, thus simplifying and increasing the availability of information to the navigator as well as automating some parts of the navigators work as an OOW. The aim of such a development is to increase the situational awareness (Endsley 1988) of the OOW while at the same time alleviating the OOW of tasks such as manual reporting, which are both work intensive as well as incorporate risks of mistakes.

The latest progress in the work on e-Navigation with regard to the process of the IMO is the approval of an e-Navigation Strategy Implementation Plan (SIP) (IMO 2014). The SIP has identified five prioritized solutions for achieving the aim of the e-Navigation concept. One of those solutions (S9) is the improved communication of VTS Service

Portfolio aiming at improving communication and efficient transfer of marine information between ships and shore stations.

1.3 *ACCSEAS & MONALISA 2.0*

There are currently two major EU-funded e-Navigation project working on various parts of the tasks and solutions identified in the e-Navigation SIP; ACCSEAS and MONALISA 2.0. ACCSEAS is a 3-year project focusing on improving access to the North Sea Region by sea transport while at the same time reducing navigational risks. Among other things, the project is studying how to harmonise and exchange maritime information using e-Navigation solutions as well as hard- and software prototypes for demonstrating the developed solutions. The latter have also been tested in simulated environments (Accseas Project Bureau 2014).

The MONALISA 2.0 project is also a 3-year project focusing on the development of e-Maritime solutions as well as e-Navigation solutions. Within the project, the concept of Sea Traffic Management (STM) has been developed. The latter in turn builds on five sub-concepts; Strategic Voyage Management, Flow Management, Dynamic Voyage Management, Port CDM (Collaborative Decision Making) and SWIM (System Wide Information Management) (Lind et al. 2014).

Both the ACCSEAS project as well as the MONALISA 2.0 project work with the development of the "route suggestion" functionality, with the purpose of developing a service to be used a part of Solution 9 of the e-Navigation SIP. Within the ACCSEAS project the "route suggestions" functionality has been augmented for use by VTS, to transmit route segments to individual vessels within its area of responsibility.

1.4 *Route suggestion*

The aim of developing the "route suggestion" functionality was originally to construct a service to be used by (in the case of ACCSEAS) VTS for sending proposals on changes to vessels' pre-planned routes while they were navigating in the area of responsibility of the VTS. The reason for proposing changes to an already pre-planned route through an area could for instance be based on changes in traffic intensity, e.g. in order to avoid expected congestion, fairway work such as dredging, or other safety related matters.

The information would be transferred between the VTS and the concerned vessel by sending a "route suggestion" from the VTS-system to the onboard ECDIS and would be in the form of waypoints. The information could be accompanied by a small text message from the VTSO, wherein the VTSO would be able to give a reason form the

proposed change to the vessel's pre-planned route, thus allowing the OOW to better understand the thinking behind the proposal. However, information containing the reason for suggesting a change to a pre-planned route could also be transferred using voice communication via VHF.

By transferring the information and presenting the "route suggestion" graphically in the onboard ECDIS and accompanying this visual information by text or voice communication, the general idea is increase the situational awareness of the OOW as compared to only using a single mean for transferring the information.

2 METHOD

The present study was part of the ACCSEAS project and was conducted in September 2014 at the Institution for Shipping and Marine Technology at Chalmers University of Technology, Sweden.

The simulated area was chosen since it was part of a bigger testbed area in the project, spanning from the Humber estuary to the Port of Rotterdam with interconnecting traffic. The Humber estuary is heavily trafficked, has several separate fairways and is monitored by VTS.

2.1 *Participants*

11 Subject Matter Experts (SMEs) were involved in the experiment. The SMEs came from the United Kingdom, Denmark and Sweden, and all had varying backgrounds as navigators, maritime pilots, and VTSOs. All of the SMEs had active or previous experience from navigating or working as VTSOs in the Humber estuary, which was a selection criterion, making it difficult to reach any larger numbers of SMEs for the present experiment. There were in all 8 SMEs operating as pilots. Of the pilots, 2 were operating in the Humber estuary, 5 were operating in Swedish waters and 1 was operating in Danish waters. 1 SME was operating as a ship's captain, holding a so called Pilot Exemption Certificate (PEC) for the Humber Estuary, calling in the area between two to three times per week. 2 of the SMEs were VTSOs, one regularly operating in the Humber estuary and the other in Gothenburg, Sweden. The latter was an experienced navigator having previously held a (PEC) for the Humber estuary. 2 of the SMEs who were currently operating as pilots, had previous experience as VTSOs, and were used in dual roles during the week.

All SMEs were male and were between 32 and 58 years old, the mean age being 47.4 and the sd=8.2. The SMEs had between 8 and 33 years of experience as navigating officers, the mean time being 22 and the sd=8.6. The 8 pilots had between 1 and 18 years of experience in their role, the mean time being 11.9

and the sd=5.8. The VTSOs had between 5 and 18 years of experience in their role, the mean being 9.3 and the sd=5.9.

All of the SMEs had previous simulator experience, both on ships simulators as well as VTS simulators for those of the SMEs who were or had been VTSOs. 2 of the SMEs were simulator instructors; 1 of the VTSOs and 1 of the pilots. The latter had previously also served as VTSO and thus had experience from VTS simulators as well.

2.2 Procedure

Five scenarios were constructed and set up in order to create five different types of events in the fairway through the Humber estuary where the navigating SMEs were to sail a 180m Ro-Pax ferry. All of the scenarios involved the use of the "route suggestion" functionality from the VTS located at Spurn Head, but were also create as to test other functionalities developed in the ACCSEAS project, such as "intended route" and "NoGo-area service" (which are both disseminated separately). The geographical area for the simulation and the five scenarios is depicted in Figure 1.

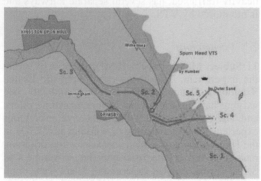

Figure 1. The Humber estuary with the five scenarios and the location of the Spurn Head VTS depicted in red.

The first scenario involved the Ro-Pax (own vessel) sailing to Hull and approaching the Sea Reach Traffic Separation Scheme (TSS) in a strong northerly winds, due to which the Ro-Pax was re-routed to the more southerly Rossy Reach TSS.

In the second scenario the Ro-Pax was asked to change its planned route by the VTS, using the "route suggestion" functionality, since the Ro-Pax was expected to meet with a very wide special transport vessel in a narrow stretch of the fairway.

In the third scenario there was an anticipated traffic congestion situation outside Immingham Oil Terminal involving several large vessels, due to which the VTS, using the "route suggestion" functionality, propose a change to the pre-planned route of the Ro-Pax in order to avoid even greater congestion (thereby creating an even greater navigational risk).

In the fourth scenario the outbound Ro-Pax is asked by the VTS, using the "route suggestion" functionality, to use a more southerly route through the TSS as an inbound deep draft vessel needed to use the more northerly TSS off Spurn Head.

In the fifth scenario the Ro-Pax is just leaving Anchorage A outside Spurn Head bound for Immingham. The area is known for a strong southbound current which is a well-known cause for accidents involving the North New Sand N-cardinal buoy. The VTS, using the "route suggestion" functionality, proposes a route to the Ro-Pax in order for the vessel to make a more northerly turn as to avoiding coming too close to the buoy.

During the simulations, two simulated ship bridges and two simulated VTSs were run in parallel worlds. The vessels interacted with one VTS respectively. The vessels were not able to interact with one another, nor were the VTSs. Both of the bridges were manned by two navigators each, as would commonly be the case in reality in the area.

The VTSs were manned by one VTSO each. Since one of the local VTSOs were unable to take part on the last day of the simulations a VTSO normally serving in VTS Gothenburg served as VTSO for one of the VTSs. The VTSO had previously worked together with a local Spurn Head VTSO during the day before as to get accustomed to the VTS-area.

The five scenarios were simulated during two days, and were run in two consecutive turns thus spanning over four days. For the first two days the scenarios were run in the normal order (1-5). For the second two days, with new SMEs, the scenarios were run in the following order: 5, 4, 1, 3, 2. Each of the two day runs included an initial familiarization for the SMEs to get used to the setup and equipment in the simulators, an initial briefing about the two day run, debriefings after each scenario, and a final discussion. At the end of the final discussions (following each of the two day simulations) the SMEs were asked to complete a questionnaire gathering demographic data as well as a survey rating their professional acceptance of the "route suggestion" functionality.

The SMEs operating as VTSOs were instructed to use the "route suggestion" functionality in their interaction with the vessels during the scenarios. They were also instructed to think aloud as they interacted with the simulation (Patton 2002). During the debriefings following each scenario, discussions between the SMEs (both bridge personnel and VTSOs) were focusing on the functionalities used during the specific scenario and the acceptance of such use. Data was gathered using video protocol and through observation, as well as by using the demographic questionnaire and the professional acceptance survey.

2.3 *Equipment*

Two simulated full mission bridges, Transas 5000, were used to simulate the vessel. The simulated environments, including target vessels, were controlled by a simulator instructor.

Instead of using the Transas ECDIS as normally installed on the bridge, an ECDIS-like test system, the e-Navigation Prototype Display (EPD) developed by the Danish Maritime Authority, was used to simulate the ECDIS. The system included the tested functionalities as described above and is so designed that it includes the major features of an ordinary ECDIS. The purpose of the EPD system is to be used as a simulated ECIDS for testing purposes, allowing system designers to develop, tweak and test new functionalities.

Two simulated VTSs were set up based on the EPD Shore system. The EPD Shor has been so designed that it includes the major features of an ordinary VTS system. The VTSOs were able to monitor the area using two displays. Each of the VTSs was equipped with a VHF, monitoring a local traffic channel as well as channel 16 for voice communication with vessels navigating the VTS-area.

The EPD system as a whole (both the ECDIS-like and the Shore system) has the possibility of exchanging route information ship-ship, ship-shore and shore-ship.

Apart from the technical installations a digital camcorder was used for recording video protocol as well as pen and paper for note-taking during observation.

2.4 *Analysis*

The video protocol and observation notes were imported into MaxQDA (Belous 2007) and analysed in two steps. The initial, macro, analysis involved playback of each of the video recordings parallel with the observation protocol from the five scenarios and the following debriefings. Data was categorised and general notes were made as to what the VTSOs were focusing their attention at on the system displays. The data was then coded and in turn micro-coded with a special focus on communication aspects, for which the recordings were played back a second time. The second part of the analysis involved a comparison between communication aspects of the information sharing process of the VTSOs.

The data from the demographic questionnaires was collated and related to the findings from the video protocol and observation notes. Similarities and differences were identified between SMEs with regard to the use and acceptance of the "route suggestion" functionality.

3 RESULTS

Results indicate that there were no differences in acceptance of the tested "route suggestion" functionality between participants of different age, nor previous experiences as navigators/VTS Operators. Further, the results strongly indicate that the acceptance of the "route suggestion" functionality was due to the VTS Operators bridging the introduction of the new functionality by voice communication.

The analysis indicated that the data from the video protocol and observation notes could be categorised as either technical or operational aspects. Technical aspects were related to system usability, whereas operational aspects were related to the use of the system in the simulated environment.

With regard to technical aspects, some traits were clearly noticeable from how the SMEs worked with the EPD Shore system and further emphasised by the SMEs specifically pointing out various aspects during the debriefs and final discussions. One such aspect was the lack of current (simulated) time in display, which made it more difficult to make estimates of vessel movements in the VTS-area. The SMEs also found it difficult to send "route suggestions" to the simulated vessel, commenting on the apparent lack of training in system use (only during the familiarisation) but also stressing that with time they would learn to handle the system more easily. The latter was also the case from a more general usability aspect, where the SMEs had obvious difficulties in remembering where to find different features in the EPD Shore system. Also, a recurring comment was the aspect of readability, where some of the SMEs commented on the small text and symbols on the display making it difficult to operate.

Operationally, the video protocol and observation notes from the scenarios indicate that the SMEs grew more accustomed to the use of the "route suggestion" functionality over time. However, one specific incident during the second scenario of the first simulation run indicates that the general acceptance of the "route suggestion" functionality became very high. In the scenario the Ro-Pax vessel was asked to change its planned route, since the Ro-Pax was expected to meet with a very wide special transport vessel in a narrow stretch of the fairway. The VTSO first suggested an alternative route to the OOW of the Ro-Pax using voice communication over VHF, also explaining the reason for suggesting an alternative route. Upon finishing the VHF communication, the VTSO said "I will send you a "route suggestion" so that you can see what I mean".

Other operational aspects, upon which the SMEs specifically commented on, was the question of workload of navigators and the acceptance of "route suggestions" from VTS onboard different types of vessels. With regard to workload in the VTS, the VTSOs agreed that the possibility to use the "route suggestion" functionality would reduce the workload, even though the use of the feature would add a new task. The reason for this, they argued, was that today they would often spend several minutes of VHF communication trying to convey to a non-native English speaker an alternative route. Using the "route suggestion" functionality, this process would be highly simplified. On the other hand, the navigating SMEs, although agreeing with the VTSOs, pointed out that a route always have to be checked for primarily safety before set into action – a formal responsibility of the ship's master – and would increase workload among navigators. Also, it was pointed out, the recurring vetting may make it difficult for an onboard OOW or Captain to change the pre-planned route due to a "route suggestion" from a VTS, since this would be a divergence from the plan and thus not be acceptable to the vetting inspectors.

The survey probing the SMEs on their "professional acceptance" of the "route suggestion" functionality was unfortunately only answered by 9 of the 11 participating SMEs, since 2 of the SMEs had to leave slightly early on the last day of the simulations. The survey was however completed by all 4 of the SMEs with current or previous experience as VTSO. The survey included three questions, two of which had predesigned answers (below) and one which asked the SMEs to rate their acceptance on a scale:

– What is your opinion about the tested suggested route concept? Very good; Good; Don't know; Bad; Very bad.
– Do you think a similar suggested routes concept will become reality in the future? Most probably; Probably; Don't know; Probably not; Most Probably not.
– What is your professional opinion about the system tested? The scale ranging from 0 to 5, where 0 indicated "Totally unacceptable", 1 indicated "Not very acceptable", 2 indicated "Neither for, nor against", 3 indicated "Acceptable", 4 indicated "Very acceptable" and 5 indicated "Extremely acceptable".

1 SME responded that the tested "route suggestion" functionality was "Very good" and 8 SMEs responded that the tested functionality was "Good". None of the SMEs responded that the "route suggestion" functionality was "Bad", "Very bad" or that they "Don't know".

3 SMEs responded that they believed that a similar concept to that of the "route suggestion" functionality "Most probably" would become a reality in the future. 6 SMEs responded that the tested functionality would "Probably" become a reality in the future. None of the SMEs responded that they did not know or did not expect ("Don't know; "Probably not"; "Most probably not") the tested functionality to become a reality in the future.

The rating of the "professional acceptance" indicated that among the 9 responding SMEs, the acceptance rating of the tested functionality was 3.7 which is between "Acceptable" and "Very acceptable".

When comparing the demographic data from the questionnaire with the SMEs operations and comments, this did not indicate any differences based neither upon age nor previous experience (as navigators or VTSOs).

4 DISCUSSION

The scenarios in the simulation were based upon a focus group meeting held during the spring of 2014. The scenarios were so constructed that they simulated normally recurring events in the Humber estuary, and were all based upon real scenarios either taken from AIS data delivered from Spurn Head VTS, or as described during the focus group. Thus, it can be argued that in the current setting, using full mission bridges with ECIDS-like displays as well as high-fidelity VTS-like system, the ecological validity was high for the present experiment.

It is difficult to fully estimate the effects on the results of the technical aspects as indicated in the data. The lack of training most certainly had an effect on the SMEs understanding for the work needed to operate the systems, thus affecting their response as to the anticipated future workload onboard. The SMEs with current or previous VTSO experience were all of the understanding that a future introduction of the "route suggestion" functionality would ease the workload of VTSOs.

The acceptance rating clearly indicates that the SMEs were generally very positive towards the tested functionality and that they expected to see the tested functionality in operation in the future. Although indicative, these responses point toward a broader acceptance from both navigators as well as VTS personnel. The lack of differences between the ages (although tested on a very small sample) further indicates the perceived usefulness of the system, despite the comments made by SMEs with no VTS experience on future acceptance onboard due to for instance vetting of tankers or the master's responsibility for the vessel's navigational safety and thus the need to check a route before putting it into action for liability reasons.

With regard to the latter, and in relation to the situation described by Chawla (2015) above, it is difficult to see how there could be a perceived difference between a VTSO verbally communicating a suggested change of route and the same VTSO under the same circumstances sending a "route suggestion" using an integrated system functionality. In the latter case, the "route suggestion" would serve as a visual tool for communicating a piece of information which would otherwise have been more complicated and time consuming to convey. Hence, it could be argued that the "route suggestion" functionality, while serving as a visual tool, would not only reduce the risk of miscommunication but also simplify the creation of a shared SA between VTSO and OOW.

5 CONCLUSIONS

The results indicate that the "route suggestion" functionality served as a graphical means of supporting voice communication between navigator and VTS Operator. It was also indicated that this is considered valuable by both navigators and VTSOs and that both groups expect to see this kind of feature in future operational use.

The results further indicate that the use of the "route suggestion" functionality could reduce the risk of miscommunication between VTSOs and navigators and that the functionality could assist in increasing the shared situational awareness between VTSOs and navigators sailing in a VTS-area.

REFERENCES

Accseas Project Bureau. (2014). "Accseas Web Site." Retrieved 2014-06-11, 2014.

Belous, I. (2007). MAXQDA2007. U. Kuckartz. Berlin, Germany, http://www.maxqda.com/.

Brodje, A., M. Lundh, J. Jenvald & J. Dahlman (2012). "Exploring non-technical miscommunication in vessel traffic service operation." Cognition, Technology & Work: 1-11.

Brodje, A., M. Lützhöft & J. Dahlman (2010). The Whats, Whens, Whys and Hows of VTS Operators use of sensor information. International Conference on Human Performance at Sea, Glasgow, University of Strathclyde.

Chawla, P. (2015). Happy talk: verbal communication and effective navigation. The Navigator. London, The Nautical Institute: 4-5.

EMSA (2008). Maritime Accident Review 2007. Lisbon, Portugal, European Maritime Safety Agency.

EMSA (2009). Maritime Accident Review 2008. Lisbon, Portugal, European Maritime Safety Agency.

EMSA (2010). Maritime Accident Review 2009. Lisbon, Portugal, European Maritime Safety Agency.

EMSA (2011). Maritime Accident Review 2010. Lisbon, Portugal, European Maritime Safety Agency.

Endsley, M. R. (1988). Situation awareness global assessment technique (SAGAT). Aerospace and Electronics Conference, 1988. NAECON 1988., Proceedings of the IEEE 1988 National.

IALA (2008). IALA - Vessel Traffic Services Manual. Saint Germain en Laye, France, International Association of Marine Aids to Navigation and Lighthouse Authorities.

IALA. (2015). "e-Navigation." Retrieved 2015-02-09, 2015.

IMO (1978). Standards of Training, Certification and Watchkeeping International Maritime Organization. London, IMO.

IMO (2001). Resolution A918(22); IMO Standard Marine Communication Phrases. A918(22). International Maritime Organization. London, IMO.

IMO. (2014, 2011). "E-navigation." Retrieved 2012-09-22, 2012.

IMO (2014). Navigation, Communications, Search and Rescue - Report of the first session of the Sub-Committee MSC94/9. International Maritime Organization. London, U.K., IMO.

Lind, M., A. Brodje, R. Watson, S. Haraldson, P.-E. Holmberg & M. Hägg (2014). Digital Infrastructures for enabling Sea Traffic Management. The 10th International Symposium ISIS 2014 "Integrated Ship's Information Systems", Hamburg.

Patton, M. Q. (2002). Qualitative Research and Evaluation Methods. Thousand Oaks, Sage Publications Inc.

Stopford, M. (2009). Maritime Economics. New York, USA, Routledge.

4M Overturned Pyramid (MOP) Model: Case Studies on Indonesian and Japanese Maritime Traffic Systems (MTS)

W. Mutmainnah
Graduate School of Maritime Sciences, Kobe University, Kobe, Japan

M. Furusho
Kobe University, Kobe, Japan

ABSTRACT: The aim of this paper is to show the characteristics of ship collision accidents that occur both in Indonesian and Japanese maritime traffic systems. There were 22 collision cases in 2008–2012 (8 cases in Indonesia and 14 cases in Japan). These cases have been analyzed using the proposed 4M overturned pyramid (MOP) model. The MOP model is adopting epidemiological model that determines causes of accidents, including not only active failures but also latent failures and barriers. The characteristics presented in this paper show failure events at every stage of the three accident development stages (the beginning of an accident, the accident itself, and the evacuation process).

1 INTRODUCTION

In the Asian region, Indonesia and Japan are the largest archipelagoes. Thus, it is important to maintain transportations at sea, which act as their lifelines and support their productivities. In the different stages of the ship's operation at sea, many errors can lead to incidents and/or accidents. Because the characteristics of each type of maritime accident are different, it is necessary to analyze each type of accident separately. In this study, we consider the ship collision accident because it involves problems associated with more than one ship. Often, ship collision results in explosion, sinking, grounding, etc. Therefore, in an effort to reduce the number of accidents caused by collisions, one of the steps could be to analyze previous collision accidents to identify their characteristics.

In Japan, the Japan Transportation Safety Board (JTSB) investigates major accidents in the Japan area, and in Indonesia, the National Transportation Safety Committee (NTSC) performs the same function as JTSB. All investigation reports can be viewed on their respective webpages.

The aim of this paper is to clarify the characteristics of ship collision accidents that occur both in Indonesian and Japanese maritime traffic systems (MTS). By understanding the characteristics of ship collision accidents, necessary countermeasures can be proposed.

In this paper, we propose a 4M overturned pyramid (MOP) model to analyze MTS, include past

accidents, and determine their characteristics. The proposed model is based on the combination of the Septigon model (society and culture, physical environment, practice, technology, individual, group, and organizational environment network) created by Grech et al. (2008) and the IM model proposed by Furusho (2000, 2013). The IM model consists of 4M factors (man, machine, media, and management) that are connected by the individual element (I) as the core of the system. The MOP model is drawn as a three-dimensional relationship that appears as a three-sided inverted pyramid, where each corner of the pyramid represent one 4M factor. Each corner (factor) is connected to and affects the other factors. The man factor should always be at the bottom of the inverted pyramid because it is the intrinsic factor that significantly affects all other factors. Because the model is drawn three-dimensionally as a three-sided inverted pyramid, it has four corners representing the 4M factors, and six edges representing interaction between the two factors that are connected by the edges. The edges, which are called line relations, show that the system is the result of interactions among the 4M Factors. Thus, to obtain a safe system, all corners and edges should be reliable and balanced.

In total, there are 22 collision cases that are analyzed in this paper; 14 from Japan and 8 from Indonesia, from 2008–2012. The analyzed cases are the investigation reports from each government. Note that, not all reports on the JTSB or NTSC website are translated in English. Therefore, only the English versions of the investigation reports were considered in this study for analyses. Thus, for Japan, there are 14 such cases from 2008–2012 and these cases involve 27 ships. In Indonesia, there are 8 collision reports that involve 16 ships from 2008–2012.

These investigation reports were analyzed in two steps, as will be explained in the section 3 There are three accident development stages: the beginning of the accident, the accident itself, and the evacuation process; these stages are labeled as Stage 1, Stage 2, and Stage 3, respectively (Nurwahyudi 2014). When characterizing the accidents in the reports, failure events at every stage of accident development are described. Several failures that occur are categorized into these stages. If we know the failures that occur at every stage until the evacuation process, we can predict the loss caused by the accident and also develop countermeasures.

The characteristics of the Indonesian and Japanese ship collision accidents can be determined using the proposed MOP model. This paper shows several analyses on the characteristics of these ship collision accidents. In several terms, ship collision accidents in Indonesia and Japan have the same characteristics, for example, the time of accidents (night) and improper lookout. Both Indonesia and Japan has a high ratio in these causative factors. However, each country also has unique characteristics that are not evident in the other country. The details of the accident characteristics are provided in Section 2.

2 MTS ACCIDENTS IN JAPAN AND INDONESIA

As explained in Section 1, a total of 22 cases are analyzed in this study. Figure 1 show the number of collision accidents that have occurred for each year from 2008 until 2012.

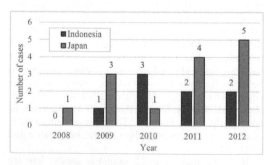

Figure 1. Number of collision cases in Indonesia and Japan from 2008–2012 (NTSC 2014, JTSB 2014)

The accident reports available for Indonesia are from 2007 to 2013. However, there were no reports of collisions in 2008. On JTSB's website, the English versions of the reports are available for accidents from 2008–2013. The analysis in this paper was carried out for the collisions in 2008–2012 because the collisions in 2013 are yet to be reported.

3 MOP MODEL

The proposed MOP model was developed in the maritime domain. As stated previously, MTS is better explained by the epidemiological model that consists of latent conditions, barriers, and active conditions. The proposed MOP model is a combination of the epidemiological, Septigon, and IM models.

The Septigon model is a concept that categorizes the MTS into seven domains: society and culture, physical environment, practice, technology, individual, group, and organizational environment network (Grech et al. 2008). All domains are connected to and affected by each other in a system. Any error in one domain can affect the system. In 2000, Furusho proposed a simpler system called the IM model. This model consists of 4M factors (man, machine, media, and management) that are connected by the individual element (I) as the core of the system (Furusho 2000, 2004).

The proposed MOP model is drawn three-dimensionally as a three-sided inverted pyramid that has four corners, representing the 4M factors, and six edges, representing an interaction between two 4M factors that are connected by the edges, as shown in Figure 2

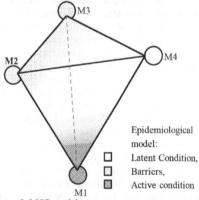

Epidemiological model:
☐ Latent Condition,
☐ Barriers,
▨ Active condition

Figure 2. MOP model

The edges, called line relations, show that the system is a result of interactions among the 4M factors. Failures that are classified into the corner of the MOP model do not occur only because of that particular corner. Often, the failure is caused by other corners. When there are failures that are caused by several corners, it implies that the line relations connecting those corners are also contributing to the instability of the system. For example, consider a failure in communication. Communication cannot be classified into one corner because communication is related to all four corners. The failure in communication among seafarers is classified as man factor (M1) because this type of communication depends on the person. Often, several seafarers do not share information with other seafarers. However, communication failure among ships and port administrations does not belong to the man factor. It can belong to either the management or the machine factor that is affected by the media factor. The classification of failure depends on the condition of the accidents. When a line relation contributes to the accident, a preventive action for the line relation has to be determined. Thus, for a safe system, all corners and edges should be reliable and balanced.

Because the MTS consists of latent conditions, barriers, and active conditions, any accident that occurs in the MTS should be traced for each of these factors separately. Each factor (corner) of the MOP model represents the epidemiological model as shown in Figure 1. In Figure 1, the individual from M1 (man factor) receives some information from M3 (media factor: environment) and from M2 (machine factor: crew complement) factors; then, this information is used for decision making. Hazard perceptions are also influenced by M4 (management factor).

Table 1. Definition and examples of each corner of the MOP Model

4M Factors	Definition (Example)
Man (M1)	All elements that affect people doing their tasks (Knowledge, skills, abilities, memory, motivation, alertness, experience, etc.)
Machine (M2)	All elements, including technology, which help people to complete their tasks (Equipment, information display, environmental design, crew complements, construction, etc.)
Media (M3)	All environments that affect the system and/or people (Climatic/weather condition (temperature, noise, sea state, vibration, wave, tide, wind, etc.), economic condition, social politics, culture, etc.)
Management (M4)	All elements that can control the system and/or people (Training scheme, communication among companies/institution, work schedule, supervising/ monitoring, regulatory activities, procedures, rules, maintenance, etc.)

Table 1 lists the definition and examples of the corners in the MOP model. By understanding the definition, it is easier to determine the causes of the accidents using the epidemiological model, and then, prevention actions can be considered. Besides, the characteristics of several accidents can be explored by analyzing several accident reports and by finding the tendency, as carried out in this paper.

4 ANALYSIS ON THE MTS ACCIDENTS USING THE MOP MODEL

The investigation reports explain all facts and causes of the accidents. We re-analyzed those reports using the MOP model. The analyses were carried out in two steps: corner analysis, which is listing causative factors for each corner of the MOP model; and line relation analysis, where the relationship between each causative factor in the corner is explored.

4.1 Corner Analysis

In this step, we traced and listed all failures that caused accidents and divided them based on the definition of each corner of the MOP model. Then, we counted the number of failures after all reports were analyzed. Tables 2–5 list the causes and the number of failures for each causative factor for each corner of the MOP model.

Table 2. Number of failures for each causative factor of the MTS accidents in Japan and Indonesia categorized as M1

Code	Causative Factors	Japan	Indonesia	Total
SLIPSHOD WORKMANSHIP				
M1-01	Inconsistency in the navigation course	3	1	4
M1-02	Wrong course decision	0	3	3
M1-03	Wrong speed decision	3	3	6
M1-04	Lack of communication among the seamen	4	3	7
M1-05	Improper Lookout	7	4	11
M1-06	Misunderstanding conditions (wrong judgment)	5	0	5
M1-07	Lighting inappropriate (navigational light)	1	0	1
INCAPABILITY OF SEAFARER				
M1-08	In utilizing AIS*	6	1	7
M1-09	In utilizing Radar	1	1	2
M1-10	In communicating by VHF	2	0	2
M1-11	In communicating with other vessel	0	4	4
M1-12	In understanding how to avoid collision	0	4	4
M1-13	In conducting abandoned ship	0	1	1

* AIS means Automatic Identification System

277

Table 3. Number of failures for each causative factor of the MTS accidents in Japan and Indonesia categorized as M2

Code	Causative Factors	Japan	Indonesia	Total
M2-01	Cannot place the object on Radar	2	0	2
M2-02	Lack of communication tools	0	1	1
M2-03	The anchor could not be placed	0	1	1

Table 4. Number of failures for each causative factor of the MTS accidents in Japan and Indonesia categorized as M3

Code	Causative Factors	Japan	Indonesia	Total
M3-01	Rain	3	0	3
M3-02	Insufficient visibility	2	0	2
M3-03	Strong wind	1	1	2
M3-04	High/Strong wave	1	0	1
M3-05	Insufficient light	9	6	15
M3-06	Tidal stream	1	1	2
M3-07	Low Tidal	0	1	1
M3-08	Traffic route was crowded	0	1	1

Table 5. Number of failures for each causative factor of the MTS accidents in Japan and Indonesia categorized as M4

Code	Causative Factors	Japan	Indonesia	Total
POOR MANAGEMENT OF PERSONNEL ON BOARD				
M4-01	Seaman has an expired certificate/ no certificate	2	2	4
M4-02	In understanding the passage plan (in new area)	0	2	2
M4-03	Lack of personnel	0	3	3
M4-04	In rotating personnel	1	3	4
POOR COMMUNICATION				
M4-05	About traffic route from the company (onshore)	3	1	4
M4-06	In monitoring and supervising from onshore	2	2	4
M4-07	Among onshore and other vessels utilizing radio	0	1	1
M4-08	Poor management of emergency drilling	0	1	1
M4-09	Poor application of safety management system	1	2	3

From Tables 2–5, the most common failures for each corner of the MOP model can be identified. They are "improper lookout," "cannot place the object on radar," and "insufficient light" for man, machine, and media factors, respectively. For the management factor, there are four failures that have the same total number, they are "seaman has an expired certificate/no certificate," "poor management of personnel on board in rotating personnel," "poor communication about traffic route from the company (onshore)," and "poor communication in monitoring and supervising from onshore."

There are some classifications in Table 2 and 5. When classifying the failures for the man and management factors, M1-01 until M1-07 can be classified again into slipshod workmanship. This classification is only for making it easy to divide and analyze the failures. Failures that caused accidents in Indonesia and Japan can be broadly classified into slipshod workmanship and incapability of seafarer. Then, in terms of the management factors (Table 6), the failures were poor management of personnel on board, poor communication, poor management of emergency drill, and poor application of safety management system.

Only observing the number of failures is not sufficient to obtain the characteristic of the accidents for each country because the number of analyzed accidents in Indonesia and Japan are different. The ratio of occurrence from among the analyzed report should be known. For example, a total of 11 failures were attributed to the improper lookout causative factor in the man factors. From the 14 reports, improper lookout was the reason for 7 cases. This means that the occurrence ratio for this causative factor is 0.5 (7 divided by 14). In the case of Indonesian collisions, improper lookout was the cause for 4 of 8 accidents. Thus, the occurrence ratio of improper lookout in Japan and Indonesia is the same. Figures 3–6 show the occurrence ratio (on the horizontal axes) of each causative factor on the vertical axes.

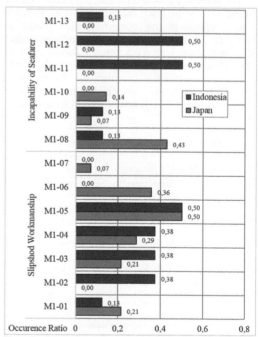

Figure 3. Comparison of occurrence ratio between Japanese and Indonesian MTS accidents for each causative factor categorized as the man Factor (M1)

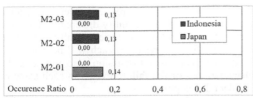

Figure 4. Comparison of occurrence ratio between Japanese and Indonesian MTS accidents for each causative factor categorized as the machine Factor (M2)

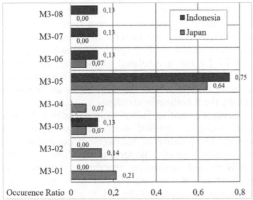

Figure 5. Comparison of occurrence ratio between Japanese and Indonesian MTS accidents for each causative factor categorized as the media Factor (M3)

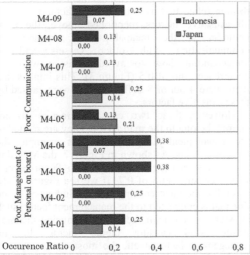

Figure 6. Comparison of occurrence ratio between Japanese and Indonesian MTS accidents for each causative factor categorized as the management Factor (M4)

4.2 Line Relation Analysis

Causative factors written in Tables 2–5 are not pure belongs to one corner. This line relation analysis step connects one corner to the other corners that are related to the causative factors that occurred. The causative factor in a corner that is related to another corner is marked to the related corner, and then, the line relation that connects these corners is obtained. The line relation that connects the man factor (M1) to the machine factor (M2) is labeled as M12. Therefore, M23 implies the line relation that connects the machine factor (M2) to the media factor (M3).

In this step, of all the causative factors listed, the relationship among the corners of the MOP model is explored. By performing line relation analysis, we can understand which line relation is the most vulnerable to failure for each country or in both countries. The causative factors listed in Figures 7–10 show the comparison of the line relations causing ratio that connects causative factors to several corners of the MOP model between Indonesia and Japan.

The causing ratio for a corner is obtained by dividing the number of causative factors that are related to other corners with the total number of causative factor in that corner. For example, in the corner man factor, there are 13 causative factors in total. However, not all causative factors caused the accidents in Indonesia or Japan. In Indonesia, for the corner man factor there were 9 causative factors (refer Table 2). Of these 9 factors, 5 were related to the management factor, i.e., M1-04, M1-05, M1-8, M1-9, and M1-10. Thus, the M14 line is causing the man factor corner with a ratio of 0.56 (5 divided by 9).

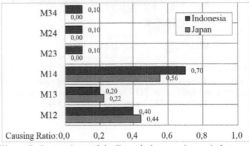

Figure 7. Comparison of the line relation causing ratio between Japanese and Indonesian MTS accidents in which the causative factors are categorized as man factors (M1)

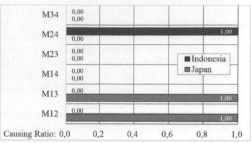

Figure 8. Comparison of line relation causing ratio between Japanese and Indonesian MTS accidents in which the causative factors are categorized as machine factors (M2)

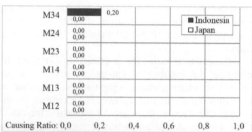

Figure 9. Comparison of line relation causing ratio between Japanese and Indonesian MTS accidents in which the causative factors are categorized as media factors (M3)

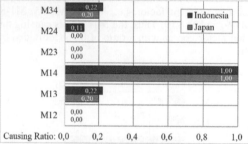

Figure 10. Comparison of line relation causing ratio between Japanese and Indonesian MTS accidents in which the causative factors are categorized as management factors (M4)

5 DISCUSSION

The characteristics of ship collision accidents in any country should be known to prepare preventive actions to decrease the number of accidents. In this research, we propose a new method, called the MOP model, for obtaining these characteristics. We analyzed accidents for two countries to demonstrate how the MOP model can explore the characteristics as well as the differences in the accident characteristics for each country or for both countries. However, this research does not aim at debating which country is better.

Figures 3–10 illustrate the comparison of the occurrence ratio for the corner and line relation analyses between Indonesia and Japan. These figures indicate that there are several causative factors that affect each country differently; there are factors that affect both countries equally as well. In subsection 5.1, the result of each will be discussed. In the last part, this discussion also provides the characteristics of the failure events in each stage of accident development.

5.1 Characteristics of MTS accidents in Indonesia

A total of 54 failures occur in the 8 cases of collisions. 25 of these failures are related to man factors with 10 types of causative factors, 2 failures are related to machine factors, 10 failures are related to media factors with 5 types of causative factors, and 17 failures are related to management factors with 9 types of causative factors. These numbers indicate that failures in the man factor are dominant among the 8 collisions accidents.

Several causative factors only occur in Indonesia. For example, slipshod workmanship in deciding course, incapability of seafarers in communicating with other vessels, understanding how to avoid collision, and conducting abandoned ship measures (categorized as man factors), lack of communication tools and the anchor could not be placed (classified as machine factors), low tidal and traffic route was crowded (classified as media factors), and poor communication among onshore and other vessels utilizing radio and poor management of emergency drilling (classified as management factors). Among all causative factors mentioned above, the most common is the incapability of seafarers in communicating with other vessels and in understanding how to avoid collision with the occurrence ratio of 0.5 (Figure 3). This implies that there were 4 out of 8 collisions that were caused by these causative factors.

However, these two causative factors do not entirely belong to the man factor. They are related to the management factor as well, for example, the term of training. As shown in Figure 7, the M14 line is the most causing line to the man factor corner, with a causing ratio of 0.7. If we see Figures 7–10 carefully, the most causing line for Indonesia is the line that is connected to the management factor, such as M14 causing man factor, M24 causing machine factor, M34 causing media factor. In the management factor itself, the most causing line is M14 with a causing ratio of 1, which means that all the causative factors in the management factor are related to the man factor.

From another perspective, the characteristics of the accident can be seen from the failure that happens in the three stages of accident development. From the 54 failures that occurred, 33 failures occurred in the beginning of accidents, 18 failures occurred in the accident or in an effort to avoid the collision, and 3 failures occurred in the evacuation process. However, there were no failures in machine and media factors in Stages 2 and 3. All failures in Stages 2 and 3 are failures in the man and management factors.

The failures that occur in the evacuation process are also dangerous because they result in an increase in the number of people who are missing, who sustain injuries, or those that are fatally wounded. Two collisions had failures in Stage 3. The failures are slipshod workmanship in communicating among the seafarers (M1-04), incapability of seafarer in conducting abandoned ship (M1-13), and poor communication among onshore and other vessels utilizing radio (M4-07).

In Stage 2, 15 out of 18 failures belong to man factors with 5 types of failures/ causative factors. These causative factors are M1-01, M1-02, M1-03, M1-11, and M1-12. All common failures that occur in all 8 collisions in Indonesia occur in Stage 2. Thus, the incapability of the seafarers in communicating with other vessels (M1-11) and in understanding how to avoid the collision (M1-12) caused the collision. If these failures were addressed, the collisions could have been averted.

In Stage 1, most failures were in the management factor, which implies latent failures, and in the media factor. However, the most common failure that occurred in this stage also occurred in the cases in Japan.

Therefore, from the discussion above, the most vulnerable condition is the man factor corner, man–management line relation and Stage 2 of the accident development process. To avoid collisions, these three points must be considered for taking preventive actions.

5.2 Characteristics of MTS accidents in Japan

Japan had 60 failures in the 14 collision cases from 2008–2012. 32 of these failures belong to the man factor with 9 types of causative factors, 2 of them belong to the machine factor, 17 belong to the media factors, with 6 types of causative factors, and 9 belong to the management factor with 5 types of causative factors. Similar to Indonesia, the man factor dominates the system. The difference is the composition of the media and the management factors. Although the management factor in Indonesia is much higher than the media factor, in Japan, the media factor is much higher than the management factor.

The number of causative factors that only occur in Japan is smaller compared to Indonesia; these factors are slipshod workmanship in misunderstanding condition/ wrong judgment (M1-06), lighting inappropriate navigational light (M1-07), and incapability of seafarers in communicating by VHF (M1-10) in man factor; cannot place the object on radar (M2-01) in machine factor; rain

(M3-01), insufficient visibility (M3-02), and high/strong wave (M3-04) in media factor. There are no management factors that occur only in Japan. From all causative factors mentioned above, failures in misunderstanding condition/wrong judgment (M1-06) has the highest causing ratio in the system. The causing ratio was 0.36 (Please refer to Figure 3). There were 5 out of 14 cases that had this causative factor. The master on board judged something wrong regarding the opposite vessel. The master assumes that the other vessel will perform a maneuver that can avoid the accident, and therefore, only the master can perform an action that can positively avoids the accident.

In the line relation analysis of Japanese accidents, there is no special characteristics that can be extracted because the high causing ratio of the line relation in Japan is the same as in Indonesia. However, the causing ratio in Japan is smaller than that in Indonesia, except in the machine factor corner. In the machine factor corner, M12 and M13 entirely cause the machine factor failures with a causing ratio of 1. This means that all failures in the machine factor are affected by M12 and M13.

Now, let us discuss how the failures are divided into stages 1 and 3. Unlike Indonesia, the 14 cases in Japan do not have any failure in the evacuation process except because of bad weather (media factor). Only 1 accident out of 14 has 13 people missing. The collision happened at night, in a rainy, wavy, and extremely dark environment, and therefore, the ship suffered from fatal damage and no one was found at sea. Six hours after the collision, an evacuation vessel reached the accident point; however, it could not find the 13 missing people. Other than that case, the number of injury and death is not as large compared to the case in Indonesia when the seafarer could not carry out the evacuation process.

From the 60 failures, 16 failures occurred in Stage 2 and the rest occurred in Stage 1. Although the most common causative factors in Indonesia occurred in Stage 2, the accidents in Japan have different characteristics. The misunderstanding condition, which is the most common causative factor in Japan, occurred in both Stages 1 and 2. In addition, there are no special characteristics failures in Stage 2. Most failures occurred in Stage 1 where the man factor corner has 21 failures and there are no machine factors.

5.3 Characteristics of MTS accidents in Both Indonesia and Japan

In this subsection, we discuss the causative factors that occur in both Indonesia and Japan. These causative factors are listed in Tables 2–5. From all causative factors, the insufficient light (M3-05) from media factors is the most common causing factor for the system in both Indonesia and Japan. The second most common factor is improper lookout (M1-05) from the man factor. Insufficient light implies that at the time of the accident, it was too dark or the accident occurred at night. Indeed, this condition is more difficult. If it we note the occurrence ratio in Figure 5, the accidents that occurred had a higher occurrence ratio of 0.75 in Indonesia, while in Japan, the ratio is 0.64. In the case of improper lookout, both Indonesia and Japan have the same occurrence ratio (0.5) as can be seen in Figure 3.

In both Indonesia and Japan, the most causing line relation is the M14 Line in the management factor, as seen in Figure 10. It has a 1.0 causing ratio, which means that all failures in the management factor that occur in both Indonesia and Japan are related to the man factor.

6 CONCLUSION

In this paper, several collision accident characteristics of Japan and Indonesia have been discussed. There are several characteristics that only occur in Indonesia and Japan separately, and those which happen both in Indonesia and Japan. From the study, we can draw the following conclusions:

- To avoid collisions, more attention needs to be paid to the capability of an Indonesian seafarer
  The most common failure of all causative failures in 8 collision cases in Indonesia occurred in Stage 2, that is, failure to avoid the accident. The most common causative factors are incapability of the seafarer in communicating with the other vessel (M1-11) and in understanding how to avoid collision (M1-12); these causative factors belong to the man factor.
- Misunderstanding condition (wrong judgment) from Masters is a characteristic of 14 Japanese collision cases from 2008–2012
  5 of the 14 collisions cases in Japan were caused by this causative factor, and it does not occur in Indonesia. In order to reduce the number of accidents that have the same causes, the seafarer needs to reconsider the judgment. It is necessary to communicate with the opposite vessel so that the right action can be performed to avoid collision.
- Most accidents in Indonesia and Japan happened at night and the seafarers did not ensure a proper look-out.
  These two causative factors have high occurrence ratios compared with other causative factors. Therefore, paying more attention to the lookout is definitely required and when a transportation occurs at night, the seafarer should be more careful.

ACKNOWLEDGEMENT

I would like to thank Mr. Aleik Nurwahyudy, one of the investigators at the Indonesian National Transportation Safety Committee, who supported me by explaining several conditions of the Indonesian MTS and by introducing the concept of the three stages of accident development.

REFERENCES

Furusho, M. 2000. IM Model for Ship Safety, proceedings of inaugural general assembly. Turkey: 26-31.

Furusho, M. 2013. Disaster of Italian Passenger Ship Costa Concordia – a Nightmare 100 Years after the Titanic-. The Mariners's Digest. Vol. 28, 31-35 (Magazine)

Grech, M. R., Horberry, T. J., and Koester, T. 2008. Human Factors in the Maritime Domain. CRC Press: France.

NTSC 2014. http://kemhubri.dephub.go.id/knkt/ntsc_maritime/maritime.htm (2015/01/15)

Nurwahyudi, A. 2014. Contemporary Issus in Domestic Ro-Ro Passenger Ferry Operation in Developing Countries: Identification of safety issues in domestic ferry operation based on accident investigation reports on ferry involved accidents in Indonesian waters, 2003–2013. World Maritime University: Malmo

JTSB 2014. http://www.mlit.go.jp/jtsb/marrep.html (2015/01/15)

AUTHOR INDEX

Alfredini, P., 147
Andreassen, N., 217
Arasaki, E., 147

Baldauf, M., 107
Başar, E., 211
Bauk, S., 253
Benedict, K., 107
Borch, O.J., 217
Borup, O., 85, 267
Brodje, A., 85, 267

Camre, D., 85, 267
Cardoso, R.M., 157
Charantonis, A.A., 67
Charou, E., 225
Chen, L., 185
Chen, X., 185

Dolzhenko, D.O., 55
Drolias, N.G., 67

Fischer, S., 107
Furusho, M., 275

Galor, W., 195
Gegenav, A., 153
Giannakopoulos, T., 225
Gluch, M., 107
Grzelakowski, A.S., 201
Guedes Soares, C., 99
Gyftakis, S., 225

Ilcev, D.S., 21, 33, 45

Jiménez-Castañeda, R., 179

Kakhidze, A., 153
Kapidani, N., 253
Karagianni, E.A., 67
Kirchhoff, M., 107
Kizielewicz, J., 243
Korcz, K., 75
Koromila, I., 225
Koshevyy, I.V., 55
Koshevyy, V.M., 13, 55, 61
Kuśmińska-Fijałkowska, A., 131, 135

Łącki, M., 123
Łebkowski, A., 117, 259
Lewiński, A., 173
Lima, D.C.A., 157
Łukasik, Z., 131, 135, 173
Luković, T., 243

Miranda, P., 157
Mitropoulos, A.P., 67
Montewka, J., 231
Moreira, A.S., 147
Mou, J., 185
Mutmainnah, W., 275

Neumann, T., 3, 9
Nivolianitou, Z., 225

Özdemir, Ü., 211

Park, J.S., 93
Park, S.W., 93
Park, Y.S., 93
Perantonis, S., 225
Perzyński, T., 173
Piniella, F., 179
Popovich, V.V., 139
Porathe, T., 85, 267

Querol, A., 179

Rijo, N., 157

Sarantopoulos, A.D., 67
Schaub, M., 107
Semedo, A., 157
Sharabidze, I., 153
Shershnova, A.A., 61
Shyshkin, O., 13
Smirnova, O.V., 139
Soares, P.M.M., 157
Sorokin, R.P., 139
Spyrou, K., 225

Thanh, N.X., 93
Tsvetkov, M.V., 139

Vettor, R., 99

Wang, Y., 165
Weber, R., 85, 267
Weintrit, A., 3, 9
Wu, B., 165

Yan, X.P., 165
Yılmaz, H., 211
Yue, X., 185

Zhang, J.F., 165
Zheng, X., 231